ELEMENTS OF
Petroleum
Geology

FRONTISPIECE *Geologist viewing the petroliferous mud volcano of Monkeytown, Trinidad. (From Wall and Sawkins, 1860.)*

ELEMENTS OF
Petroleum
Geology

Richard C. Selley

W. H. FREEMAN AND COMPANY
NEW YORK

Library of Congress Cataloging in Publication Data

Selley, Richard C., 1939–
    Elements of petroleum geology.

    Includes bibliographies and index.
    1. Petroleum—Geology.   2. Gas, Natural—Geology.
I. Title.
TN870.5.S425 1985   553.2′82   84-24718
ISBN 0-7167-1630-5

Printed in the United States of America

1 2 3 4 5 6 7 8 9 0   MP   3 2 1 0 8 9 8 7 6 5

Contents

3

Methods of Exploration   37

6

The Reservoir   219

7

The Trap   276

8

Sedimentary Basins

Preface

Geology is one of several disciplines involved in the exploration and production of oil and gas. It merges into petroleum engineering on one side, and geophysics on the other. Petroleum exploration also involves engineering, law, accounting, and economics. This text thus covers only one aspect of petroleum exploration.

I have written this book primarily for those who are either studying petroleum exploration in the terminal phases of university work or just beginning a career in the oil industry. Basically, it is written for geologists, rather than engineers or geophysicists. I anticipate, therefore, that the reader is familiar with the basic concepts of geology and has, in particular, attended courses in stratigraphy, petrography, and sedimentology. Although the main thrust of the book is toward petroleum geology, it also includes material on geophysics and petroleum reservoir engineering. Both these branches are presented at a very basic level, and petroleum geologists need more knowledge than this book contains. Textbooks on both of these topics are readily available, and the novice petroleum geologist will attend courses on seismic interpretation and log analysis early in his or her professional career.

I also hope that this book may be useful for petroleum engineers and geophysicists, especially if they have acquired some familiarity of geology from courses or experience. Those that have will already appreciate how different geology is from their own disciplines.

Although geology may, to a large extent, be based on numerical data, it is less susceptible to the rigorous mathematical analysis used in engineering and geophysics. Furthermore, it can only be approached through a blizzard of jargon. To engineers I would say: when you read this book, keep a geology dictionary handy. Ignore the sections on log analysis and petroleum reservoir engineering. To geophysicists I would say, again, keep a geology dictionary handy, and ignore the section on geophysical methods.

R. C. SELLEY
September 1984

Acknowledgments

There are two main problems to overcome when writing a book on petroleum geology. The subject is a vast one, ranging from arcane aspects of molecular biochemistry to the mathematical mysteries of seismic data processing. The subject is also evolving very fast as new data become available and new concepts are developed. Most innovations are made within oil company research laboratories and do not get published; they may not become known to the general public for several years.

I am very grateful to the many people who have read drafts of the manuscript, pointed out errors of fact or emphasis, and suggested improvements. Much of this load was borne by staff of Imperial College, London University. The geophysical sections were dealt with by Dr. Thomas-Betts and Messrs. Weildon and Williamson, geochemistry by Dr. Kinghorn, petroleum engineering by Professor Wall, and most of the remaining topics by Professor Stoneley. Mr. Maret of Schlumberger reviewed the formation evaluation section.

For permission to use previously published illustrations I am grateful to the following: Academic Press, the American Association of Petroleum Geologists, Applied Science Publishers, Blackwell Scientific Publications, Gebruder Borntraeger, the Canadian Association of Petroleum Geologists, Chapman and Hall, W. H. Freeman and Company, The Geologists Association of London, the Geological Society of London, the Geological Society of South Africa, Gulf Coast Association of Geological Societies, the Institute of Petroleum, the *Journal of Geochemical Exploration*, the *Journal of Petroleum Geology*, McGraw-Hill, the Norwegian Petroleum Society, the Offshore Technology Conference, Princeton University Press, Schlumberger Wireline Logging Services, the Society of Petroleum Engineers, the Society of Professional Well Log Analysts, Springer-Verlag, John Wiley & Sons, and World Oil.

ELEMENTS OF
Petroleum Geology

Introduction

*And God said unto Noah .... Make thee an
ark of gopher wood; rooms shalt thou
make in the ark, and shalt pitch it within
and without with pitch.*

GENESIS 6:13–14

HISTORICAL REVIEW OF PETROLEUM EXPLORATION

Petroleum from Noah to OPEC

Petroleum exploration is a very old pursuit, as the preceding quotation illustrates. The Bible contains many references to the use of pitch or asphalt
collected from the natural seepages with which the Middle East abounds.

Herodotus, writing in about 450 B.C., described oil seeps in Carthage
(Tunisia) and the Greek island Zachynthus (Herodotus, c. 450 B.C.). He
gave details of oil extraction from wells near Ardericca in modern Iran,
although the wells could not have been very deep, as fluid was extracted in
a wineskin on the end of a long pole mounted on a fulcrum. Oil, salt, and
bitumen were produced simultaneously from these wells. Throughout the
first millenium A.D. oil and asphalt were gathered from natural seepages in
many parts of the world.

The early uses of oil were for medication, waterproofing, and warfare.
Oil was applied externally for wounds and rheumatism and administered
internally as a laxative. From the time of Noah pitch has been used to make
boats watertight. Pitch, asphalt, and oil have long been employed in warfare. When Alexander the Great invaded India in 326 B.C., he scattered the
Indian elephant corps by charging them with horsemen waving pots of
burning pitch. Nadir Shah employed a similar device, impregnating the
humps of camels with oil and sending them ablaze against the Indian elephant corps in 1739 (Pratt and Good, 1950). *Greek fire* was invented by

Callinicus of Heliopolis in A.D. 668. Its recipe is unknown, but it is believed to have included quicklime, sulfur, and naphtha and it ignited when wet. It was a potent weapon in Byzantine naval warfare.

Up until the mid-nineteenth century, asphalt, oil, and their by-products were only produced from seepages, shallow pits, and hand-dug shafts. In 1694 the British Crown issued a patent to Masters Eele, Hancock, and Portlock to "make great quantities of pitch, tarr, and oyle out of a kind of stone." The stone in question was of Carboniferous age and occurred at the appropriately named Pitchford in Shropshire (Eele, 1697).

The first well specifically sunk to search for oil (as opposed to water or brine) appears to have been at Pechelbronn, France, in 1745. Outcrops of oil sand were noted in this region, and Louis XV granted a license to M. de la Sorbonniere, who sank several borings and built a refinery in the same year (Redwood, 1913). The birth of the oil shale industry is credited to James Young, who began retorting oil from the Carboniferous shales at Torban, Scotland, in 1847 (see p. 401).

The resultant products of these early refineries included ammonia, solid paraffin wax, and liquid paraffin (kerosene or coal oil). The wax was used for candles and the kerosene for lamps. Kerosene became cheaper than whale oil, and therefore the market for liquid hydrocarbons expanded rapidly in the mid-nineteenth century. Initially the demand was satisfied by oil shales and from oil in natural seeps, pits, hand-dug shafts, and galleries.

Before oil exploration, cable-tool drilling was an established technique in many parts of the world in the quest for water and brine (Fig. 1.1). The first well to actually *produce* oil was drilled at Oil Creek, Pennsylvania, by Colonel Drake in 1859 (Owen, 1975). A rapid growth in oil production from subsurface wells soon followed, both in North America and around the world. A major stimulus to oil production was the development of the internal combustion engine in the 1870s and 1880s. Gradually the demand for lighter petroleum fractions overtook that for kerosene. However, uses were found for all the refined products, from the light gases, via petrol, paraffin, diesel oil, tar, and sulfur, to the heavy residue.

Demand for oil products was increased greatly by the First World War (1914–1918). By the 1920s the oil industry was dominated by seven major companies, termed the *seven sisters* by Enrico Mattei (Sampson, 1975). These companies included:

$$
\left.\begin{array}{l}
\text{British Petroleum} \\
\text{Shell}
\end{array}\right\} \text{European}
$$

$$
\left.\begin{array}{l}
\text{Exxon (formerly Esso)} \\
\text{Gulf} \\
\text{Texaco} \\
\text{Mobil} \\
\text{Socal (or Chevron)}
\end{array}\right\} \text{American}
$$

FIGURE 1.1 *Early cable-tool rig used in America. The motive power was provided by one man and a spring-pole. (Courtesy of British Petroleum.)*

British Petroleum and Shell found their oil reserves abroad from their parent countries, principally in the Middle and Far East, respectively. They were thus involved early in long-distance transport by sea, measuring their oil by the seagoing tonne. The American companies, by contrast, with shorter transportation distances, used the barrel as their unit of measurement.

The American companies began overseas ventures, mainly in Central and South America, in the 1920s. In the 1930s the Arabian-American Oil Company (Aramco) evolved from a consortium of Socal, Texaco, Mobil, and Exxon.

Following the Second World War and the postwar economic boom, the consortia principle became established over much of the free world. Oil companies risked the profits from one productive area to explore for oil in new areas. To take on all the risks in a new venture is unwise, so companies would invest in several joint ventures, or consortia. Table 1.1

3

TABLE 1.1   Partners of some of the major overseas oil consortia[a]

Companies	Iran The Consortium	Iraq I.P.C.	Saudi Arabia Aramco	Kuwait Kuwait Oil Co.	Abu Dhabi Admar	A.D.P.C.	Libya Oasis
B.P.	X	X		X	X	X	
Shell	X	X				X	X
Exxon	X	X	X			X	
Mobil	X	X	X			X	
Gulf	X			X			
Texaco	X		X				
Socal	X		X				
C.F.P.	X					X	
Conoco							X
Amerada							X
Marathon							X

[a]Note that partners and their percentage interest varied over the lifetime of the various consortia.

shows some of the major consortia, demonstrating the stately dance of the seven sisters as they changed their partners around the world. In this process the major consortia shared a mutual love-hate relationship. The object of any business is to maximize profit. Thus it was to their mutual benefit to export oil from the producing countries as cheaply as possible and to sell it in the world market for the highest price possible. The advantage of a cartel is offset by the desire of every company to enhance its sales at the expense of its competitors by selling its products cheaper.

In 1960 the Organization of Petroleum Exporting Countries (OPEC) was founded in Baghdad, consisting initially of Iraq, Iran, Kuwait, Saudi Arabia, and Venezuela (Martines, 1969). It later expanded to include Algeria, Dubai, Ecuador, Gabon, Indonesia, Libya, Nigeria, Qatar, and the United Arab Emirates. To qualify for membership, a country's economy must be predominantly based on oil exports; therefore the United States and the United Kingdom do not qualify.

The object of OPEC is to control the power of the independent oil companies by a combination of price control and appropriation of company assets. For many OPEC countries oil is their only natural resource. Once it is depleted, they will have no assets unless they can maximize their oil revenues and spend them in the development of other industries. The OPEC objective has been notably successful, although its large price increases in the early 1970s contributed to a global recession, which affected both the developed and the poorer third world countries alike.

The idea of the producing state controlling the oil company's activities has now been exported from OPEC. Not only have state oil companies been formed in countries that formerly lacked indigenous oil expertise (e.g., Statoil in Norway and Petronas in Malaysia), but they have also been formed in those that had the expertise (Petrocan in Canada and Britoil in the United Kingdom). Formerly the profit that the oil companies made in one country was the risk capital to be invested in the next country. Now the risk capital is provided by the taxpayer.

Evolution of Petroleum Exploration Concepts and Techniques

From the days of Noah to OPEC the role of the petroleum geologist has become more and more skilled and demanding. In the early days oil was found by wandering about the countryside with a naked flame, optimism, and a sense of adventure. One major U.S. company, which will remain nameless, once employed a chief geologist whose exploration philosophy was to drill on old Indian graves. Another oil finder used to put on an old hat, gallop about the prairie until his hat dropped off, and start drilling where it landed. History records that he was very successful (Cunningham-Craig, 1912).

One of the earliest exploration tools was *creekology*. It gradually dawned on the early drillers that oil was more often found by wells located on river bottoms than by those on the hills (Fig. 1.2). The anticlinal theory of oil entrapment, which explained this phenomenon, was expounded by Hunt (1861). Up to the present day the quest for anticlines has been one of the most successful exploration concepts.

Experience soon showed, however, that oil could also occur off-structure. Carll (1880) noted that the oil-bearing marine Venango sands of Pennsylvania occurred in trends that reflected not structure but paleo-shorelines. Thus was borne the concept that oil could be trapped strati-

FIGURE 1.2 *Creekology—the ease of finding oil in the old days.*

graphically as well as structurally. Stratigraphic traps are caused by variations in deposition, erosion, or diagenesis within the reservoir.

Through the latter part of the nineteenth century and the early part of the twentieth, oil exploration was based on the surface mapping of anticlines. Stratigraphic traps were found accidentally by serendipity or by subsurface mapping and extrapolation of data gathered from wells drilled to test structural anomalies.

Unconformities and disharmonic folding limited the depth to which surface mapping could be used to predict subsurface structure. The solution to this problem began to emerge in the mid-1920s, when seismic (refraction), gravity, and magnetic methods were all applied to petroleum exploration. Magnetic surveys seldom proved to be effective oil finders, whereas gravity and seismic methods proved to be effective in finding salt dome traps in the Gulf of Mexico coastal province of the United States. In the same period geophysical methods were also applied to borehole logging, with the first electric log run at Pechelbronn, France, in 1927. Further electric, sonic, and radioactive logging techniques followed.

Aerial surveying began in the 1920s, but photogeology, which employs stereophotos, only became widely used after the Second World War. At this time aerial surveys were cheap enough to allow the rapid reconnaissance of large concessions, and photogeology was notably effective in the deserts of North Africa and the Middle East, where vegetation does not cover surface geology.

Pure geological exploration methods advanced slowly but steadily during the first half of the twentieth century. One of the main applications to oil exploration was the development of micropaleontology. The classic biostratigraphic zones, which are based on macrofossils such as ammonites, could not be identified in the subsurface because of the destructive effect of drilling. New zones had to be defined by microfossils, which were calibrated at the surface with macrofossil zones.

The study of modern sedimentary environments in the late 1950s and early 1960s, notably on Galveston Island (Texas), the Mississippi delta, the Bahama Bank, and the Dutch Wadden Sea, gave a new insight into ancient sedimentary facies and their interpretation. This insight provided improved prediction of the geometry and internal porosity and permeability variation of reservoirs.

The 1970s saw major advances on two fronts: geophysics and geochemistry. The advent of the computer resulted in a major quantum jump in seismic processing; instead of seismologists poring painfully over a few bunched galvanometer traces, vast amounts of data could be displayed on continuous seismic sections. Reflecting horizons could be picked out in bright colors, first by geophysicists and later even by geologists. As techniques improved, seismic lines became more and more like geological cross-sections, until stratigraphic and environmental concepts were directly applicable.

From the earliest days of scientific investigation, the formation of petroleum had been attributed to two origins: inorganic and organic. Chemists, such as Mendele'ev in the nineteenth century, and astronomers, such as Gold and Hoyle in the twentieth, argued for an inorganic origin—sometimes igneous, sometimes extraterrestrial, or a mixture of both. Most petroleum geologists believe that petroleum forms from the diagenesis of buried organic matter and note that it is indigenous to sedimentary rocks rather than igneous ones. The advent of cheap and accurate chemical-analytical techniques allowed petroleum source rocks to be studied. It is now possible to match petroleum with its parent shale and to identify potential source rocks, their tendency to generate oil or gas, and their level of thermal maturation.

It remains to be seen what the main advance in petroleum exploration technology during the 1980s will be. All techniques may be expected to improve. Remote sensing from satellites may be one major new tool, as may direct sensing from surface geochemical or geophysical methods. These latter methods generally involve the identification of gas microseeps and fluctuations in electrical conductivity of rocks above petroleum accumulations. Such methods have been around for half a century, but have yet to be widely accepted. This unacceptance can only please geologists, who, like lawyers, have a vested interest in incompetence. The advent of a "black box" for finding petroleum would result in mass unemployment, such as would occur if lawyers could settle court cases within days or weeks rather than months or years.

For a commercial oil accumulation to occur, five conditions must be fulfilled:

1. There must be an organic-rich source rock to generate the oil and/or gas.
2. The source rock must have been heated sufficiently to yield its petroleum.
3. There must be a reservoir to contain the expelled hydrocarbons. This reservoir must have:
   a. Porosity: to contain the oil and/or gas.
   b. Permeability: to permit fluid flow.
4. The reservoir must be sealed by an impermeable cap rock to prevent the upward escape of petroleum to the earth's surface.
5. Source, reservoir, and seal must be arranged in such a way as to trap the petroleum.

Chapter 5 deals with the generation and migration of petroleum from source rocks. Chapter 6 discusses the nature of reservoirs, and Chapter 7 deals with the different types of traps.

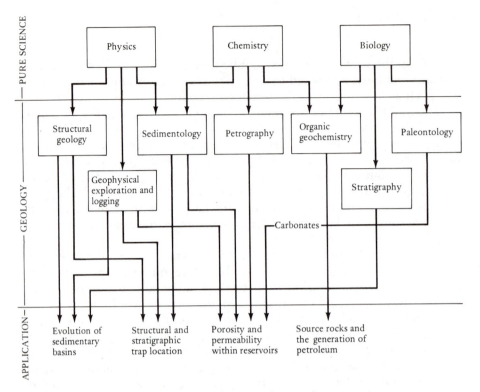

Physics **Chemistry** **Biology**

Structural geology Sedimentology Petrography Organic geochemistry Paleontology

Geophysical exploration and logging Stratigraphy

—Carbonates—

Evolution of sedimentary basins

Structural and stratigraphic trap location

Porosity and permeability within reservoirs

Source rocks and the generation of petroleum

FIGURE 1.3 *The relationship of petroleum geology to the pure sciences.*

THE CONTEXT OF PETROLEUM GEOLOGY

Relationship of Petroleum Geology to Science

Petroleum geology is the application of geology (the study of rocks) to the exploration for and production of oil and gas. Geology itself is firmly based on chemistry, physics, and biology, involving the application of essentially abstract concepts to observed data. In the past these data were basically observational and subjective, but they are now increasingly physical and chemical, and therefore more objective. Geology, in general, and petroleum geology, in particular, still rely on value judgements based on experience and an assessment of validity among the data presented.

The preceding section showed how petroleum exploration had advanced over the years as various geological techniques were developed. It is now appropriate to consider in more detail the roles of chemistry, physics, and biology in petroleum exploration (Fig. 1.3).

Chemistry and Petroleum Geology

The application of chemistry to the study of rocks (geochemistry) has many uses in petroleum geology. Detailed knowledge of the mineralogical com-

8

position of rocks is important at many levels. In the early stages of exploration certain general conclusions as to the distribution and quality of potential reservoirs could be made from their gross lithology. For example, the porosity of sandstones tends to be facies related, whereas in carbonate rocks this is generally not so. Detailed knowledge of the mineralogy of reservoirs enables estimates to be made of the rate at which they may lose porosity during burial, and this detailed mineralogical information is essential for the accurate interpretation of geophysical well logs through reservoirs. Knowledge of the chemistry of pore fluids and their effect on the stability of minerals can be used to predict where porosity may be destroyed by cementation, preserved in its original form, or enhanced by solution of minerals by formation waters.

Organic chemistry is involved both in the analysis of oil and gas and in the study of the diagenesis of plant and animal tissues in sediments and the way in which the resultant organic compound, kerogen, generates petroleum.

Physics and Petroleum Geology

The application of physics to the study of rocks (geophysics) is very important in petroleum geology. In its broadest application geophysics makes a major contribution to understanding the earth's crust and, especially through the application of modern plate tectonic theory, to the genesis and petroleum potential of sedimentary basins. More specifically, physical concepts are required to understand folds, faults, and diapirs, and hence their roles in petroleum entrapment.

Modern petroleum exploration is unthinkable without the aid of magnetic, gravity, and seismic surveys in finding potential petroleum traps. Nor could any finds be evaluated effectively without geophysical wireline well logs to measure the lithology, porosity, and petroleum content of a reservoir.

Biology and Petroleum Geology

Biology is applied to geology in several ways, notably through the study of fossils (paleontology), and is especially significant in establishing biostratigraphic zones for regional stratigraphical correlation. The shift in emphasis from the use of macrofossils to microfossils for zonation, caused by oil exploration, has already been noted. Ecology, the study of the relationship between living organisms and their environment, is also important in petroleum geology. Carbonate sediments, in general, and reefs, in particular, can only be studied profitably with the aid of a detailed knowledge of the ecology of modern marine fauna and flora. Biology, and especially biochemistry, is important in studying the transformation of plant and animal tissues into kerogen during burial and the generation of oil or gas that may be caused by this transformation.

Relationship of Petroleum Geology
to Petroleum Exploration and Production

Geologists, in contrast to some nongeologists, believe that knowledge of the concepts of geology can help to find petroleum and, furthermore, often think that petroleum geology and petroleum exploration are synonymous, which they are not. Theories that petroleum is not formed by the transformation of organic matter in sediments have already been noted and are examined in more detail in Chapter 5. If the petroleum geologists' view of oil generation and migration are not accepted, then present exploration methods would need extensive modification.

Some petroleum explorationists still do not admit to a need for geologists to aid them in their search. In 1982 a successful oil finder from Midland, Texas, admitted to not using geologists because when his competitors hired them, all it did was increase their costs per barrel of oil found. The South African state oil company (SOEKOR) is under a statutory obligation imposed by its government to put to the test every claim to an oil-finding method, be it dowsing or some sophisticated scientific technique. These examples are not isolated cases, and it has been argued that oil may better be found by random drilling than by the appliance of scientific principles (see p. 411).

Petroleum geology is only one aspect of petroleum exploration and production. Leaving aside atypical enterprises, petroleum exploration now involves integrated teams of people possessing a wide range of professional skills (Fig. 1.4). These skills include political and social expertise,

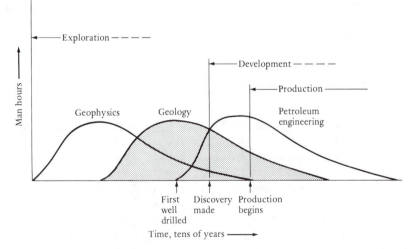

FIGURE 1.4 *Graph showing how petroleum geology is part of a continuum of disciplines employed in the exploration and production of oil and gas.*

which is involved in the acquisition of prospective acreage. Geophysical surveying is involved in preparing the initial data on which leasing and, later, drilling recommendations are based. Geological concepts are applied to the interpretation of the geophysical data once it has been acquired and processed. As soon as an oil well has been drilled, the engineering aspects of the discovery need appraisal. Petroleum engineering is concerned with establishing the reserves of a field, the distribution of petroleum within the reservoir, and the most effective way of producing it. Thus petroleum geology lies within a continuum of disciplines, beginning with geophysics and ending with petroleum engineering, but overlapping both in time and subject matter (Fig. 1.5).

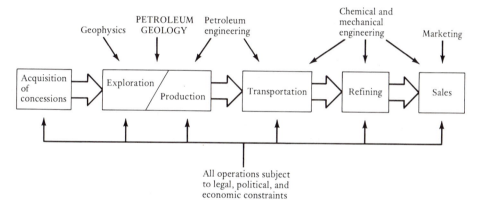

FIGURE 1.5 *Flowchart showing how petroleum geology is only one aspect of petroleum exploration and production, and how these enterprises themselves are part of a continuum of events subjected to various constraints and expedited by many disciplines.*

Underlying this sequence of events is the fundamental control of economics. Oil companies exist not only to find oil and gas but, like any business enterprise, to make money. Thus every step of the journey, from leasing to drilling, to production, and finally to enhanced recovery, is monitored by accountants and economists. Activity in petroleum exploration and production accelerates when the world price of petroleum products increases, and it decreases or even terminates when the price drops.

Petroleum geologists are in the unusual position of being subject to firing for either technical incompetence or excellence, but not for mediocrity. If they find no oil, they may be fired; and, similarly, if they find too much oil, then their presence on the company payroll is unnecessary.

Competent geologists are more important to small companies, for whom a string of dry holes spells catastrophe. Major companies can tolerate a fair degree of incompetence because they have the financial resources to withstand a string of disasters. Nowhere is this more true than in state oil companies. With an endless supply of taxpayers' money to sustain them, the political expediency of searching for indigenous petroleum reserves may outweigh any economic consideration.

SELECTED BIBLIOGRAPHY

For an account of the early historical evolution of the oil industry, see:

REDWOOD, SIR B. 1913.   *A Treatise on Petroleum*, vol. 1. London: Griffin & Co., 367 pp.

For the evolution of geological concepts in petroleum exploration, see:

DOTT, R. H. and REYNOLDS, M. J. 1969.   *Source Book for Petroleum Geology*. Tulsa: Am. Assoc. Petrol. Geol., Mem. No. 5, 471 pp.

For accounts of the evolution of the oil industry over the last century, see:

PRATT, W. E. and GOOD, D. 1950.   *World Geography of Petroleum*. Princeton: Princeton Univ. Press, 464 pp.

OWEN, E. W. 1975.   *Trek of the Oil Finders: A History of Exploration for Petroleum*. Tulsa: Am. Assoc. Petrol. Geol., 1647 pp.

SAMPSON, A. 1975.   *The Seven Sisters*. London: Hodden and Stoughton, 334 pp.

REFERENCES

CARLL, J. F. 1880.   The geology of the oil regions of Warren, Venango, Clarion and Butler Counties. *Penn. Geol. Surv.*, 3, 482.

CUNNINGHAM-CRAIG, E. H. 1912.   *Oil-Finding*. London: Edward Arnold, 195 pp.

EELE, M. 1697.   On making pitch, tar and oil out of a blackish stone in Shropshire. *Phil. Trans. Roy. Soc.*, xix, 544.

HERODOTUS, H. ?450 B.C.   *The Histories*. 9 books.

HUNT, T. S. 1861.   Bitumens and mineral oils. *Montreal Gazette*, 1 March.

LYMAN, E. J. 1877.   Tokyo: *Report on the Geological Survey of the Oil Lands of Japan*.

MARTINEZ, A. R. 1969.   *Chronology of Venezuelan Oil*. London: Allen & Unwin, 207 pp.

OWEN, E. W. 1975.   *Trek of the Oil Finders: A History of Exploration for Petroleum*. Tulsa: Am. Assoc. Petrol. Geol., 1647 pp.

PRATT, W. E. and GOOD, D. 1950.   *World Geography of Petroleum*. Princeton: Princeton Univ. Press, 464 pp.

REDWOOD, SIR B. 1913.   *A Treatise on Petroleum*, vol. 1. London: Griffin & Co., 367 pp.

SAMPSON, A. 1975.   *The Seven Sisters*. London: Hodden and Stoughton, 334 pp.

SHIRLEY, T. 1667.   The description of a well and earth in Lancashire taking fire by a candle approached to it. *Phil. Trans. Roy. Soc.*, 2, 482.

2

Physical and Chemical Properties of Oil and Gas

Petroleum exploration is largely concerned with the search for oil and gas, two of the chemically and physically diverse group of compounds termed the *hydrocarbons*. Physically, hydrocarbons grade from gases, via liquids and plastic substances, to solids. The hydrocarbon gases include *dry* gas (methane) and the *wet* gases (ethane, propane, butane, etc.). Condensates are hydrocarbons that are gaseous in the subsurface, but condense to liquid when they are cooled at the surface. Liquid hydrocarbons are termed oil, crude oil, or just *crude*. The plastic hydrocarbons include asphalt and related substances. Solid hydrocarbons include coal and kerogen. Gas hydrates are ice crystals with peculiarly structured atomic lattices, which contain molecules of methane and other gases.

This chapter describes the physical and chemical properties of natural gas, oil, and the gas hydrates; it is a necessary prerequisite to Chapter 5, which deals with petroleum generation and migration. The plastic and solid hydrocarbons are discussed in Chapter 9, which covers the tar sands and oil shales.

The earth's atmosphere is composed of "natural gas." In the oil industry, however, natural gas is defined as "a mixture of hydrocarbons and varying quantities of nonhydrocarbons that exists either in the gaseous phase or in solution with crude oil in natural underground reservoirs." The foregoing is the definition adopted by the American Petroleum Institute, the American Association of Petroleum Geologists, and the Society of Petroleum Engineers. The same authorities subclassify natural gas into dissolved, associated, and nonassociated gas. Dissolved gas is in solution in crude oil in the reservoir. Associated gas, commonly known as gas cap gas, overlies and is in contact with crude oil in the reservoir. Nonassociated gas is in reservoirs that do not contain significant quantities of crude oil. Natural gas liquids, or NGLs, are the portions of the reservoir gas that are liquefied at the surface in lease operations, field facilities, or gas-processing plants. Natural gas liquids include, but are not limited to, ethane, propane, butane, pentane, natural gasoline, and condensate. Basically, natural gases encountered in the subsurface can be classified into two groups: those of organic origin and those of inorganic origin (Table 2.1).

TABLE 2.1   Natural gases and their dominant modes of formation

Gas		Dominant source
Inert gases	Helium	
	Argon	
	Krypton	Inorganic
	Radon	
	Nitrogen	
	Carbon dioxide	Mixed
	Hydrogen sulfide	
	Hydrogen	
Hydrocarbons	Methane—dry gas	
	Ethane	Mainly organic
	Propane wet gases	
	Butane	

Gases are classified as *dry* or *wet* according to the amount of liquid vapor that they contain. A dry gas may be arbitrarily defined as one with less than 0.1 gal/1000 ft^3 of condensate; chemically, dry gas is largely methane. A wet gas is one with over 0.3 gal/1000 ft^3 of condensate; chemically, these gases contain ethane, butane, and propane. Gases are also described as *sweet* or *sour*, based on the absence or presence, respectively, of hydrogen sulfide.

NATURAL GASES

Hydrocarbon Gases

The major constituents of natural gas are the hydrocarbons of the paraffin series (Table 2.2). The heavier members of the series decline in abundance with increasing molecular weight. Methane is the most abundant; ethane,

TABLE 2.2   Significant data of the paraffin series

Name	Formula	Molecular weight	Boiling point at atmospheric pressure, °C	Solubility g/10°g water
Methane	CH_4	16.04	−162	24.4
Ethane	C_2H_6	30.07	−89	60.4
Propane	C_3H_8	44.09	−42	62.4
Isobutane	C_4H_{10}	58.12	−12	48.9
n-Butane	C_4H_{10}	58.12	−1	61.4
Isopentane	C_5H_{12}	72.15	30	47.8
n-Pentane	C_5H_{12}	72.15	36	38.5
n-Hexane	C_6H_{14}	86.17	69	9.5

butane, and propane are quite common; and paraffins with a molecular weight greater than pentane are the least common.

Methane (CH_4) is also known as *marsh gas* if found at the surface or *fire damp* if present down a coal mine. Traces of methane are commonly recorded as *shale gas* or *background gas* during the drilling of all but the driest of dry wells. Methane is a colorless, flammable gas, which is produced (along with other fluids) by the destructive distillation of coal. As such, it was commonly used for domestic purposes in Europe until replaced by natural gas, itself largely composed of methane.

Methane is the first member of the paraffin series. It is chemically nonreactive, sparingly soluble in water, and lighter than air (0.554 relative density). Methane is known to occur as a by-product of bacterial decay of organic matter at normal temperatures and pressures. This biogenic methane has considerable potential as a source of energy. In the nineteenth century eminent Victorians debated the possibility of lighting the streets of London from methane from the sewers. Today's avant-garde agriculturalists acquire much of the thermal energy needed for their farms by collecting the gas generated by the maturation of manure.

Biogenic methane is commonly formed in the shallow subsurface by the bacterial decay of organic-rich sediments. As the burial depth and temperature increase, however, this process diminishes and the bacterial action is extinguished. The methane encountered in deep reservoirs is produced by thermal maturation of organic matter. This process will be discussed in detail later (p. 190).

The other major hydrocarbons that occur in natural gas are ethane, propane, butane, and occasionally pentane. Their chemical formulas and molecular structure are shown in Figure 2.1. Their occurrence in various gas reservoirs are given in Table 2.3. Unlike methane these heavier members of the paraffin series do not form biogenically. They are produced

TABLE 2.3 Chemical composition of various gas fields

Field and area	Composition									Reference
	Methane	Ethane	Propane	Butane	Pentane+	CO_2	N	H_2S	He	
Southern N. Sea basin										
Groningen	81.3	2.9	0.4	0.1	0.1	0.9	14.3	Tr[a]	Tr	Cooper (1977) (1975)
Hewett	83.2	5.3	2.1	0.4	0.5	0.1	8.4	Tr	Tr	Cumming and Wyndham (1975)
West Sole	94.4	3.1	0.5	0.2	0.2	0.5	1.1			Butler (1975)
Algeria										
Hassi-R'Mel	83.5	7.0	2.0	0.8	0.4	0.2	6.1			Cooper (1977)
France										
Lacq	69.3	3.1	1.1	0.6	0.7	9.6	0.4	15.2		Cooper (1977)
Canada										
Turner Valley	92.6	4.1	2.5	0.7	0.13					Slipper (1935)
U.S.A.										
Panhandle	91.3	3.2	1.7	0.9	0.56	0.1			0.5	Cotner and Crum (1935)
Hugoton	74.3	5.8	3.5	1.5	0.6				Av.[b]	Keplinger et al. (1949)
Trinidad										
Barrackpore	95.65	2.25	1.55	0.80	0.25					Suter (1952)
Venezuela										
La Concepcion	70.9	8.2	8.2	6.2	3.7	2.8				Anon (1948)
Cumarebo	63.89	9.49	14.41	8.8	5.4					Payne (1951)
New Zealand										
Kapuni	46.2	5.2	2.0	0.6	44.9	1.0				Cooper (1975)
Russia										
Baku	88.0	2.26	0.7	0.5	6.5					Shimkin (1950)
Abu Dhabi										
Zakum	76.0	11.4	5.4	2.2	1.3	2.3	1.1	0.3		Cooper (1975)
Iran										
Agha Jari	66.0	14.0	10.5	5.0	2.0	1.5	1.0			Cooper (1975)

[a] Tr = trace
[b] Av. = average

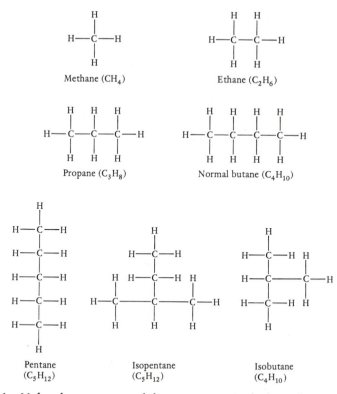

FIGURE 2.1 *Molecular structures of the more common hydrocarbon gases.*

only by the thermal maturation of organic matter. If they are recorded by a gas detector during the drilling of a well, it often indicates proximity to a significant hydrocarbon reservoir or source rock.

Nonhydrocarbon Gases

Inert Gases

Helium is a common minor accessory in many natural gases, and traces of argon and radon have also been found in the subsurface. Except for helium these gases are of no economic significance.

Helium occurs in the atmosphere at 5 ppm and has also been recorded in mines, hot springs, and fumaroles. It has been found in oil field gases in amounts of up to 8 percent (Dobbin, 1935). In North America helium-enriched natural gases occur in the Four Corners area and Texas panhandle of the United States and in Alberta and Saskatchewan, Canada (Lee, 1963; Hitchon, 1963). In Canada the major concentrations occur along areas of crustal tension, such as the Peace River and Sweetwater arches, and the foothills of the Rocky Mountains. Other regions containing helium-enriched natural gases include Poland, Alsace, and Queensland in Australia.

Helium is known to be produced by the decay of various radioactive elements, principally uranium, thorium, and radium. The rates of helium production for these elements are shown in Table 2.4. Rogers (1921) calculated that between 282 to 1.06 billion ft^3 of helium are generated annually. Based on these data, the helium found in natural gas is widely believed to have emanated from deep-seated basement rocks, especially granite. Although the actual rate of production is slow and steady, the expulsion of gas into the overlying sediment cover may occur rapidly when the basement is subjected to thermal activity or fracturing by crustal arching.

TABLE 2.4   Production rates of helium from various radioactive elements

Radioactive substance	Helium production g/(year)(mm^3)
Uranium	2.75×10^{-5}
Uranium in equilibrium with its products	11.0×10^{-5}
Thorium in equilibrium with its products	3.1×10^{-5}
Radium in equilibrium with emanation, radium A, and radium C	158

From Levorsen, 1967. Reprinted with permission.

Ideally it would be useful to demonstrate a correlation between helium-enriched natural gases and radioactive basement rocks. This correlation is generally difficult to establish because helium tends to occur in deep, rather than shallow, wells, and there is seldom sufficient well control to map the geology of the basement.

The major source of helium in the United States is the Panhandle-Hugoton field in Texas. This field locally contains up to 1.86 percent helium, and an extraction plant has been working since 1929 (Pippin, 1970). It is significant that this field produces helium from sediments pinching out over a major granitic fault block.

Helium also occurs from the breakdown of uranium ore bodies within sedimentary sequences; for example, at Castlegate in central Utah. Apart from helium, traces of other inert gases have been found in the subsurface. Argon occurs in the Panhandle-Hugoton field (as does radon) and in Japan. Argon and radon are by-products of the radioactive disintegration of potassium and radium, respectively, and are believed to have an origin similar to that of helium.

Nitrogen

Nitrogen is another nonhydrocarbon gas that frequently occurs naturally in the earth's crust. It is commonly associated both with the inert gases just described and with hydrocarbons. Nitrogen is found in North America in a belt stretching from New Mexico to Alberta and Saskatchewan. The Rattlesnake gas field of New Mexico is a noted example (Hinson, 1947) and has the following composition:

Nitrogen	72.6%
Methane	14.2%
Helium	7.6%
Ethane	2.8%
Carbon dioxide	2.8%
	100.0%

Note the association of nitrogen with helium. Nitrogen is also a common constituent of the Rotliegende gases of the southern North Sea basin. Published accounts show ranges from 1 to 14 percent in the West Sole and Groningen fields, respectively (Butler, 1975; Stauble and Milius, 1970). Individual wells in the German offshore sector are reported to have encountered far higher quantities.

The origin of nitrogen is less straightforward than that of the inert gases. Nitrogen has been recorded from volcanic emanations, but some studies have suggested that it may form organically, for example, by the bacterial degradation of nitrates via ammonia. Although chemically feasible, such biogenic nitrogen is likely to be produced only in shallow conditions, whereas in nature it occurs in deep hydrocarbon reservoirs. It has also been suggested that the thermal metamorphism of bituminous carbonates could generate both nitrogen and carbon dioxide. This mechanism has been postulated for the Beaverhill Lake Formation of Alberta and Saskatchewan (Hitchon, 1963).

Several pieces of information suggest that the major source of nitrogen is basement, or more specifically, igneous rock. First, the association of nitrogen with basement-derived inert gases has been already noted. Second, it has been stated that in the southern North Sea basin nitrogen content is unrelated to the variation of the coal rank of the Carboniferous beds that created the hydrocarbon gases (Eames, 1975). The amount of nitrogen in Rotliegendes reservoirs appears to increase eastward across the basin in the direction of increased igneous intrusives. Another significant case has been described from southwestern Saskatchewan by Burwash and Cumming (1974). Natural gas produced from basal Paleozoic clastics contains 97 percent nitrogen, 2 percent helium, and 1 percent carbon dioxide. No associated hydrocarbons are found. The accumulation overlies an up-

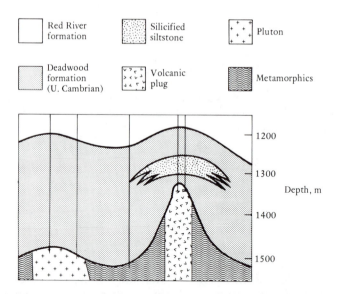

FIGURE 2.2 *Cross-section of the accumulation of nitrogen, helium, and carbon dioxide gases over a buried volcanic plug in southwestern Saskatchewan, Canada. (From Burwash and Cumming, 1974. Reprinted with permission.)*

standing volcanic plug in metamorphic basement (Fig. 2.2). Taken all together, these various lines of evidence suggest that nitrogen natural gases are predominantly of inorganic origin, although organic processes may be significant generators of atmospheric and shallow nitrogen. Furthermore, some atmospheric nitrogen may have been trapped in sediments during deposition, occurring now as *connate gas.*

Hydrogen

Free hydrogen gas rarely occurs in the subsurface, partly because of its reactivity and partly because of its mobility. Nonetheless, one or two instances are known. Over 1.36 trillion ft^3 of hydrogen gas have been discovered in Mississippian rocks of the Forest City basin, Kansas. Analysis of the gas discovered revealed a composition of 40 percent hydrogen, 60 percent nitrogen, and traces of carbon dioxide, argon, and methane (Anon, 1984).

Hydrogen is commonly dissolved in subsurface waters and in petroleum as traces, but it is seldom recorded in conventional analyses (Hunt, 1979). Subsurface hydrogen is probably produced by the thermal maturation of organic matter.

Carbon Dioxide

Carbon dioxide (CO_2) is often found as a minor accessory in hydrocarbon natural gases. It is also associated with nitrogen and helium, as previously presented data show. Natural gas accumulations in which carbon dioxide is the major constituent are common in areas of extensive volcanic activity, such as Sicily, Japan, New Zealand, and the Cordilleran chain of North America, from Alaska to Mexico. Levorsen (1967) cites a specific example from the Tensleep Sandstone (Pennsylvanian) of the Wertz Dome field, Wyoming:

Carbon dioxide	42.00%
Hydrocarbons	52.80%
Nitrogen	4.09%
Hydrogen sulfide	1.11%
	100.00%

Levorsen also quotes wells in New Mexico capable of producing up to 26 million ft^3/day of 99 percent pure carbon dioxide.

Both organic and inorganic processes can undoubtedly generate significant amounts of carbon dioxide in the earth's crust. Carbon dioxide is commonly recorded in natural gases associated with volcanic eruptions and earthquakes (Sugisaki et al., 1983). It may also be generated where igneous intrusives metamorphose carbonate sediments. Permeable limestones and dolomites can also yield carbon dioxide when they are invaded and leached by acid waters of either meteoric or connate origin. As we shall see in Chapter 5 (p. 190), carbon dioxide is a normal product of the thermal maturation of kerogen, generally being expelled in advance of the petroleum.

Carbon dioxide is also given off during the fermentation of organic matter (see p. 186), as well as by the oxidation of mature organic matter due to either fluid invasion or bacterial degradation. Specifically, methane in the presence of oxygenated water may yield carbon dioxide and water:

$$3CH_4 + 6O_2 = 3CO_2 + 6H_2O$$

Much of the carbon dioxide will remain in solution as carbonic acid. A duplex origin for carbon dioxide thus seems highly probable.

In the past, carbon dioxide was generally of limited economic value, being used principally for dry ice. The recent application of carbon dioxide to enhance oil recovery has increased its usefulness (Taylor, 1983).

Hydrogen Sulfide

Hydrogen sulfide (H_2S) occurs in the subsurface both as free gas and, because of its high solubility, in solution with oil and brine. It is a poisonous, evil-smelling gas, whose presence causes operational problems in both oil and gas fields. It is highly corrosive to steel, quickly attacking production pipes, valves, and flowlines. Gas or oil containing significant traces of hydrogen sulfide are referred to as *sour*—in contrast to *sweet*, which refers to oil or gas without hydrogen sulfide. Small amounts of hydrogen sulfide are economically deleterious in oil or gas because a washing plant must be installed to remove them: both to prevent corrosion and to render the residual gas safe for domestic combustion. Extensive reserves of sour gas can be turned to an advantage, however, since it may be processed as a source of free sulfur.

One analysis of a hydrogen-sulfide-rich gas from Emory, Texas, yielded the following composition (Anon, 1951):

Hydrogen sulfide	42.40%
Hydrocarbons	53.10%
Carbon dioxide	4.50%
	100.00%

Hydrogen sulfide is commonly expelled together with sulfur dioxide from volcanic eruptions. It is also produced in modern sediments in euxinic environments, such as the Dead Sea and the Black Sea. This process is achieved by sulfate-reducing bacteria working on metallic sulfates, principally iron, according to the following reaction:

$$2C + MeSO_4 + H_2O \rightarrow MeCO_3 + CO_2 + H_2S$$

where Me = metal

Note that sour gas generally occurs in hydrocarbon provinces where large amounts of evaporites are present. Notable examples include the Devonian basin of Alberta, the Paradox basin of Colorado, southwest Mexico, and extensive areas of the Middle East.

Anhydrite may be converted to calcite in the presence of organic matter, the reaction leading to the generation of hydrogen sulfide according to the following equation:

$$CaSO_4 + 2CH_2O = CaCO_3 + H_2O + CO_2 + H_2S$$
Organic matter

There is a further interesting point about sour gas. Not only is it associated with evaporites but it is frequently associated with carbonates, generally reefal, and lead-zinc sulfide ore bodies. A genetic link has long been postulated between these telethermal ores and evaporites (Davidson, 1965).

Dunsmore (1973) has shown that the reduction of anhydrite to calcite and sour gas can occur inorganically, without the aid of bacteria. The reaction is an exothermic one, which may generate the high temperatures needed to mobilize the metallic sulfides.

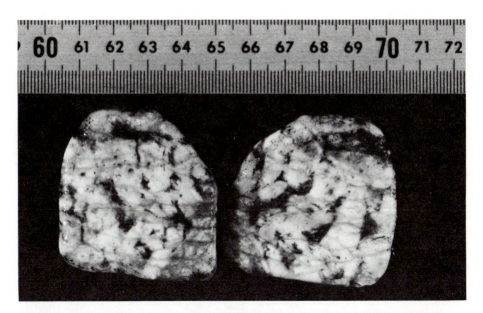

FIGURE 2.3 *Samples of massive gas hydrate cored by the Deep Sea Drilling Project. (Courtesy of Drs. R. von Huene and J. Aubouin and the Deep Sea Drilling Project, Scripps Institute of Oceanography.)*

GAS HYDRATES

Composition and Occurrence

Gas hydrates are compounds of frozen water that contain gas molecules. The ice molecules themselves are referred to as clathrates. Physically, hydrates look similar to white, powdery snow (Fig. 2.3) and have two types of unit structure. The small structure holds up to 8 methane molecules within 46 water molecules. This clathrate may contain not only methane but also ethane, hydrogen sulfide, and carbon dioxide. The larger clathrate consists of unit cells with 136 water molecules. This clathrate can hold the larger hydrocarbon molecules of the pentanes and *n*-butanes.

Gas hydrates occur only in very specific pressure-temperature conditions. Figure 2.4 shows that they are stable at high pressures and low temperatures, the pressure required for stability increasing logarithmically for a linear thermal gradient. Gas hydrates occur in shallow arctic sediments and in deep oceanic deposits. Gas hydrates in arctic permafrost have been described from Alaska and Siberia. In Alaska they occur between about 750 to 3500 m (Holder et al., 1976). In Siberia, where the geothermal gradient is lower, they extend down to about 3100 m (Makogon et al., 1971).

Hydrates have been found in the sediments of many of the oceans around the world. They have been recovered in the DSDP borehole cores, and their presence has been inferred from seismic data (e.g., Moore and Watkins, 1979; MacLeod, 1982). Specifically, they have been recognized

23

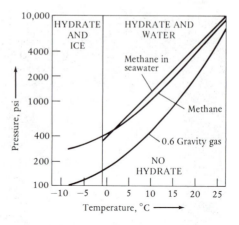

FIGURE 2.4 *Pressure-temperature graph showing the stability field of gas hydrates. (After Hunt, 1979.)*

from *bright spots* (see p. 109) on seismic lines in water depths of 1000 to 2500 m off the eastern coast of North Island, New Zealand (Katz, 1981), and in water depths of 1000 to 4000 m in the western North Atlantic (Dillon et al., 1980).

Gas hydrates have generally been attributed to a shallow biogenic origin (e.g., Hunt, 1979). Based on their carbon and helium isotope ratios, MacDonald (1983) has postulated a crustal inorganic origin for hydrates.

Identification and Economic Significance

As just noted, the presence of gas hydrates can be suspected, but not proved, from seismic data. The lower limit of hydrate-cemented sediment is often concordant with bathymetry, because the velocity contrast between the gas-hydrate-cemented sediment and underlying noncemented sediment is large enough to generate a detectable reflecting horizon. The bottom simulating reflector may appear as a *bright spot* (see p. 109), which cross-cuts bedding-related reflectors. Care must be taken to distinguish gas hydrate bottom reflectors from ordinary seabed multiples (Fig. 2.5). Basal hydrate reflectors commonly increase in sub-bottom depth with increasing water depth because of the decreasing temperature of the water above the sea floor.

The presence of gas hydrates can only be proved, however, by engineering data. The penetration rate of the bit is low when drilling clathrate-cemented sediments. Pressure core barrels provide the ultimate proof, since gas hydrates show a characteristic pressure decline when pressure cells are brought to the surface and opened (Stoll et al., 1971). Gas hydrates also show certain characteristic log responses (Bily and Dick, 1974). They have high resistivity and acoustic velocity, coupled with low density (Fig. 2.6).

FIGURE 2.5 *Seismic line from offshore New Zealand showing* bright spot, *which is interpreted as the lower surface of a gas hydrate accumulation. (From Katz, 1981. Reprinted with permission.)*

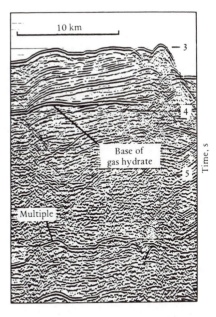

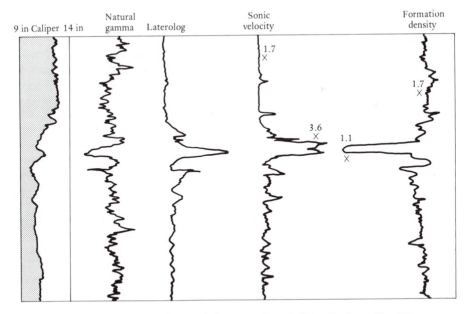

FIGURE 2.6 *Wireline log of part of the Deep Sea Drilling Project, Site 570, showing a gas hydrate zone. It is characterized by high resistivity and acoustic velocity, and low radioactivity and density. Crosses show velocities and densities in hydrate-bearing and normal intervals. (Courtesy of Drs. R. von Huene and J. Aubouin and the Deep Sea Drilling Project, Scripps Institute of Oceanography.)*

Large areas of the arctic permafrost and of the ocean floors contain vast reserves of hydrocarbon gas locked up in clathrate deposits. Clathrates can hold six times as much gas as can an open, free, gas-filled pore system. Unfortunately, gas hydrates present considerable production problems, which have yet to be overcome. These problems are due, in part, to the low permeability of the reservoir and, in part, to chemical problems concerning the release of gas from the crystals. Clathrate deposits may be of indirect economic significance, however, by acting as cap rocks. Because of their low permeability, they form seals that prevent the upward movement of free gas.

CRUDE OIL

Crude oil is defined as "a mixture of hydrocarbons that existed in the liquid phase in natural underground reservoirs and remains liquid at atmospheric pressure after passing through surface separating facilities" (joint API, AAPG, and SPE definition). In appearance crude oils vary from straw yellow, green, and brown to dark brown or black in color. Oils are naturally oily in texture and have widely varying viscosities. Oil on the surface tend to be more viscous than oils in warm subsurface reservoirs. Surface viscosity values range from 1.4 to 19,400 centistokes and vary not only with temperature but also with the age and depth of the oil (see p. 32).

Most oils are lighter than water. Although the density of oil may be measured as the difference between its specific gravity and that of water, it is often expressed in gravity units defined by the American Petroleum Institute according to the following formula:

$$°\text{API} = \frac{141.5}{\text{specific gravity 60/60F}} - 131.5$$

where 60/60F is the specific gravity of the oil
at 60°F compared with that of water at 60°F.

Note that API degrees are inversely proportional to density. Thus light oils have API gravities of over 40° (0.83 specific gravity), whereas heavy oils have API gravities of less than 10° (1.0 specific gravity). Heavy oils are thus defined as those oils that are more dense than water. Variations of oil gravity with age and depth are discussed on page 374. Oil viscosity and API gravity are generally inversely proportional to one another.

Chemistry

In terms of elemental chemistry, oil consists largely of carbon and hydrogen with minor amounts of oxygen, nitrogen, and sulfur. Oil also contains

TABLE 2.5   Elemental composition of crude oils by weight %

Element	Minimum	Maximum
Carbon	82.2	87.1
Hydrogen	11.8	14.7
Sulfur	0.1	5.5
Oxygen	0.1	4.5
Nitrogen	0.1	1.5
Other	Trace	0.1

From Levorsen, 1967. Reprinted with permission.

traces of vanadium, nickel, and other elements (see Table 2.5). Although the elemental composition of oils is relatively straightforward, there may be an immense number of molecular compounds. No two oils are identical either in the compounds contained or in the various proportions present. However, certain compositional trends are related to the age, depth, source, and geographical location of the oil. Concoco's Ponca City crude oil from Oklahoma, for example, contains at least 234 compounds (Rossini, 1960). For convenience the compounds found in oil may be divided into two major groups: (1) the hydrocarbons, which contain three major subgroups, and (2) the heterocompounds, which contain other elements. These various compounds are described in the following sections.

Paraffins

The paraffins, often called alkanes, are saturated hydrocarbons, with a general formula C_nH_{2n+2}. For values of $n < 5$ the paraffins are gaseous at normal temperatures and pressures. These compounds (methane, ethane, propane, and butane) are discussed in the section on natural gas (p. 15). For values of $n = 5$ (pentane, C_5H_{12}) through to $n = 15$ (pentadecane, $C_{15}H_{32}$) the paraffins are liquid at normal temperatures and pressures; and for values of $n > 15$ they grade from viscous liquids to solid waxes.

Two types of paraffin molecules are present within the series, both having similar atomic compositions (isomers), which increase in molecular weight along the series by the addition of CH_2 molecules. One series consists of straight-chain molecules, the other of branched-chain molecules. These structural differences are illustrated in Figure 2.7.

The paraffins occur abundantly in crude oil, the normal straight-chain varieties dominating over the branched-chain structures. Individual members of the series have been recorded up to $C_{78}H_{158}$. For a given molecular weight the normal paraffins have higher boiling points than do equivalent weight isoparaffins.

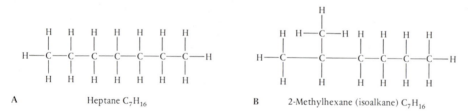

A   Heptane C₇H₁₆   B   2-Methylhexane (isoalkane) C₇H₁₆

FIGURE 2.7 *The isomeric structure of the paraffin series. Octane may have a (A) normal chain structure or a (B) branched-chain isomer.*

The paraffins are the major constituents of hydrocarbon gases. They are also quantitatively significant in light gasoline and kerosene oils, making up to 30 percent and 25 percent of the oil, respectively. With increasing boiling point the paraffin fraction of oils gradually decreases.

Naphthenes

The second major group of hydrocarbons found in crude oils is the naphthenes, or cycloalkanes. This group has a general formula C_nH_{2n}. Like the paraffins they occur in a homologous series consisting of five- and six-membered carbon rings termed the cyclopentanes and cyclohexanes, respectively (Fig. 2.8). Unlike the paraffins all the naphthenes are liquid at normal temperatures and pressures. They make up about 40 percent of both light and heavy crude oils.

FIGURE 2.8 *Examples of the molecular structures of five- and six-ring naphthenes (cycloalkanes). For convenience the accompanying hydrocarbon atoms are not shown.*

FIVE-RING SERIES   SIX-RING SERIES

Cyclopentane   Cyclohexane
(C_5H_{10}) (C_6H_{12})

Cyclohexane   Ethyl cyclohexane
(C_6H_{12}) (C_8H_{16})

Aromatics

Aromatic compounds are the third major group of hydrocarbons commonly found in crude oil. Their molecular structure is based on a ring of six carbon atoms. The simplest member of the family is benzene (C_6H_6), whose structure is shown in Figure 2.9. One major series of the aromatic compounds is formed by substituting hydrogen atoms by alkane (C_nH_{2n+2}) molecules. This alkyl benzene series includes ethyl benzene (C_6H_5, C_2H_5) and toluene ($C_6H_5CH_3$). Another series is formed by straight- or branched-chain carbon rings. This series includes naphthalene ($C_{10}H_8$) and anthracene ($C_{14}H_{10}$). The aromatic hydrocarbons include asphaltic compounds. These compounds are divided into the resins, which are soluble in *n*-pentane, and the asphaltenes, which are not.

The aromatic hydrocarbons are liquid at normal temperatures and pressures (the boiling point of benzene is 80.5°C). They are present in relatively minor amounts (about 10 percent) in light oils, but increase in quantity with decreasing API gravity to over 30 percent in heavy oils. Toluene ($C_6H_5CH_3$) is the most common aromatic component of crude oil, followed by the xylenes ($C_6H_4(CH_3)_2$) and benzene.

FIGURE 2.9 *Molecular structure of aromatic hydrocarbons commonly found in crude oil.*

Usual representation of benzene ring

Benzene (C_6H_6)

Toluene ($C_6H_5CH_3$)

Dimethyl benzene
(one of three isomers of xylene)

Heterocompounds

Crude oil contains many different heterocompounds that contain elements other than hydrogen and carbon. The principal ones are oxygen, nitrogen, and sulfur, together with rare metal atoms, commonly nickel and vanadium.

The oxygen compounds range between 0.06 and 0.4 percent by weight of most crudes. They include acids, esters, ketones, phenols, and alcohols. The acids are especially common in young, immature oils and include fatty acids, isoprenoids, and naphthenic and carboxylic acids. The presence of steranes in some crudes is an important indication of their organic origin. Nitrogen compounds range between 0.01 and 0.9 percent by weight of most crudes. They include amides, pyridines, indoles, and pyroles.

Sulfur compounds range from 0.1 to 7.0 percent by weight in crude oils. This figure represents genuine sulfur-containing carbon molecules; it does not include hydrogen sulfide gas (H_2S) or native sulfur. Elemental sulfur occurs in young, shallow oils (above 100°C sulfur combines with hydrogen to form hydrogen sulfide). Five main series of sulfur-bearing compounds are in crude oil: alkane thiols (mercaptans), the thio alkanes (sulfides), thio cycloalkanes, dithio alkanes, and cyclic sulfides.

Traces of numerous other elements have been found in crude oils, but determining whether they occur in genuine organic compounds or whether they are contaminants from the reservoir rock, formation waters, or production equipment is difficult. Hobson and Tiratsoo (1975) have tabulated many of these trace elements. Their lists include common rock-forming elements, such as silicon, calcium, and magnesium, together with numerous metals, such as iron, aluminum, copper, lead, tin, gold, and silver. Of particular interest is the almost ubiquitous presence of nickel and vanadium. Unlike with the other metals it can be proved that nickel and vanadium occur not as contaminants but as actual organometallic compounds, generally in porphyrin molecules. The porphyrins contain carbon, nitrogen, and oxygen, as well as a metal radical. The presence of porphyrins in crude oil is of particular genetic interest because they may be derived either from the chlorophyll of plants or the hemoglobin of blood. Data compiled by Tissot and Welte (1978) show that the average vanadium and nickel content of 64 crude oils is 63 ppm and 18 ppm, respectively. The known maximum values are 1200 ppm vanadium and 150 ppm nickel in the Boscan crude of Venezuela.

Metals tend to be associated with resin, sulfur, and asphaltene fractions of crude oils. Metals are rare in old, deep marine oils and are relatively abundant in shallow, young, or degraded crudes. Figure 2.10 illustrates the abundance of the various compounds occurring in crude oils.

Classification

Many schemes have been proposed to classify the various types of crude oils. Broadly speaking, the classifications fall into two categories: (1) those pro-

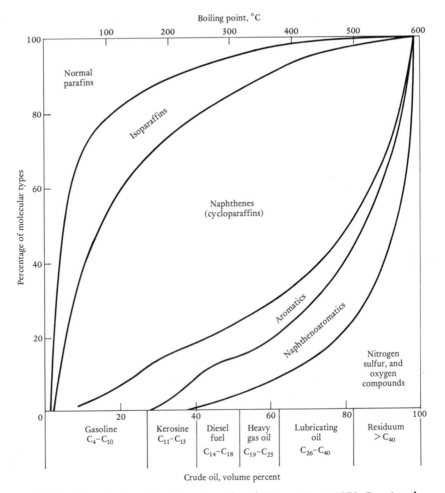

FIGURE 2.10 *Chemical composition of crude oil. (From Hunt, 1979. Reprinted with permission.)*

posed by chemical engineers interested in refining crude oil and (2) those devised by geologists and geochemists as an aid to understanding the source, maturation, history, or other geological parameters of crude oil occurrence.

The first type of classification is concerned with the quantities of the various hydrocarbons present in a crude and their physical properties, such as viscosity and boiling point. For example, the *n.d.M.* scheme is based on refractive index, density, and molecular weight (*n* = refractive index, *d* = density, and *M* = molecular weight).

Classificatory schemes of interest to geologists are concerned with the molecular structures of oils, because these may be keys to their source and geological history. One of the first schemes was developed in the U.S.

TABLE 2.6   Classification of crude oils

Concentration in crude oil ($> 210°C$)					
S = saturates AA = aromatics + resins + asphaltenes	P = paraffins N = naphthenes	Crude oil type	Sulfur content in crude oil (approximate)	Number of samples per class (total = 541)	
S > 50% AA < 50%	P > N and P > 40%	Paraffinic		100	
	P ⩽ 40% and N ⩽ 40%	Paraffinic- naphthenic	< 1%	217	
	N > P and N > 40%	Naphthenic		21	
S ⩽ 50% AA ⩾ 50%	P > 10%	Aromatic intermediate	> 1%	126	
	P ⩾ 10% N ⩽ 25%	Aromatic asphaltic		41	
	N ⩾ 25%	Aromatic naphthenic	Generally S < 1%	36	

From Tissot and Welte, 1978. Reprinted with permission.

Bureau of Mines (Smith, 1927; Lane and Garton, 1935). It classifies oils into paraffinic, naphthenic, and intermediate types according to their distillate fractions at different temperatures and pressures. Later, Sachenen (1945) devised a scheme that also included asphaltic and aromatic types of crude.

A more recent scheme by Tissot and Welte (1978) is based on the ratio between paraffins, naphthenes, and aromatics, including asphaltic compounds (Table 2.6 and Fig. 2.11). The great advantage of this classification is that it can also be used to demonstrate the maturation and degradation paths of oil in the subsurface (see p. 374).

A scheme based more on geological occurrence was proposed by Biederman (1965). This empirical classification is based on an arbitrarily defined depth and age:

1. Mesozoic and Cenozoic oil at less than 600 m.
2. Mesozoic and Cenozoic oil at over 3000 m.
3. Paleozoic oil at less than 600 m.
4. Paleozoic oil at more than 3000 m.

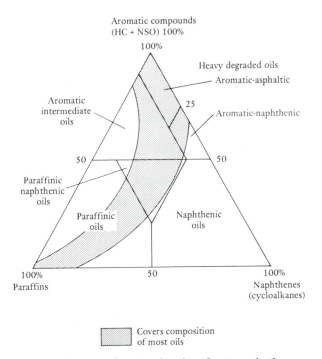

FIGURE 2.11 *Ternary diagram showing the classification of oils proposed by Tissot and Welte (1978).*

Essentially, this scheme recognized four classes of oil: young-shallow, young-deep, old-shallow, and old-deep.

A detailed study of many oils by Martin et al. (1963) showed a number of statistically significant chemical and physical differences between these groups. Young-shallow oils tend to be heavy and viscous. They are generally sulfurous and relatively low in paraffins and rich in aromatics. Young-deep oils, on the other hand, are less viscous, of higher API gravity, more paraffinic, and low in sulfur content. Old-shallow oils are broadly comparable to young-deep crudes in purity, viscosity, and paraffinic nature. Like young-shallow oil, however, they tend to be relatively sulfurous. Old-deep oils tend to have the lowest viscosity, density, and sulfur content of the four groups.

This system is very much a rule-of-thumb classification, and the variations of crude oils based on geological parameters are discussed further in Chapter 8. In particular, note that oils vary not only with age and depth but also with variations in their source rock and the degree of degradation to which they have been subjected. This degradation causes very wide variations, especially in shallow oils.

SELECTED BIBLIOGRAPHY

HUNT, J. M. 1979. *Petroleum Geochemistry and Geology*. San Francisco: Freeman. Chapter 3 (pp. 28–66) gives a detailed account of the chemistry of oil and gas.

KINGHORN, R. F. K. 1983. *An Introduction to the Physics and Chemistry of Petroleum*. New York: Wiley, 432 pp.

NEUMANN, I. B., PAZYNSKA-LAHME, and SEVERIN, D. 1981. *Composition and Properties of Petroleum*. London: Wiley, 148 pp.

TIRATSOO, E. N. 1972. *Natural Gas*, Second Edition. Beaconsfield: Scientific Press. Chapter 1 (pp. 1–38) gives a good account of the occurrence of natural gases, hydrocarbons, and others.

TISSOT, B. P. and WELTE, D. H. 1978. *Petroleum Formation and Occurrence*. Berlin: Springer-Verlag. Part IV, Chapter 1 (pp. 333–368) gives a detailed account of the chemistry of oil and gas.

REFERENCES

ANON. 1948. Oil fields of Royal Dutch-Shell Group in Western Venezuela. *Am. Assoc. Petrol. Geol. Bull.*, 32, 595.

ANON. 1951. Sour gas discovery at Emory field, Texas. *Oil and Gas J.*, 24 May, 82.

BARKER, C., HINCH, H. H., JONES, R. W., McAULIFFE, C., MOMBER, J., and PRICE, L. 1978. *Physical and Chemical Constraints on Petroleum Migration*. Continuing Education Course Note Series, No. 8. Tulsa: Am. Assoc. Petrol. Geol., 293 pp.

BIEDERMAN, E. W. 1965. Crude oil composition—a clue to migration. *World Oil*, No. 161, 78–82.

BILY, C. and DICK, J. W. L. 1974. Naturally occurring gas-hydrates in the Mackenzie Delta, N.W.T. *Can. Petrol. Geol. Bull.*, 22, 340–352.

BURWASH, R. A. and CUMMING, G. L. 1974. Helium source-rock in southwestern Saskatchewan. *Can. Petrol. Geol. Bull.*, 22, 405–412.

BUTLER, J. B. 1975. The West Sole Gas Field. In: *Petroleum and the Continental Shelf of Northwest Europe*, vol. 1. Geology. A. W. Woodland (ed.). London: Applied Science Publishers, 213–222.

COTNER, V. and CRUM, H. E. 1935. Natural gas in Amarillo District, Texas. In: *Geology of Natural Gas*. Tulsa: Am. Assoc. Petrol. Geol., 409.

CUMMING, A. D. and WYNDHAM, C. L. 1975. The geology and development of the Hewett gas field. In: *Petroleum and the Continental Shelf of Northwest Europe*. A. W. Woodland (ed.). London: Applied Science Publishers, 313–326.

DAVIDSON, C. F. 1965. A possible mode of strata-bound copper ores. *Econ. Geol.*, 60, 942–954.

DILLON, W. P., GROW, J. A., and PAULL, C. K. 1980. Unconventional gas hydrate seals may trap gas off Southeast U.S.A. *Oil & Gas J.*, 7 Jan., 124–130.

DOBBIN, C. E. 1935. Geology of natural gases rich in helium, nitrogen, carbon dioxide and hydrogen sulphide. In: *Geology of Natural Gas*. Tulsa: Am. Assoc. Petrol. Geol., 1053–1064.

DUNSMORE, H. E. 1973. Diagenetic processes of lead-zinc emplacement in carbonates. *Trans. Inst. Miner. Metall.*, Section B, *82*, B. 168–173.

EAMES, T. D. 1975. Coal rank and gas source relationships—Rotliegendes reservoirs. In: *Petroleum and the Continental Shelf of Northwest Europe*, vol. 1. *Geology*. A. W. Woodland (ed.). London: Applied Science Publishers, 191–204.

HINSON, H. H. 1947. Reservoir characteristics of Rattlesnake oil and gas field, San Juan County, N. Mexico. *Am. Assoc. Petrol. Geol. Bull.*, *31*, 731–771.

HITCHON, B. 1963. Geochemical studies of natural gas, part III. Inert gases in western Canada natural gases. *J. Can. Petrol. Technol.*, 2, 165–174.

HOLDER, G. D., KATZ, D. L., and HAND, J. H. 1976. Hydrate formation in subsurface environments. *Am. Assoc. Petrol. Geol. Bull.*, 60, 981–988.

HUNT, J. M. 1979. *Petroleum Geochemistry and Geology*. San Francisco: Freeman, 617 pp.

IKORSKIY, S. V. 1967. *Organic substances in minerals of igneous rocks in the Khibina massif.* (Translated from the Russian by L. Shapiro and I. A. Breger). McLean, VA: Clark Co., 155 pp.

KATZ, D. L. D., CORNELL, P., KOBAYASHI, R., POETTMAN, F. H., VARY, J. A., ELENBAAS, J. R., and WEINANG, C. F. 1959. Water-hydrocarbon systems. In: *Handbook of Natural Gas Engineering*. D. L. Katz et al. (eds.). New York: McGraw-Hill, 189–221.

KATZ, H. R. 1981. Probable gas hydrate in continental slope east of the North Island, New Zealand. *J. Petrol. Geol.*, 3, 315–324.

KEPLINGER, C. H., WANEMACHER, C. H., and BURNS, K. R. 1949. Hugoton, worlds largest dry gas field in amazing development. *Oil and Gas J.*, 6 Jan., 84–88.

LANE, E. C. and GARTON, E. L. 1935. 'Base' of a crude oil. *U.S. Bur. Mines Rep.*, No. 3279.

LEE, H. 1963. The technical and economic aspects of helium production in Saskatchewan. *J. Can. Petrol. Technol.*, 2, 16–27.

LEVORSEN, A. I. 1967. *Geology of Petroleum*, Second Edition. San Francisco: Freeman, 724 pp.

MacDONALD, G. J. 1983. The many origins of natural gas. *J. Petrol. Geol.*, 5, 31–362.

MacLEOD, M. K. 1982. Gas hydrates in ocean bottom sediments. *Am. Assoc. Petrol. Geol. Bull.*, 66, 2649–2662.

MAGLIONE, P. R. 1970. Triassic gas field of Hassi er R'Mel, Algeria. In: *Geology of Giant Petroleum Fields*. M. T. Halbouty (ed.). Am. Assoc. Petrol. Geol., Mem. No. 14, 489–501.

MAKOGON, Y. F., TREBIN, F. A., TROFIMUK, A. A., TSAREV, V. P., and CHERSKIY, N. V. 1971. Detection of a pool of natural gas in a solid (hydrated gas) state. *Dokl. Akad. Nauk. SSSR.*, *196*, 197–200.

MARTIN, R. L., WINTERS, J. C., and WILLIAMS, J. A. 1963. Composition of crude oil by gas chromatography: geological significance of hydrocarbon distribution. Frankfurt: *Proc. 6th World Petrol. Cong.*, Section 5, Paper 13.

MASSA, D., RUHLAND, M., and THOUVENIN, J. 1972. Structure et fracturation du Champ d'Hassi Messaoud (Algerie) Pt. 1. Observation des phenomenes tectoniques au Tassili des Ajjers. *Rev. Institut Francais du Petrol.*, 27, 489–534.

MOORE, J. C. and WATKINS, J. S. 1979. Middle America trench. *Geotimes*, 24, 20–22.

PAYNE, A. L. 1951. Cumarebo oil field, Falcon, Venezuela. *Am. Assoc. Petrol. Geol. Bull.*, *35*, 1870.

PIPPIN, L. 1970.   Panhandle-Hugoton field, Texas–Oklahoma–Kansas—the first fifty years. In: *Geology of Giant Petroleum Fields*. M. T. Halbouty (ed.). Tulsa: Am. Assoc. Petrol. Geol., Mem. No. 14, 204–222.

PORFIR'EV, V. B. 1974.   Inorganic origin of petroleum. *Am. Assoc. Petrol. Geol. Bull.*, *58*, 3–33.

ROGERS, G. S. 1921.   *Helium-Bearing Natural Gas*. U.S. Geol. Surv., 113 pp.

ROSSINI, F. D. 1960.   Hydrocarbons in petroleum. *J. Chem. Ed.*, *37*, 554–561.

SACHENEN, A. N. 1945.   *Chemical Constituents of Petroleum*. New York: Rheinhold, 414 pp.

SHIMKIN, D. B. 1950.   Is petroleum a Soviet weakness? *Oil & Gas J.*, 21 Dec., 214–226.

SLIPPER, S. E. 1935.   Natural Gas in Alberta. In: *Geology of Natural Gas*. Tulsa: Am. Assoc. Petrol. Geol., 51.

SMITH, N. A. C. 1927.   The interpretation of crude oil analyses. *U.S. Bur. Mines Rep.*, No. 2806.

STAUBLE, A. J. and MILIUS, B. 1970.   Geology of Groningen gas field, Netherlands. In: *Geology of Giant Petroleum Fields*. M. T. Halbouty (ed.). Tulsa: Am. Assoc. Petrol. Geol., Mem. No. 14, 359–369.

STOLL, R. D., EWING, J., and BRYAN, G. M. 1971.   Anomalous wave velocities in sediments containing gas hydrates. *J. Geophys. Res.*, *76*, 2090–2094.

SUGISAKI, R., ID, M. T., TAKEDA, H., and ISOBE, Y. 1983.   Origin of hydrogen and carbon dioxide in fault gases and its relation to fault activity. *J. Geol.*, *91*, 239–258.

SUTER, H. H. 1952.   The general and economic geology of Trinidad. In: *Colonial Geol. & Miner. Res.*, *4*, 28 pp.

TAYLOR, G. 1983. CO_2 projects to test recovery theories. *AAPG Explorer*, June, 1, 20–21.

TIRATSOO, E. N. 1972.   *Natural Gas—A Study*, Second Edition. Beaconsfield: Scientific Press, 352 pp.

Methods of Exploration

WELL DRILLING AND COMPLETION

In the earliest days of oil exploration, oil was collected from surface seepages. In China, Burma, and Romania mine shafts were dug to produce shallow oil. Access was gained by ladders or hoist, and air was pumped down to the mines through pipes. Oil seeped into the shaft and was lifted to the surface in buckets. This type of technology was not conducive to a healthy life and long retirement for the miners.

Oil has also been mined successfully in several parts of the world by driving horizontal adits into reservoirs. Oil dribbles down the walls onto the floor and flows down to the mine entrance. This technique has recently been reintroduced by Conoco in the North Tisdale field of Wyoming (Dobson and Seelye, 1981). Conventionally, however, oil and gas are located and produced by drilling boreholes. The two methods, cable-tool and rotary-tool drilling, are briefly described in the following sections.

Cable-Tool Drilling

Cable-tool drilling seems to have developed spontaneously in several parts of the world. In the early nineteenth century the Chinese were sinking shafts to depths of some 700 m using an 1800-kg bit suspended from a rattan cord. Hole diameters were of the order of 10 to 15 cm, and the rate of penetration was about 60 to 70 cm/day. The wells were cased with bamboo or hollow cypress trunks (Imbert, 1828; Coldre, 1891). These wells were sunk in the search for fresh water and for brines from which salt could be extracted.

In modern cable-tool drilling a heavy piece of metal, termed the *bit*, is banged up and down at the end of a cable on the bottom of the hole. The bit is generally chisel-shaped. This repeated percussion gradually chips away the rock on the bottom of the hole. Every now and then the bit is withdrawn to the surface and a bailer is fitted to the end of the cable. This bailer is a steel cylinder with a one-way flap at the bottom. As the bailer is dropped on the floor of the hole, chips of rock are forced into it through the trap flap. When the bailer is lifted prior to the next drop, the rock cuttings are retained in the trap (Fig. 3.1). When the cuttings have been removed from the borehole, the bailer is drawn out and emptied. The bit is then put back on the end of the cable, and percussion recommences. As the hole is gradually deepened, the sides have a natural tendency to cave in. This tendency is counteracted by lining the hole with steel casing.

In the early days of oil exploration the percussive power of the cable

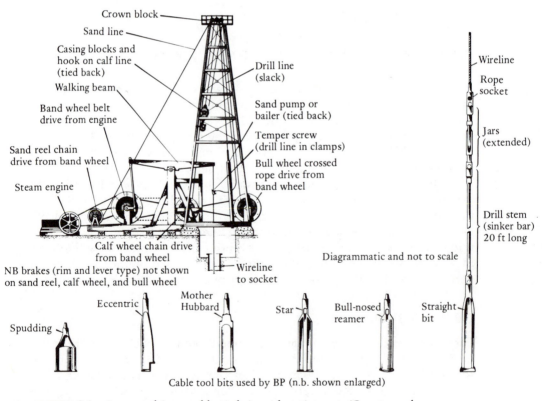

FIGURE 3.1 *A steam-driven cable-tool rig and equipment. (Courtesy of British Petroleum.)*

tool was provided by a man or men pulling on the rope or, later, aided by a spring-pole. In more recent times, however, motive power was provided by a steam or internal combustion engine. The cable-tool drilling method evolved toward the end of the eighteenth century. It was then used primarily for the sinking of water wells. Occasionally such wells would find water contaminated with oil, to the displeasure of the driller.

When the economic uses of oil were discovered in the mid-nineteenth century, however, the cable-tool rig became the prime method of sinking oil wells and it remained so for some 80 years. Cable-tool drilling has several major mechanical constraints. First, the depth to which one may drill is severely limited. The deeper the hole, the heavier is the cable. There comes a point, therefore, when the cable at the well head is not strong enough to take the combined weight of the bit and the downhole cable. Although cable-tool rigs have drilled to over 3000 m, the average capability is about 1000 m. This capability is adequate for most water wells, but too shallow for the increased depths required for oil exploration.

A further limitation of the cable-tool method is that it can only work in an open hole. The cable must be free to move, so it is not possible to keep the hole full of fluid. The bailer removes water that oozes into the hole. Thus when the bit breaks through into a high-pressure formation, the oil or gas shoots up to the surface as a *gusher*.

Because of these limitations of penetration depth and safety, cable-tool drilling is of limited use in petroleum exploration.

Rotary Drilling

Because of the greater safety and depth penetration of rotary drilling, it has largely superseded the cable-tool method for deep drilling in the oil industry (Fig. 3.2). In this technique the bit is rotated at the end of a hollow steel tube called the *drill string*. Many types of bit are used, but the most common consists of three rotating cones set with teeth (Fig. 3.3). The bit is rotated and the teeth gouge or chip away the rock at the bottom of the borehole. Simultaneously, mud or water is pumped down the drill string, squirting out through nozzles in the bit and flowing up to the surface between the drill string and the wall of the hole.

This circulation of the drilling mud has many functions: It removes the rock cuttings from the bit; it removes cavings from the borehole wall; it keeps the bit cool; and, most importantly, it keeps the hole safe. The hydrostatic pressure of the mud generally prevents fluid from moving into the hole, and if the bit penetrates a formation with a high-pore pressure, the weight of the mud may prevent a gusher. A gusher can also be prevented by sealing the well head with a series of valves termed the *blowout preventers* or *BOPs*.

As the bit deepens the hole, new joints of drill pipe are screwed on to the drill string at the surface. The last length of drill pipe is screwed to a

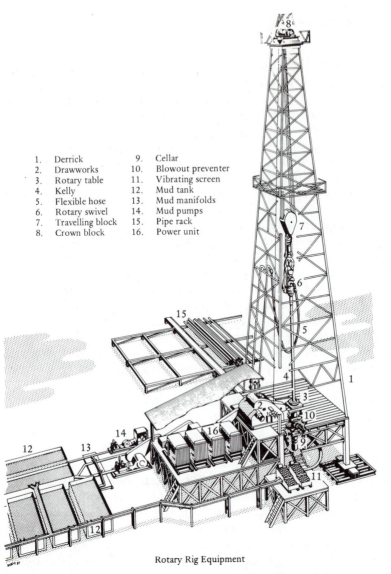

1.	Derrick	9.	Cellar
2.	Drawworks	10.	Blowout preventer
3.	Rotary table	11.	Vibrating screen
4.	Kelly	12.	Mud tank
5.	Flexible hose	13.	Mud manifolds
6.	Rotary swivel	14.	Mud pumps
7.	Travelling block	15.	Pipe rack
8.	Crown block	16.	Power unit

Rotary Rig Equipment

FIGURE 3.2 *Simplified sketch of an onshore derrick for rotary drilling. (Courtesy of British Petroleum.)*

square-section steel member called the *kelly*, which is suspended vertically in the *kelly bushing*, a square hole in the center of the rotary table. Thus rotation of the table by the rig motors imparts a rotary movement down the drill string to the bit at the bottom of the hole. As the hole deepens, the kelly slides down through the rotary table until it is time to attach another length of drill pipe (Fig. 3.4). When the bit is worn out, which depends on the type of bit and the hardness of the rock, the drill pipe is drawn out of the hole and stacked in the derrick. When the bit is brought to the surface, it is removed and a new one fitted.

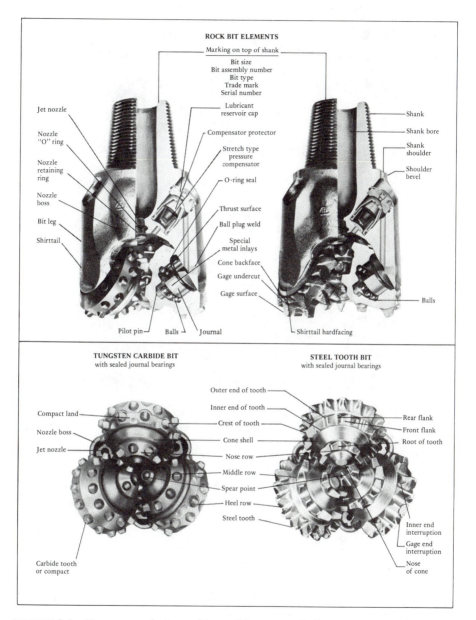

FIGURE 3.3 *Two types of tricone bit used for rotary drilling. (Courtesy of Hughes Tool Co.)*

FIGURE 3.4   *Applying the tong to make up the drill pipe on the rig floor of British Petroleum's rig, Sea Conquest. (Courtesy of British Petroleum.)*

After drilling for some depth, the borehole is lined with steel casing, and cement is set between the casing and the borehole wall. Drilling may then recommence with a narrower gauge of bit. The diameters of bits and casing are internationally standardized. Depending on the final depth of the hole, several diameters of bit will be used with the appropriate casing (Fig. 3.5). The average depth of an oil well is between 1 to 3 km, but depths of up to 11.5 km can be penetrated.

When drilling into a reservoir, an ordinary bit may be removed and replaced by a core barrel. A core barrel is a hollow steel tube with teeth, commonly diamonds, at the downhole end. As the core barrel rotates, it cuts a cylinder of rock and descends over it as the hole deepens. When the core barrel is withdrawn to the surface, the core of rock is retained in the core barrel by upward-pointing steel springs. Coring is slower than drilling with an ordinary bit and is thus more expensive. It is only used sparingly in hydrocarbon exploration to collect large, intact rock samples for geological and engineering information.

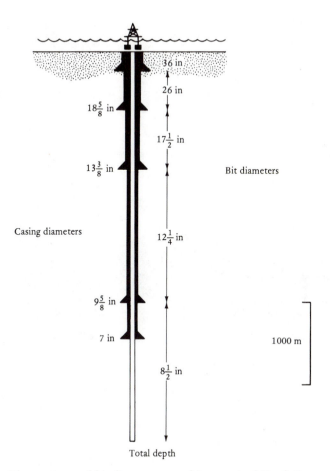

FIGURE 3.5 *The casing and bit diameters used in a typical North Sea well.*

Geologists are involved to varying degrees in drilling wells. In routine oil field development wells, where the depth and characteristics of the reservoir are already well known, geologists may not be present at the well site, although they will monitor the progress of the well from the office. On an offshore wildcat well, however, there may be five or more geologists. The oil company for whom the well is being drilled will have one of their own well site geologists on board. His or her duties include advising both the driller and the operational headquarters on the formations, fluids, and pressures to be anticipated; picking casing and coring points; deciding when to run wireline logs; supervising logging and interpreting the end result; and, most importantly, identifying and evaluating hydrocarbon shows in the well.

The other geologists are *mudloggers*, working in pairs on 12-hr shifts. The mudlogging company is contracted by the oil company to carry out a

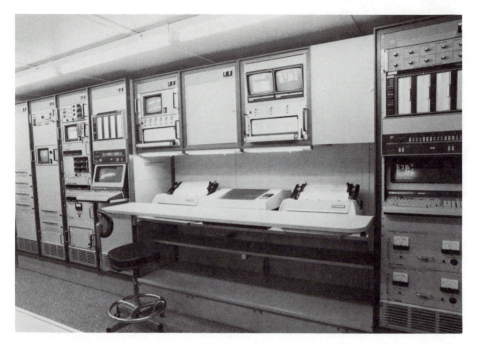

FIGURE 3.6   *The inside of a mudlogging unit. (Courtesy of Exploration Logging Services, Ltd.)*

complex and continuous evaluation of the well as it is drilled. This evaluation is done from a mudlogging unit, which is a cabin or caravan-trailer packed with computers and electronic monitors (Fig. 3.6). The mudloggers record many drilling variables, including the rate of penetration, mud temperature, pore pressure, and shale density. Gas chromatographs monitor the presence of hydrocarbon gases in the mud. Samples of the well cuttings are collected at specified depth intervals, checked with ultraviolet light and other tests for the presence of oil, identified, and described. These data are continuously plotted on a mudlog (Fig. 3.7). A useful account of the geological aspects of mudlogging and associated rig activities is given by Dickey (1979).

Various Types of Drilling Unit

The derrick for rotary drilling may be used on land or at sea. On land, prefabricated rigs are used; they can be transported by skidding, vehicle, or, in the case of light rigs, by helicopter. Once a well has been drilled, the derrick is dismantled and moved to the next location, whether the well is productive or barren of hydrocarbons (colloquially termed a *dry hole* or *duster*). Thus modern oil fields are not marked by a forest of derricks as shown in old photographs and films.

EXLOG

COMPANY ABC OIL COMPANY OF SPAIN

WELL DESMOND 1X

FIELD ANDORRA

REGION OFFSHORE SPAIN

COORDINATES 39 deg 44 min 54 s N
03 deg 23 min 33 s E

API WELL INDEX NO.

SPUD DATE 5/4/81

ELEVATION RKB to MSL : 84.5 ft
RKB to SF : 174.2 ft

TOTAL DEPTH 5345 ft

CONTRACTOR DEEP DRILLING COMPANY

RIG / TYPE CHARLIE JONES / JACKUP

LOG INTERVAL
DEPTH FROM 400 ft TO 5345 ft
DATE FROM 5/4/81 TO 21/5/81
SCALE 1:500 **UNIT** 501
LOG PREPARED BY A.EVANS, G.JONES,
A.EDWARDS

HOLE SIZE
26 in	TO 750 ft	8 1/2 in	TO 5345 ft	
17 1/2 in	TO 2300 ft		TO	
12 1/4 in	TO 4190 ft		TO	

CASING RECORD
30 in	AT 400 ft	9 5/8 in	AT 4185 ft
20 in	AT 735 ft	7 in	AT 5340 ft
13 3/8 in	AT 2295 ft		AT

MUD TYPES
SEAWATER/GEL	TO	2300 ft
KCL/POLYME	TO	5345 ft
	TO	

LITHOLOGY SYMBOLS

LIMESTONE　DOLOMITE　ANHYDRITE AND GYPSUM　HALITE

COAL AND LIGNITE　CLAY　SHALE　SILT AND SILTSTONE

SILTY SANDSTONE　SAND AND SANDSTONE　CONGLOMERATE　CHERT

IGNEOUS UNDIFFERENTIATED

EXLOG SUITE
FORMATION EVALUATION LOG	☒	WIRELINE DATA PRESSURE LOG	☒
PRESSURE EVALUATION LOG	☒	TEMPERATURE DATA LOG	☒
DRILLING DATA PRESSURE LOG	☒	GEMDAS COMPUTER LOGS	☐

ABBREVIATIONS
NB	NEW BIT	SVG	SURVEY GAS
RRB	RERUN BIT	C	CARBIDE TEST
CB	CORE BIT	W	MUD DENSITY — ppg
WOB	WEIGHT ON BIT	V	FUNNEL VISCOSITY
RPM	REVS PER MINUTE	F	FILTRATE — API
FLC	FLOW CHECK	FC	FILTER CAKE
CR	CIRCULATE RETURNS	PV	PLASTIC VISCOSITY
PR	POOR RETURNS	YP	YIELD POINT
NR	NO RETURNS	SOL	SOLIDS — %
LAT	LOGGED AFTER TRIP	SD	SAND — %
BG	BACKGROUND GAS	S	SALINITY — PPM CI
TG	TRIP GAS	RM	MUD RESISTIVITY
STG	SHORT TRIP GAS	RMF	FILTRATE RESISTIVITY
CG	CONNECTION GAS		
SWG	SWAB GAS		

CASING SEAT / WIRELINE LOG RUN

CORED INTERVAL / TEST INTERVAL

NO RECOVERY + WIRELINE TEST

SIDEWALL CORE

GAS — 100 Total gas units is equivalent to 2% methane-in-air.

OIL — Based on live oil in unwashed cuttings and percentage staining of washed cuttings.

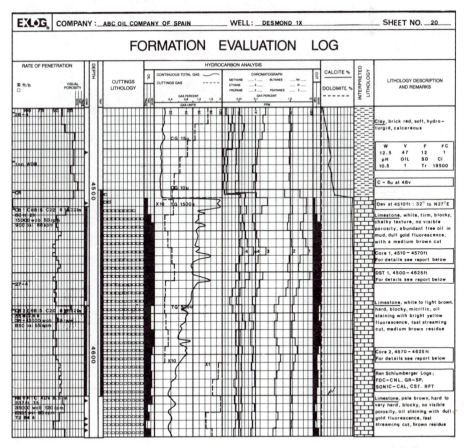

FIGURE 3.7 *Example of part of a typical mudlog, showing the range of drilling and geological data recorded. (Courtesy of Exploration Logging Services, Ltd.)*

FIGURE 3.8 *Jack-up drilling rig, Key Gibralter (on charter to British Petroleum) in British Petroleum's West Sole gas field in the North Sea. (Courtesy of British Petroleum.)*

In offshore drilling the derrick can be mounted in various ways. For sheltered inland waterways it may be rigged on a flat-bottomed barge. This technique has been employed, for example, in the bayous of the Mississippi delta for over 50 years. In water depths of up to about 100 m, jack-up rigs are used. In jack-up rigs the derrick is mounted on a flat-bottomed barge fitted with legs that can be raised or lowered. When the rig is to be moved, the legs are raised (i.e., the barge is lowered) until the barge is floating. Some barges are self-propelled, but the majority are moved by tugs. On reaching the new location the legs are lowered, thus raising the barge clear of all but the highest waves (Fig. 3.8).

Submersible units are platforms mounted on hollow caissons, which can be flooded with seawater. The platform supports can be sunk onto the seabed, leaving a sufficient clearance between the sea surface and the underside of the platform. Like jack-ups, submersible rigs are only suitable for shallow water.

For deeper waters drill ships or semisubmersible units are used. On drill ships the derrick is mounted amidships and the ship kept on location either by anchors or by a specially designed system of propellers that automatically keeps the ship in the same position regardless of wind, waves, and currents. Semisubmersibles are floating platforms having three or more floodable caisson legs. With the aid of anchors and judicious flooding of the legs, the unit can be stabilized, although still floating, with the rotary table 30 m or so above the sea. Some semisubmersibles are self-propelled, but the majority are towed by tugs from one location to another (Fig. 3.9).

In shallow-water arctic conditions, where there is extensive pack ice, drill ships, jack-ups, and semisubmersibles are unsuitable. In very shallow water, artificial islands are made from gravel and ice. Monopod rigs are also used, which, as their name implies, balance on one leg around which the pack ice can move with minimum obstruction.

FIGURE 3.9 *British Petroleum's semisubmersible drilling rig, Sea Conquest. (Courtesy of British Petroleum.)*

Various Types of Production Unit

As previously noted, when a well has been drilled, the derrick is moved to the next location. Oil and gas are produced from wells in several ways. Details of the various drive mechanisms of reservoirs are outlined in Chapter 6 (p. 263). Actual production is achieved in one of two ways, depending on whether or not the oil flows to the surface without the use of a pump.

In most cases the reservoir has sufficient pressure for oil or gas to flow to the surface. In this situation casing is generally run below the producing zone, and perforations are shot through it by explosive charges opposite the hydrocarbon pay interval. Steel tubing is hung from the well head to the producing zone, and the annulus between tubing and casing sealed off with a packer (Fig. 3.10). A system of valves, termed the *Christmas tree*, is installed at the well head, from which a flowline leads to a tank in which produced oil and gas are separated at atmospheric pressure.

FIGURE 3.10 *A typical well completion through perforated casing. Details of the various casing strings are omitted (see Fig. 3.5).*

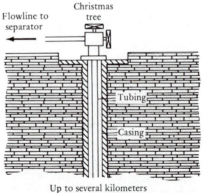

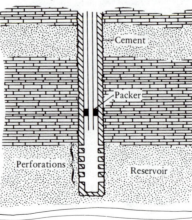

FIGURE 3.11   *Nodding donkey well head pump on British Petroleum's Kimmeridge Bay field, Dorset.*

If the reservoir pressure is too low for oil to flow to the surface, a pumping device is used: either a *nodding donkey* at the well head or a downhole pump installation. Figure 3.11 illustrates a well head pump. A diesel or electrically powered beam engine raises and lowers a connected string of *sucker rods*, which are connected to a piston at the base of the borehole.

Alternatively, an electrically driven centrifugal pump may be installed at the bottom of the hole. This device is more effective than the nodding donkey and less ecologically shocking. Maintenance costs may be higher because of difficulties of access, and reliability may be lower.

The productivity of many wells may be stimulated, either initially or worked over later during their life. Typical stimulation techniques involve fracturing the formation, generally by high-pressure pumping of metallic or plastic shrapnel into fractures to wedge them open. For a finale hydrochloric or some other acid may then be injected to enlarge the fracture and hence increase permeability and productivity. These techniques are largely, but not exclusively, applied to carbonate reservoirs (Langnes, Robertson, and Chilingarian, 1972).

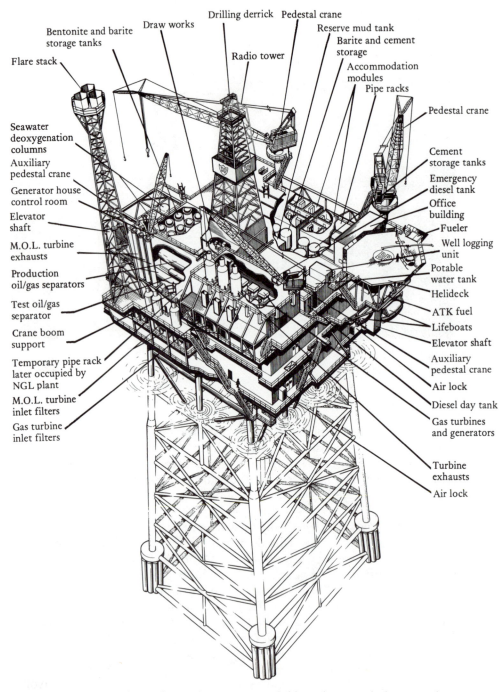

FIGURE 3.12   *One of British Petroleum's Forties field production platforms in the North Sea. (Courtesy of British Petroleum.)*

Onshore development wells are drilled on various types of geometric arrangements at spacings determined by the permeability of the reservoir and radius of drainage anticipated for each well. Offshore, however, this procedure is not feasible. Central production platforms are installed, from which many producing wells are deviated to penetrate the reservoir at the desired depth and spacing (Figs 3.12 and 3.13).

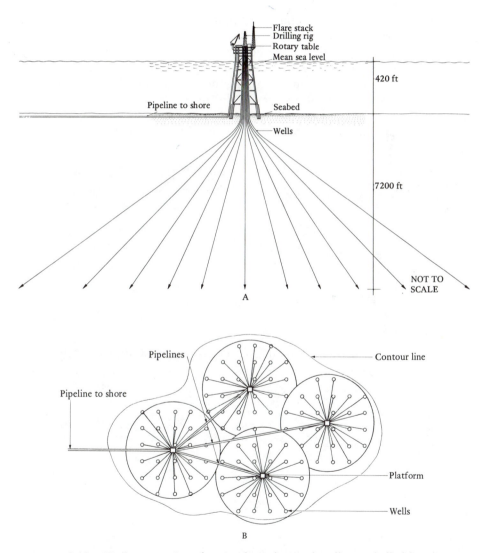

FIGURE 3.13   *(A) Cross-section showing how deviated wells are drilled from a platform to penetrate the reservoir at the required depth and spacing (B). (Courtesy of British Petroleum.)*

FORMATION EVALUATION

Boreholes yield much geological and engineering information, particularly when extensive lengths of core are recovered. Because coring is so expensive, holes are usually drilled with an ordinary rotary bit, and afterward further information is gained about the penetrated formations by measuring their geophysical properties with the aid of wireline logs. This process is generally completed before setting casing, so several log runs are necessary during the drilling of a single well. The equipment that measures the geophysical properties of the penetrated rocks is housed in a cylindrical *sonde*, which is lowered down the borehole on a multiconductor electric cable (Fig. 3.14). On reaching the bottom of the hole, the recording circuits are switched on and the various properties measured as the sonde is drawn to the surface. The measurements are transmitted up the cable and recorded on film or magnetic tape in the logging unit. In onshore operations the recorders are generally mounted on a truck (Fig. 3.15).

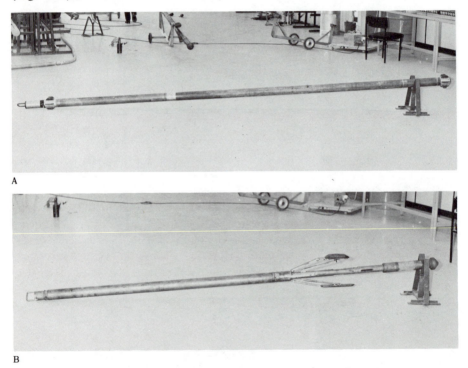

A

B

FIGURE 3.14   *Two sondes for measuring electric properties of formations.*
(A) Induction resistivity tool, which combines the S.P., a 16-in. normal to measure
R_{xo}, *and a deep induction log to measure* R_t. *(B) A microfocused log, which*
records S.P., R_{xo}, *and hole diameter. (Courtesy of Tesel Services, Ltd.)*

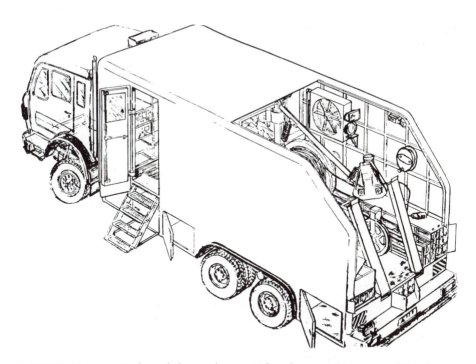

FIGURE 3.15 *A typical truck for onshore wireline logging. (Courtesy of Tesel Services, Ltd.)*

Many different parameters of the rocks can be recorded, such as formation resistivity, sonic velocity, density, and radioactivity. The recorded data can be interpreted to determine the lithology and porosity of the penetrated formations and the type and quantity of fluids (oil, gas, or water) within the pores.

Formation evaluation is a large and complex subject. Many oil companies employ full-time log analysts. The following account is only intended as an introduction to the topic. The various types of log are described and the principles of log analysis reviewed. Petroleum geologists generally attend courses on log analysis early in their professional career, which provide up-to-date and detailed information on new tools and analytical methods.

Electric Logs

The Spontaneous Potential, or S.P., Log

The spontaneous potential, or self potential, log is the oldest type of geophysical log in use (Fig. 3.14). The first one was run more than 50 years ago. The spontaneous potential log records the electric potential set up

FIGURE 3.16 *Basic arrangement for*
the spontaneous potential log.

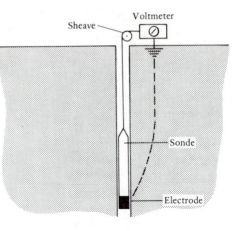

between an electrode in a sonde drawn up the borehole and a fixed electrode at the earth's surface (Fig. 3.16). It can only be used in open (i.e.,
uncased) holes filled with conductive mud. Provided that there is a minimum amount of permeability, the S.P. response is dependent primarily on
the difference in salinity between drilling mud and the formation water.

The electric charge of the S.P. is caused by the flow of ions (largely
Na^+ and Cl^-) from concentrated to more dilute solutions. Generally this
flow is from salty formation water to fresher drilling mud (Fig. 3.17). This

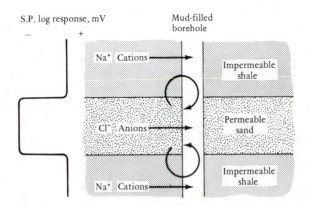

FIGURE 3.17 *Diagram showing how ionic diffusion causes the spontaneous*
potential effect. Looped arrows show the direction of positive current flow. Log
response refers to the situation in which the salinity of the formation water is
greater than that of the drilling mud.

naturally occurring electric potential (measured in millivolts) is basically related to the permeability of the formation. Deflection of the log from an arbitrarily determined shale baseline indicates porous sandstones or carbonates. In most cases this deflection, termed a *normal* or *negative* S.P. deflection, is to the left of the baseline. Deflection to the right of the baseline, termed *reversed* or *positive* S.P., occurs when formation waters are fresher than the mud filtrate. A poorly defined or absent S.P. deflection occurs in uniformly impermeable formations or where the salinities of mud and formation water are comparable (Fig. 3.18). In most cases, with a normal S.P. the curve can be used to differentiate between interbedded impermeable shales and permeable sandstones or carbonates.

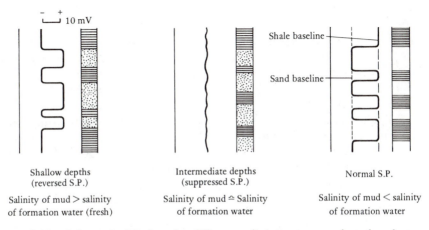

Shallow depths
(reversed S.P.)

Salinity of mud > salinity
of formation water (fresh)

Intermediate depths
(suppressed S.P.)

Salinity of mud ≏ Salinity
of formation water

Normal S.P.

Salinity of mud < salinity
of formation water

FIGURE 3.18 *Schematic S.P. logs for different salinity contrasts of mud and formation water. Reversed S.P. logs are very rare. Suppressed S.P. logs occur where salt muds are used. The usual response is shown in the lower part of the log.*

Note that the S.P. curve has no *absolute* value. The logging engineer shifts the baseline as the curve gradually drifts across the scale during the log run, and the S.P. deflections of adjacent wells are not comparable. Similarly, although local deflections on the curve are caused by vertical variations in permeability, no actual millidarcy values can be measured.

The amount of the current, and hence the amplitude of deflection on the S.P. curve, is related not only to permeability but also to the contrast between the salinity of the drilling mud and the formation water. Specifically, it is the contrast between the resistivity of the two fluids. Empirically, it has been found that:

$$E = K \log \frac{R_{mf}}{R_w}$$

where E = S.P. charge (mV)
        K = a constant, which is generally taken as
             $65 + 0.24T(°C)$ or $61 + 0.133T(°F)$
        R_{mf} = resistivity of mud filtrate (ohm/m)
        R_w = resistivity of formation water (ohm/m)

Details of this relationship are found in Wyllie (1957) and the various
wireline-logging service company manuals. Note that since the resistivity
of salty water varies with temperature, it is necessary to allow for that in
solving the equation. The resistivity of the filtrate of the drilling mud may
be measured at the surface and, if the bottom hole temperature is known,
can be recalculated for the depth of the zone at which the S.P. charge is
measured. The equation may then be solved, and R_w, the resistivity, and
hence the salinity of the formation water can be determined.

In summary the S.P. log may be used to delineate permeable zones,
and hence it aids lithological identification and well-to-well correlation.
The S.P. log can also be used to calculate R_w, the resistivity of the formation water. The S.P. is limited by the fact that it cannot be run in cased hole
and is ineffective when R_{mf} is approximately equal to R_w, which occurs
with many offshore wells using saltwater-based drilling muds.

Resistivity Logs

At present the three main ways of measuring the electrical resistivity of
formations penetrated by boreholes are the *normal log, laterolog,* and *induction log* techniques. With the normal, or *conventional resistivity*, log an
electric potential and flow of current is set up between an electrode on the
sonde and an electrode at the surface. A pair of electrodes on the sonde is
used to measure the variation in formation resistivity as the sonde is raised
to the surface. The spacing between the current electrode and the recording electrode can be varied, as shown in Figure 3.19 (top left). The three
electrode spacings usually employed are 16 in (*short normal*), 64 in (*long
normal*), and 8 ft 8 in (*long lateral*). They can generally be run simultaneously with an S.P. log.

Normal resistivity devices, although largely superseded by more sophisticated types, may be encountered on old well logs. For low-resistivity
salty muds, laterologs, or guard logs, are now generally used. In these
systems single electrodes cause focused current to flow horizontally into
the formation. This horizontal flow is achieved by placing two guard
electrodes above and below the current electrode. By balancing the guard

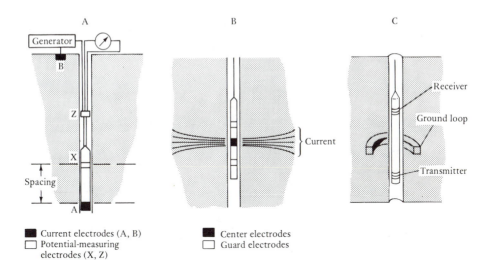

A   B   C

Generator

Receiver

Ground loop

Current

Spacing

Transmitter

■ Current electrodes (A, B)   ■ Center electrodes
☐ Potential-measuring   ☐ Guard electrodes
   electrodes (X, Z)

FIGURE 3.19 *(Top left) Schematic arrangement for a normal resistivity logging device. Variation of the spacing between A and X determines the distance away from the borehole at which the resistivity is measured. (Top right) Diagram of the laterolog method. (Bottom) Diagram of the induction principle. For explanations see text.*

electrode current with that of the central generating electrode, a sheet of current penetrates the formation. The potential of the guard and central electrodes is measured as the sonde is raised, as shown in Figure 3.19 (top right). As with conventional resistivity logs, various types of laterologs can be used to measure resistivity at different distances from the borehole. This measurement is achieved by changing the geometry of the focusing electrodes.

For freshwater or oil-based muds, which have a hi~~low~~ resistivity, a third type of device is used. This device is the induction log, in which transmitter and receiver coils are placed at two ends of a sonde and are used to transmit a high-frequency alternating current. This current creates a magnetic field, which, in turn, generates currents in the formation. These currents fluctuate according to the formation resistivity and are measured at the receiver coil, as shown in Figure 3.19 (bottom).

The electrical resistivity of formations varies greatly. Solid rock is highly resistive, as is porous rock saturated in fresh water, oil, or gas. Shale, on the other hand, and porous formations saturated with salty water or brine have very low resistivities. When run simultaneously, S.P. and resistivity logs enable qualitative interpretations of lithology and the nature of pore fluids to be made (Fig. 3.20).

One of the functions of drilling mud is to prevent the flow of fluids into the borehole from permeable formations. At such intervals a cake of

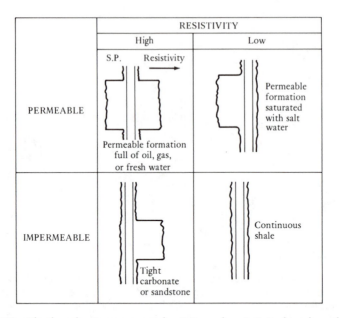

	RESISTIVITY	
	High	Low
PERMEABLE	*(S.P. Resistivity curves)* Permeable formation full of oil, gas, or fresh water	Permeable formation saturated with salt water
IMPERMEABLE	Tight carbonate or sandstone	Continuous shale

FIGURE 3.20   *The four basic responses for S.P. and resistivity logs for a bed between impermeable formations.*

mud builds up around the borehole wall, and mud filtrate is squeezed into the formation. Thus the original pore fluid is displaced, be it connate water, oil, or gas. Thus a circular invaded, or flushed, zone is created around the borehole with a resistivity (referred to as R_{xo}) that may be very different from the resistivity of the uninvaded zone (R_t). A transition zone separates the two. These concepts are shown in Figure 3.21. As already noted, various types of resistivity log are not only adapted for different types of mud but also for measuring resistivity of both the uninvaded zone (R_t) and the

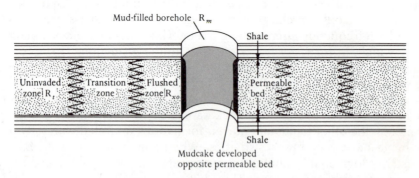

FIGURE 3.21   *The situation around a borehole adjacent to a permeable bed. Resistivity of flushed zone (R_{xo}); resistivity of uninvaded zone (R_t); resistivity of mud filtrate (R_{mf}); and resistivity of formation water (R_w). (Courtesy of Schlumberger.)*

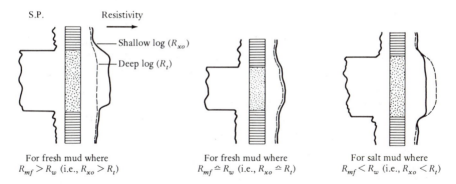

FIGURE 3.22   *S.P. and resistivity logs showing the responses for a water-bearing reservoir for cases of various differences between the resistivity of the drilling mud (R_m), and that of the formation water (R_w).*

flushed zone (R_{xo}). The latter are generally referred to as microresistivity logs, of which there are many different types with different trade names. Figure 3.22 shows the responses of R_t and R_{xo} (deep and shallow) resistivity logs in various water-bearing reservoirs. Note the convention that the deep-penetrating log is shown by a dashed curve and the shallow log by a continuous line. Where formations are impermeable, there is no separation between the two logs, since there is no flushed zone. A single sonde can simultaneously record the S.P. and more than one resistivity curve (Fig. 3.14).

How hydrocarbon zones can be identified from the resistivity log has already been shown. Where two curves are run together, the response is as shown in Figure 3.23. This figure is, however, only a qualitative interpretation. No absolute resistivity cut-off reading can be used to determine oil saturation.

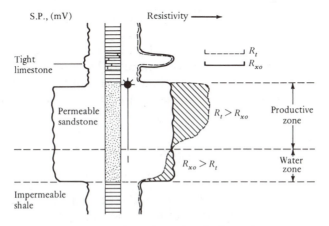

FIGURE 3.23   *S.P. and resistivity logs through a hydrocarbon reservoir showing typical responses for the situation $R_{mf} > R_w$.*

Quantitative Calculation of Hydrocarbon Saturation

Having reviewed the various resistivity logs and their qualitative interpretation, it is appropriate to discuss the quantitative calculation of the hydrocarbon content of a reservoir. Generally this calculation is approached in reverse by first calculating the water saturation (S_w). A reservoir whose pores are totally filled by oil or gas has an S_w of 0 percent, or 0.00; a reservoir devoid of hydrocarbons has an S_w of 100 percent, or 1.0.

Consider now a block of rock whose resistivity is measured as shown in Figure 3.24. Then:

$$\text{Resistance (ohm)} = R \times \frac{L}{A} \, [\text{ohm}/(\text{m})(\text{m}^2)]$$

where R = resistivity (ohm/m)
L = length of conductive specimen
A = cross-sectional area

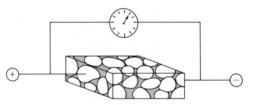

FIGURE 3.24   *Sketch showing how the resistivity of a rock is measured.*

Note also that:

$$\text{Resistivity} = \frac{1}{\text{conductivity (mho's)}}$$

Most reservoirs consist of mineral grains that are themselves highly resistive, but between them are pores saturated by fluids whose conductivity varies with composition and temperature. An important number in log interpretation therefore is the particular formation resistivity factor (F), defined as:

$$\text{Formation resistivity factor } (F) = \frac{R_o}{R_w}$$

where R_o = resistivity of rock 100 percent saturated
with water of resistivity R_w.

Note that R_w decreases with increasing salinity and temperature, as, therefore, does R_o. The value of F increases with decreasing porosity.

Archie (1942) empirically derived what is now termed the *Archie formula:*

$$F = \frac{a}{\phi^m}$$

where a = a constant
ϕ = porosity
m = cementation factor

The values of a and m depend on formation parameters that are difficult to measure, being influenced by the tortuosities among interconnected pores.

Several values for a and m have been found that empirically match various reservoirs:

For sands:

$$F = \frac{0.81}{\phi^2}$$

For compacted formations (notably carbonates):

$$F = \frac{1}{\phi^2}$$

The Shell formula for low-porosity nonfractured carbonates:

$$F = \frac{1}{\phi^m}$$

where $m = 1.87 + \dfrac{0.019}{\phi}$

The Humble formula (suitable for soft formations):

$$F = \frac{0.62}{\phi^{2.15}}$$

Again it has been found empirically that in clean clay-free formations water saturation can be calculated from the following equation:

$$S_w^n = \frac{FR_w}{R_t}$$

where n = the saturation exponent (generally taken as 2).

But, by definition:

$$F = \frac{R_o}{R_w}$$

or

$$R_o = FR_w$$

Thus:

$$S_w^2 = \frac{R_o}{R_t}$$

or

$$S_w = \frac{\sqrt{R_o}}{R_t}$$

Using a log reading deep resistivity, R_t can be measured in the suspected oil zone and R_o in one that can be reasonably assumed to be 100 percent water saturated. Thus S_w may be calculated throughout a suspected reservoir. This method is the simplest and is only valid for clean (i.e., clay-free) reservoirs. If a clay matrix is present, the resistivity of the reservoir is reduced and oil-saturated shaley sands may be missed. This method also assumes that the value for F is the same in the oil and water zones, which is seldom true because oil inhibits cementation, which may continue in the underlying water zone (see p. 247).

To overcome some of these problems, the ratio method may be used. This method is based on the assumption that:

$$S_{xo} = S_w^{0.2}$$

From this equation the following relationship may be derived:

$$S_w = \left(\frac{R_{xo}/R_t}{R_{mf}/R_w} \right)^{5/8}$$

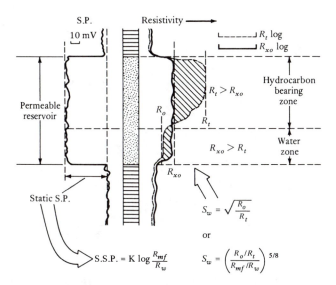

FIGURE 3.25 *S.P. and resistivity logs for a hydrocarbon zone showing where readings are taken to solve the R_w and S_w calculations. For explanation see text.*

R_{xo} can be measured from a microlog; R_t from a deep resistivity log; R_{mf} from the drilling mud at the surface (corrected for temperature at the zone of interest); and R_w from the S.P. log, as shown (Fig. 3.25).

Radioactivity Logs

The Gamma-Ray Log (and Caliper)

Three types of log that measure radioactivity are commonly used: the gamma-ray, neutron, and density log. The gamma-ray log, or gamma log, uses a scintillation counter to measure the natural radioactivity of formations as the sonde is drawn up the borehole. The main radioactive element in rocks is potassium, which is commonly found in illitic clays and to a lesser extent in feldspars, mica, and glauconite. Zirconium (in detrital zircon), monazite, and various phosphate minerals are also radioactive. Organic matter commonly scavenges uranium and thorium, and thus oil source rocks, oil shales, sapropelites, and algal coals are radioactive. Humic coals, on the other hand, are not radioactive. The gamma log is thus an important aid to lithological identification (Fig. 3.26). The radioactivity is measured in API (American Petroleum Institute) units and generally plotted on a scale of 0–100 or 0–120 API.

Conventionally, the natural gamma reading is presented on the left-hand column of the log in a similar manner to, and often simultaneously with, the S.P. log. The gamma log can be used in much the same way as an

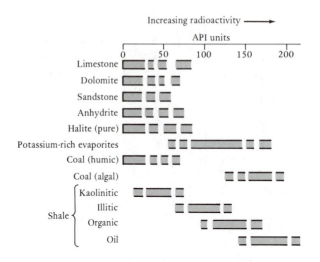

FIGURE 3.26 *The approximate gamma log ranges for various rocks. Note that small quantities of radioactive clay, for example, can increase the reading of any lithology.*

S.P., with a shale baseline being drawn. Deflection to the left of this line does not indicate permeability but rather a change from shale to clean lithology, generally sandstone or carbonate. The gamma reading is affected by hole diameter, so it is generally run together with a caliper log, a mechanical device that records the diameter of the borehole. The caliper log shows where the hole may be locally enlarged by washing out or caving and hence deviating the expected gamma-ray and other log responses. The hole may also be narrower than the gauge of the bit where *bridging* occurs. Bridging is caused by either sloughing of the side of the hole and incipient collapse or a build-up of mudcake opposite permeable zones. Although the gamma log is affected by hole diameter, it has the important advantage that it can be run through casing. The gamma log is important for identifying lithology, calculating the shaliness of reservoirs, and correlating between adjacent wells.

The Natural Gamma-Ray Spectrometry Tool

One of the limitations of the standard gamma-ray log is that it is unable to differentiate among various radioactive minerals causing the gamma response. This lack of differentiation causes a severe problem when the log is used to measure the clay content of a reservoir in which either the clay is kaolin (potassium-free and nonradioactive) or other radioactive minerals are present, such as mica, glauconite, zircon, monazite, or uranium adsorbed on organic matter.

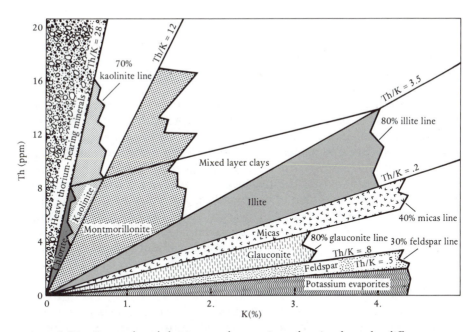

FIGURE 3.27 *Cross-plot of thorium and potassium showing how the different types of clay (and other minerals) can be identified by the gamma-ray spectrometry log. (Courtesy of Schlumberger.)*

By analyzing the energy wavelength spectrum of detected gamma radiation, the refined gamma-ray spectrometry tool measures the presence of three commonly occurring radioactive decay series, whose parent elements are thorium, uranium, and potassium (Hassan, Hossin, and Combaz, 1976). This information can be used for detailed mineral identification, and, in particular, it allows a study of clay types to be made. An example of such an application is shown in Figure 3.27, where the potassium-thorium ratio indicates the trend from potassic feldspar to kaolinite clays.

Gamma-ray spectrometry logging is also important in source rock evaluation because it can differentiate detrital radioactive minerals containing potassium and thorium from organic matter with adsorbed uranium.

The Neutron Log

The neutron log, as its name suggests, is produced by a device that bombards the formation with neutrons from an americium beryllium or other radioactive source. Neutron bombardment causes rocks to emit gamma rays in proportion to their hydrogen content. This gamma radiation is recorded by the sonde. Hydrogen occurs in all formation fluids (oil, gas, or water) in reservoirs, but not in the minerals. Thus the response of the neutron log is *essentially* correlative with porosity. Hole size, lithology, and hydrocarbons, however, all affect the neutron log response. The effect of variation in hole size is overcome by simultaneously running a caliper with an automatic correction for bit gauge. In the early days, the neutron log was recorded in API units. Because it is so accurate for clean reservoirs, the neutron log is now directly recorded in either limestone or sandstone porosity units (LPUs and SPUs, respectively). As with all porosity logs, the log curve is presented to the right of the depth scale, with porosity increasing to the left. Because shales always contain some bonded water, the neutron log will always give a higher apparent porosity reading in dirty reservoirs than actually exists. The hydrogen content of oil and water is about equal, but is lower than that of hydrocarbon gas. Thus the neutron log may give too low a porosity reading in gas reservoirs. As shown in the following section, this fact can be turned to an advantage. The neutron log can be run in cased hole because its neutron bombardment penetrates steel.

The Density Log

The third type of radioactivity tool measures formation density by emitting gamma radiation from the tool and recording the amount of gamma radiation returning from the formation (Fig. 3.28). For this reason the device is often called the *gamma-gamma* tool. Corrections are automatically made within the sonde for the effects of borehole diameter and mudcake thickness. The corrected gamma radiation reading can be related to the electron density of the atoms in the formation, which is, in turn, directly related to the bulk density of the formation. Bulk density of a rock is a function of lithology and porosity. Porosity may be calculated from the following equation:

$$\text{Porosity } (\phi) = \frac{\rho_{ma} - \rho_b}{\rho_{ma} - \rho_f}$$

where ρ_{ma} = density of the dry rock (g/cm^3)
ρ_b = bulk density recorded by the log
ρ_f = density of the fluid

Density values commonly taken for different lithologies are:

LITHOLOGY	DENSITY (g/cm^3)
Sandstone	2.65
Limestone	2.71
Dolomite	2.87

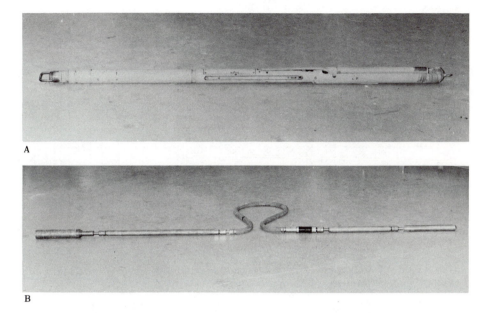

FIGURE 3.28 *(A) A density tool, which records the porosity of the formation, making automatic corrections for hole diameter and mudcake thickness. (B) A sonic sonde for measuring the acoustic velocity of formations. (Courtesy of Tesel Services, Ltd.)*

The fluid in the pores near the borehole wall is generally the mud filtrate. Because the tool has only a shallow depth of investigation and effectively "sees" only that part of the formation invaded by filtrate from the drilling mud, it reads this value for the porosity. Thus the density of the fluid may vary from 1.0 g/cm^3 for freshwater mud to 1.1 g/cm^3 for salty mud. Shale also affects the accuracy of the density-derived porosity of the reservoir. Also, several minerals have anomalous densities, the log traces of which may affect porosity values. Notable among these minerals are mica, siderite, and pyrite. The presence of oil has little effect on porosity values, but gas lowers the density of a rock and thus has the effect of causing the log to give too high a porosity. This effect can be turned to an advantage, however, when combined with the information derived from the neutron log.

The Lithodensity Log

Recent improvements to density logging techniques include the addition of a new parameter: the photoelectric cross-section, commonly denoted *Pe*. *Pe* is less dependent on porosity than is the formation density, and is particularly useful in analyzing the effects of heavy minerals on log inter-

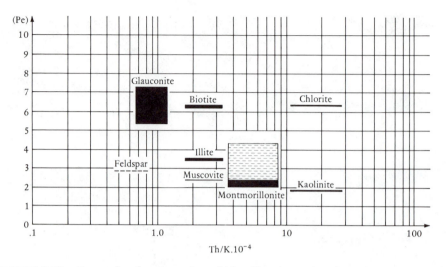

FIGURE 3.29   *Cross-plot showing mineral identification using the spectrometry and lithodensity logs. (Courtesy of Schlumberger.)*

pretation. An application that is particularly useful involves combining *Pe* with the thorium-potassium ratio from the gamma-ray spectrometry device, as indicated in Figure 3.29.

The Sonic, or Acoustic, Log

A third way of establishing the porosity of a rock is by measuring its acoustic velocity by the sonic, or acoustic, log (Fig. 3.28). In this technique the sonde emits from one end a series of *clicks* whose interval transit times are recorded on one or more receivers at the other end. The sound waves generally travel faster through the formation than through the borehole mud. The interval transit time (Δt), which is measured in microseconds/foot, can then be used to calculate porosity according to the following equation (Wyllie et al., 1956, 1958):

$$\phi = \frac{\Delta t_{\log} - \Delta t_{ma}}{\Delta t_f - \Delta t_m}$$

where $\Delta t_{\log}$ is the interval transit time recorded on the log
Δt_{ma} is the velocity of the rock at $\phi = 0$
Δt_f is the velocity of the pore fluid

Some commonly used velocities are as follows:

Lithology (i.e., pure mineral, $\phi = 0$)	Velocity	
	ft/s	μs/ft
Sandstone (quartz)	18,000–21,000	55.5–51.3
Limestone (calcite)	21,000–23,000	47.5
Dolomite (dolomite)	23,000	43.5
Anhydrite (calcium sulfate)	20,000	50.0
Halite (sodium chloride)	15,000	67.0
Fluid (fresh water or oil)	5300	189.0

The sonic log can be used only in open, uncased hole. The circuitry associated with the receiver has to be carefully adjusted for sensitivity so that it does not trigger on spurious noise, yet picks up the first arrival from the signal sound wave. A rapid to-and-fro log trace, termed *cycle skipping*, occurs if the sensitivity is insufficient or if the returned signal is very weak. It results from triggering on a later part of the sound wave pulse, which causes an erroneously long computed transit time. Cycle skip occurs in undercompacted formations (especially if gas filled), fractured intervals, and areas where the hole is enlarged and out of gauge so that a wider than normal width of borehole mud is traversed before the signal pulse enters the formation. Recent advances with computer-controlled logging devices greatly reduce the problem of balancing trigger sensitivity, which previously required continuous monitoring by the logging engineer.

The sonic method is the least accurate of the three porosity logs because it is the most affected by lithology. On the other hand, for this very reason, it is widely employed as a means of lithology identification and hence for correlation from well to well. The sonic log is also extremely useful to geophysicists because it can be used to determine the interval velocities of formations and thus relate timing of seismic reflectors to actual rocks around a borehole by means of computed time-depth conversions. For this reason the sonic log also records the total travel time in milliseconds along its length by a process of integration, the result being recorded as a series of *pips* along the length of the log. The cumulative number of these constant time interval pips can be counted between formation boundaries, and hence formation velocities can be related in time and depth.

Porosity Logs in Combination

The three porosity logs are influenced by formation characteristics other than porosity, notably by their lithology, clay content, and the presence of gas. When used in combination rather than individually, the logs give a more accurate indication of porosity and extract much other useful information.

It has already been noted that in gas zones the neutron log indicates too low a porosity and the density log, too high a porosity. These different porosity values can be turned to an advantage if the two curves are calibrated so that they track each other on the log. In limestones the curves are calibrated so that 0 LPU = 2.71 g/cm³; in sandstones 0 SPU = 2.65 g/cm³. If this calibration is done, separation between the log traces indicates the presence of gas (Fig. 3.30). Note that this phenomenon only applies in reservoir zones. Separation will commonly be seen in the shale sections, but in the reverse direction.

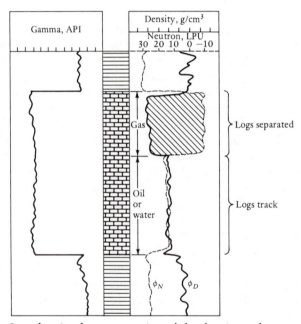

FIGURE 3.30 *Log showing how separation of the density and neutron logs may indicate gas-bearing intervals.*

Accurate identification of porosity and lithology can be determined by cross-plotting the readings of two porosity logs, using the chart books provided by the particular service company that ran the logs. Figure 3.31 shows one such graph for Schlumberger's formation density and compensated neutron logs run in freshwater holes. These cross-plots are satisfactory only in the presence of limestone, sandstone, or dolomite, but are inaccurate in the presence of clay or other anomalous minerals.

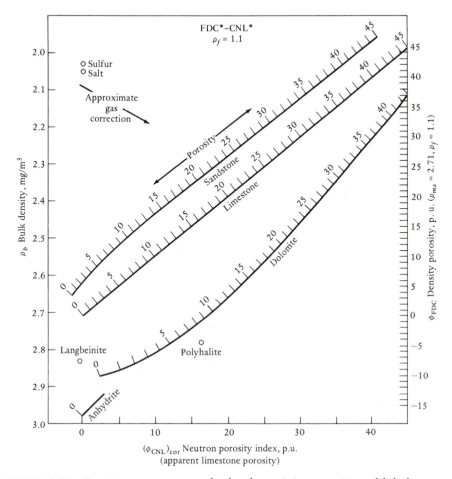

FIGURE 3.31 *Density-neutron cross-plot for determining porosity and lithology. For explanation see text. (Courtesy of Schlumberger.)*

A more sophisticated method of lithology identification is the M–N cross-plot (Burke et al., 1969). The two formulas used in this method take the readings from the three porosity logs and remove the effect of porosity, thereby leaving only the lithological effect:

For water-saturated formations:

$$M = \frac{\Delta t_f - \Delta t}{\rho_b - \rho_f} \times 0.01$$

$$N = \frac{(\phi_N)f - \phi_N}{\rho_b - \rho_f}$$

For freshwater muds:

$$\Delta t = 189 \ \mu s/ft$$

$$\phi_N = 1.0$$

$$\rho_f = 1.0$$

The M and N values are then plotted on the graph in Figure 3.32. This method has an advantage over the simple two-log cross-plot in that it can

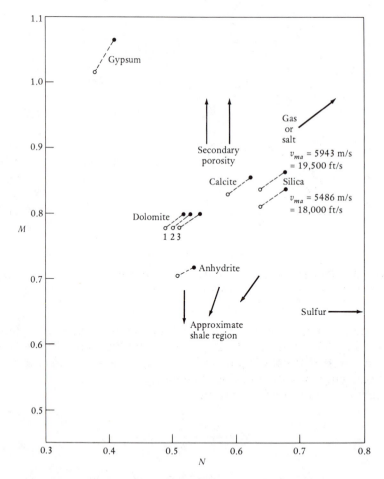

FIGURE 3.32   M–N *cross-plot for complex lithology identification. For explanation see text. (Courtesy of Schlumberger.)*

differentiate the proportions of multimineralic rocks. Note, however, that the computation is still susceptible to the effects of gas, clay, and anomalous minerals.

The Dipmeter Log

The dipmeter is a device for measuring the direction of dip of beds adjacent to the borehole. It is essentially a multiarm microresistivity log (Fig. 3.33). Three or four spring-loaded arms record separate microresistivity tracks, while, within the sonde, a magnetic compass records the orientation of the tool as it is drawn up the hole. A computer correlates deviations, or *kicks*, on the logs and calculates the amount and direction of bedding dip and assesses dip reliability (Fig. 3.34). Many different computer programs can be applied. Some programs essentially smooth the dips over large intervals; others use statistical techniques to discriminate against minor bedding features so as to display only large-scale dip trends. Both of these types are suitable for determining structural dip. Some programs calculate dips over intervals of only a few centimeters. This type of program (after removal of structural dip) can be used to discover small-scale sedimentary dips, such as cross-bedding.

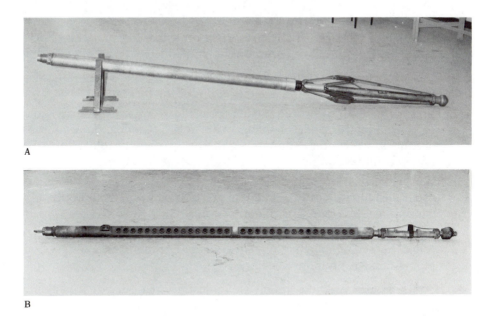

A

B

FIGURE 3.33 *(A) Three-arm dipmeter sonde. (B) Sidewall core gun. This device fires cylindrical steel bullets, which are attached to the gun by short cables, into the side of a borehole. Small samples of rock may thus be collected from known depths. (Courtesy of Tesel Services, Ltd.)*

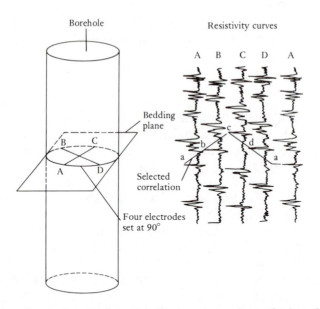

FIGURE 3.34   *Sketch showing how the dipmeter is used to calculate the direction of dip around a borehole.*

Calculated dips are usually presented on a tadpole plot (Fig. 3.35). Four basic types of motif are commonly identifiable:

1.   Uniformly low dips (referred to as green patterns) are generally seen in shales and indicate the structural dip of the formation.

2.   Upward declining dip sequences (referred to as red patterns) may be caused by the drape of shales over reefs or sandbars; the infilling of sandstones within channels; or the occurrence of folds, faults, or unconformities.

3.   Upward increasing dip sequences may be caused by sedimentary progrades in reefs, submarine fans, or delta lobes. They may also be caused by folds, faults, or unconformities.

4.   Random *bag o' nails* motifs can reflect poor hole conditions or they might be geologically significant, indicating fractures, slumps, conglomerates, or grainflows.

The dipmeter provides much valuable information, but it can only be interpreted fully in the light of other logs and geological data. For additional discussion see the service company manuals and Gilreath and Maricelli (1964), Campbell (1968), McDaniel (1968), Jageler and Matuszak (1972), Goetz et al. (1977), and Selley (1978a, 1978b).

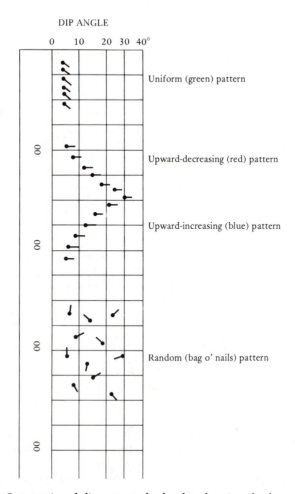

FIGURE 3.35   *Conventional dipmeter tadpole plot showing the four common motifs. Each motif can be produced by several quite different geological phenomena. The head of the tadpole shows the amount of dip. The tail of the tadpole points in the direction of dip.*

Uses of Logs in Petrophysical Analysis: Summary

The previous sections provide a short, and therefore simplified, review of a major group of borehole logging techniques that are crucial to the exploration for and production of hydrocarbons. New tools and techniques are continuously being introduced, but the basic principles remain constant. Figure 3.36 provides a summary of the more common geophysical logs and their principal uses. Figures 3.37 and 3.38 provide typical examples

ELECTRIC LOGS	Lithology	Hydrocarbons	Porosity	Pressure prediction	Structural and sedimentary dip	OTHER
Spontaneous potential	✓					Calculation of R_w, and bed shaliness, Qualitative identification of permeability
Resistivity	✓	✓		✓		Calculation of R_{xo}, R_t, and hence S_w
RADIOACTIVE LOGS						
Gamma	✓					Calculation of bed shaliness and organic content
Neutron		✓ } Gas effect	✓			} Lithology identification by cross-plots
Density		✓ } Gas effect	✓			} Lithology identification by cross-plots
SONIC	✓	✓	✓	✓		Calibration of formations with seismic data
DIPMETER					✓	

FIGURE 3.36 *Summary of the main types of wireline logs and their major applications.*

of suites of logs with their responses for parts of two boreholes. Most of the calculations of R_w, S_w, and so forth can be done using service company chart books or calculators. For lengthy runs, however, computer-processed interpretations, which give continuous readings of lithology, porosity, and the percentages of the various fluids, are preferable (Fig. 3.39). Although computer outputs look very convincing, especially when brightly colored, the old adage "rubbish in, rubbish out" must be remembered. Failure to check the quality of the original logs or failure to note the presence of an unusual or aberrant mineral will invalidate the whole procedure. For example, abundant mica flakes in a sandstone will read as "clay" on logs, giving a false value of porosity.

FIGURE 3.37 *Typical suite of logs through a sand-shale sequence in the Niger delta. Note high-resistivity hydrocarbon-bearing zones at 160 to 210 ft and below. (Courtesy of Schlumberger.)*

Spontaneous potential	Depths	Resistivity	Conductivity	Proximity log	Caliper	Gamma ray	Simultaneous compensated neutron formation density		
mV 10		ohm, m²/m	ohm, m²/m	ohm, m²/m	in	API units	Bulk density, g/cm³		
		0 16 SN 4	1000 IL 0	2 1 10 100 1000	10 20	0 150	1.8 2.3 2.8		
		0 20	2000 1000		Bit size		CNL Limestone porosity		
		0 200					47 17 −13		
		0 Rec. cond. 20					51 21 −9		
		0 200					Sandstone porosity		
							−.10 Δρ +.15		

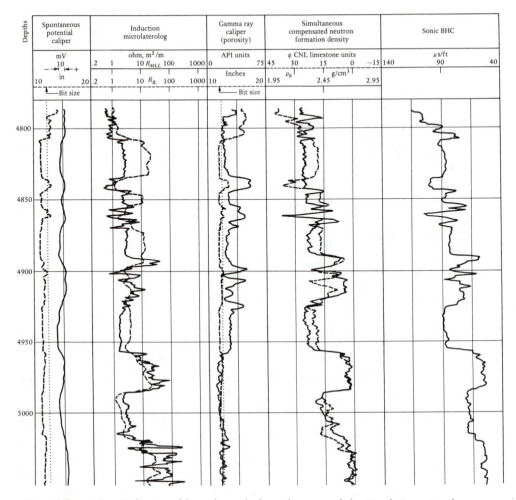

FIGURE 3.38 *Typical suite of logs through the Paleocene of the North Sea. Sands and shales in the upper part pass down into limestones. High resistivity in sands between 4805 and 4895 m suggest hydrocarbons; separation of density and neutron curves suggest that these hydrocarbons may be gas. (Courtesy of Schlumberger.)*

Applications of Logs in Facies Analysis

The previous sections reviewed the various types of geophysical logs and their application in identifying lithology, porosity, and hydrocarbon saturation. Logs are also used in other ways. When the stratigraphy of a well has been worked out, whether it be the lithostratigraphy or the biostratigraphy (in which paleontological zones are introduced), adjacent wells can be stratigraphically correlated. Logs can also be used to determine the facies and depositional environment of a reservoir. From such studies the geometry and orientation of reservoirs may be predicted.

Carbonate facies are largely defined, and their depositional environments deduced from petrography. Geophysical logs show porosity distribution in carbonates, but, because it is commonly of secondary origin, little correlation exists between log response and original facies (see p. 248). In sandstones, however, porosity is mainly of primary origin. Studies of modern sedimentary environments show that grainsize has certain characteristic vertical profiles. Notably, the fill of channels fines up, often from a basal conglomerate, via sand, to silt and clay. Conversely, prograding deltas and barrier islands deposit upward-coarsening grainsize profiles. Grainsize profiles may thus be of use in facies analysis. Both the S.P. and gamma logs may indicate grainsize profiles in sand-shale sequences. As discussed earlier, the deflection on the S.P. is locally controlled by permeability, with the maximum leftward deflection occurring in the most permeable interval. Permeability, however, increases with grainsize (see p. 235). Therefore the S.P. log is generally a vertical grainsize log in all but the most cemented sand-shale sections.

The gamma log may be used in a similar way, since the clay content (and hence radioactivity) of sands increases with declining grainsize. Exceptions to this general statement may be caused by intraformational clay clast conglomerates and the presence of anomalous radioactive minerals, such as glauconite, mica, and zircon.

Gamma and S.P. logs often show three basic motifs: (1) sands that fine upward gradually from a sharp base (bell motifs); (2) sands that coarsen up gradually toward a sharp top (funnel motifs); and (3) clean sands with sharp upper and lower boundaries (boxcar motifs). These three patterns may show variations from smooth to serrated curves when considered in detail (Fig. 3.40).

No log motif is specific to a particular sedimentary environment, but by combining an analysis of log profile with the composition of well samples, an interpretation of environment can be attempted. Constituents to look for in the cuttings are glauconite, shell debris, carbonaceous detritus, and mica. Glauconite forms during the early diagenesis of shallow marine sediments. Once formed, it is stable in the marine realm, but can be transported landward on to beaches or basinward on to deep-sea fans. Glauconite is readily oxidized at outcrop, however, so reworked second-cycle glauconite is virtually unknown. Thus the presence of glauconite grains in

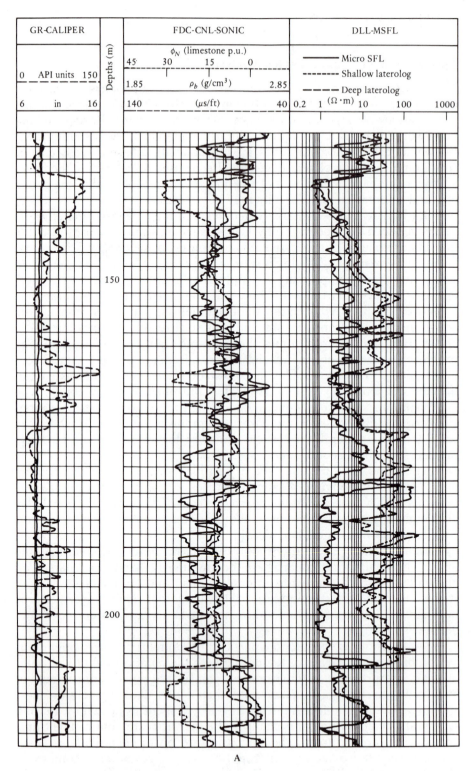

GR-CALIPER	Depths (m)	FDC-CNL-SONIC	DLL-MSFL

FDC-CNL-SONIC

ϕ_N (limestone p.u.)

45 30 15 0

1.85 ρ_b (g/cm³) 2.85

140 (μs/ft) 40

DLL-MSFL

——— Micro SFL

------- Shallow laterolog

— — — Deep laterolog

0.2 1 $(\Omega \cdot m)$ 10 100 1000

GR-CALIPER

0 API units 150

6 in 16

A

FIGURE 3.39 *(A) Suite of logs in Triassic gas-bearing sands from Algeria. (Courtesy of Schlumberger.) (Cont'd on p. 80.)*

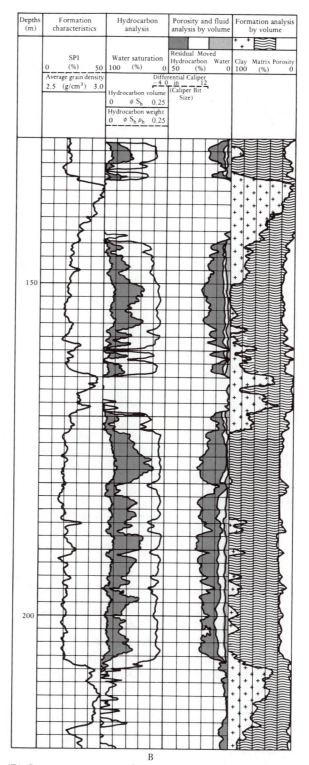

B

FIGURE 3.39 *(B) Computer-processed interpretation of logs shown in (A).
(Courtesy of Schlumberger.)*

FIGURE 3.40 *Diagram showing the three common grainsize motifs seen on S.P. and gamma logs and how, by combining these with the analysis of well samples, environmental interpretations of sand bodies can be made.*

Log motif	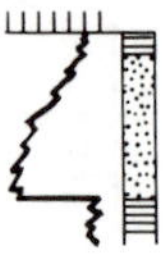	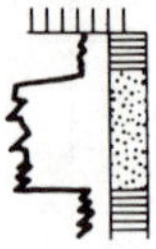	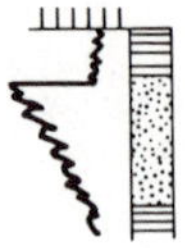
Glauconite and shell debris (high-energy marine)	Tidal channel	Tidal sand wave	Regressive barrier bar
Glauconite, shell debris, carbonaceous detritus, and mica (dumped marine)	Submarine channel		Prograding submarine fan
	Turbidite fill	Grain flow fill	
Carbonaceous detritus and mica (dumped)	Fluvial or deltaic channel	Delta distributary channel	Prograding delta or crevasse splay

a sandstone indicates a marine environment, although its absence indicates nothing.

Shell-secreting invertebrates live in freshwater and seawater environments, but shelly sands tend to be marine rather than nonmarine. Diagnosis is obviously enhanced if fragments of specifically marine fossils can be identified.

Carbonaceous detritus includes coal, plant fragments, and kerogen. These substances may be of continental or marine origin, but the preservation of organic matter generally indicates rapid deposition with minimal reworking and oxidation. Similarly, the presence of mica indicates rapid sedimentation in either marine or continental environments.

These four constituents are commonly recorded in well sample descriptions. Coupled with a study of log motifs, their presence (but not their absence) may aid the identification of the depositional environment of sand bodies and hence the prediction of reservoir geometry and trend (Fig. 3.40). This technique was first proposed by Selley (1976) and later refined (Selley 1978a, 1978b). In an ideal world facies analysis should be based on a detailed sedimentological and petrographic study of cores, but this method may be used to extend interpretation beyond cored wells or in regions where cores are not available. Figures 3.41 and 3.42 show the sort of facies synthesis that can be produced by integrating log and rock data. Real examples of some of these motifs are illustrated in this book, especially in Chapters 6 and 7.

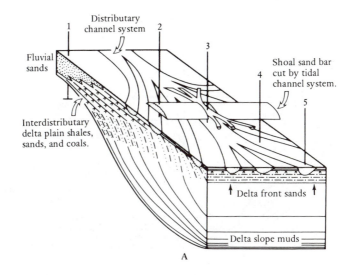

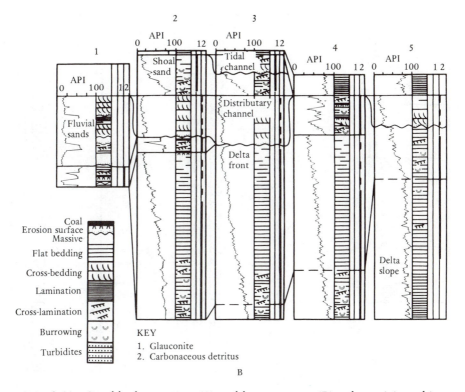

FIGURE 3.41   *Sand body geometry (A) and log responses (B) to be anticipated in a
prograding delta sequence. (From Selley, 1978a. Reprinted with permission.)*

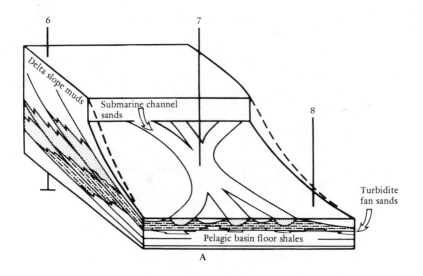

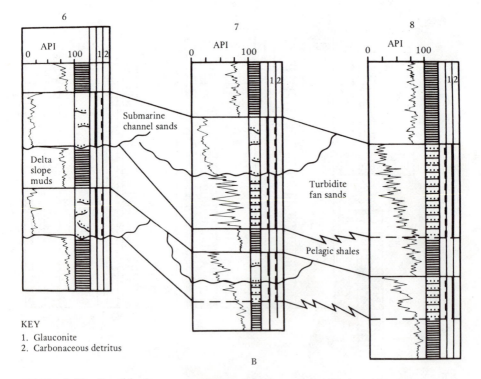

FIGURE 3.42 *Sand body geometry (A) and log responses (B) for submarine channel and fan deposits. (From Selley, 1978a. Reprinted with permission.)*

GEOPHYSICAL METHODS OF EXPLORATION

Petroleum exploration is largely concerned with the geological interpretation of geophysical data, especially in offshore areas. Petroleum geologists need to be well acquainted with the methods of geophysics. For many years a large communication barrier existed in many oil companies between the two groups, which were usually organized in different departments. This separation has now largely disappeared, as a new breed of petroleum explorationist appears—half geologist and half seismic interpreter. Unfortunately, a new division has developed within the geophysical world between the mathematicians, physicists, and computer programmers (who acquire and process geophysical information) and those who use geological concepts to interpret this information.

The following account of geophysical methods of petroleum exploration is very basic. It is probably sufficient for petroleum engineers seeking to know the origins of the maps by which fields are found and reserves assessed. For petroleum geologists, however, it is only an introduction and assumably will be followed by courses in geophysics and, early in their industrial career, a course in seismic interpretation.

Three main geophysical methods are used in petroleum exploration: magnetic, gravity, and seismic. The first two of these methods are used only in the predrilling exploration phase. Seismic surveying is used in both exploration and development phases and is by far the most important of the three methods (Fig. 3.43).

Magnetic Surveying

Methodology

The earth generates a magnetic field as if it were a dipole magnet. Lines of force radiate from one magnetic pole and converge at the other. The inclination of the magnetic field varies from vertical at the magnetic poles to

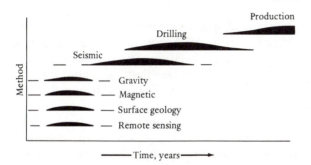

FIGURE 3.43 *Sketch showing the times at which different geophysical methods are used in petroleum exploration.*

horizontal at the magnetic equator. The magnetic axis of the earth moves about within a circle some 10 to 15 degrees of arc from the rotational axis.

The force between two magnetic poles may be expressed as follows:

$$F = \frac{AM_1M_2}{r^2}$$

where F is the force
A is a constant, generally unity
M_1 and M_2 are the strengths of the respective poles
r is the distance between the poles.

Magnetic field strength is measured either in gammas or in oersteds. One oersted is the field that exerts a force of one dyne on a unit magnetic pole. One gamma is 10^{-5} oersted. The magnetic field of the earth varies from about 60,000 gamma (0.6 oersted) at the poles to about 35,000 gamma (0.35 oersted) at the equator. By way of contrast, a schoolboy's horseshoe magnet has a field of about 350 oersteds.

The magnetic field of the earth shows considerable time-varying fluctuations, with periods ranging from hundreds of years to about a second. Diurnal variations are the most significant for magnetic surveys and have to be corrected before interpreting the results. Less easy to allow for, however, are large amplitude fluctuations in the earth's magnetic field during magnetic storms, which last for several hours and are caused by sunspot activity.

The intensity of magnetization of a magnetic mineral will normally be related to the regional field strength according to the following relation:

$$J = kH$$

where J is the intensity of magnetization, defined to be
the magnetic moment per unit volume of the rock
k is the magnetic susceptibility of the rock
H is the intensity of the magnetic field inducing
the magnetism

The intensity of magnetization may also have increments that have been acquired and retained at earlier stages in the history of the rock and may have to be taken into account in the interpretation. The magnetic susceptibility of rocks is very variable, ranging from less than 10^{-4} emu/cm^3 for sedimentary rocks to between 10^{-3} and 10^{-2} emu/cm^3 for iron-rich basic igneous rocks.

At any point above the earth, the measured geomagnetic field will be the sum of the regional field and the local field produced by the magnetic rocks in the vicinity. The purpose of magnetic surveys, therefore, is to measure the field strength over the area of interest. The recorded variations of

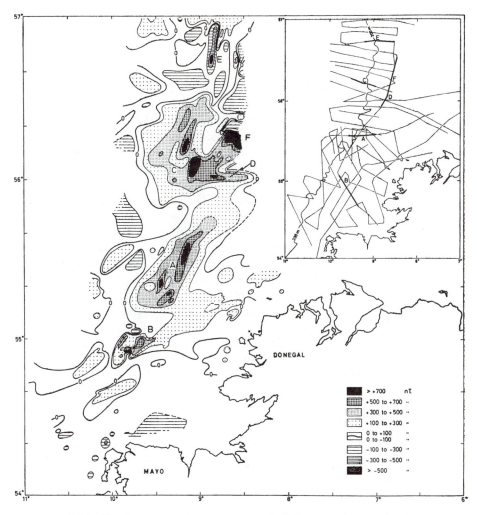

FIGURE 3.44 *Total magnetic intensity map of offshore northern Ireland. Location of survey lines shown in inset map. (From R. J. Bailey et al., 1975. Reprinted with permission.)*

the field strength will be due to changes in the earth's magnetic field and to the volume and magnetic susceptibility of the underlying rocks. These variations can be eliminated by removing the fluctuations recorded by a fixed station in the survey area. The residual values are then directly related to the rocks beneath.

Magnetic surveys can be carried out on the ground or on board ship, but taking the readings from the air is far more efficient. In aeromagnetic surveys a plane trails a magnetometer in a sonde attached to a cable and flies in a grid pattern over the survey area. The spacing of the survey lines will vary according to the nature of the survey and the budget, but the lines may be as close as 2 km with cross-tie lines on a 10-km spacing. By this method an aeromagnetic map, which contours anomalies in the earth's magnetic field in gamma units (Fig. 3.44), may be constructed.

Interpretation

Magnetic surveys are used in the early stages of petroleum exploration. Magnetic anomaly maps show the overall geological grain of an area, with the orientation of basement highs and lows generally apparent. Faults may show up by the close spacing or abrupt changes in orientation of contours. Sometimes magnetic maps may be used to differentiate basement (i.e., igneous and metamorphic rocks devoid of porosity) from sedimentary cover that may be prospective. However, this differentiation is only possible where there is a sharp contrast between the magnetic susceptibility of basement and cover and where this contrast is laterally persistent. An original magnetic survey can be interpreted and presented as a *depth-to-magnetic basement* map (Fig. 3.45). This procedure can be done when there is some well control on the basement surface, backed by measured rock parameters, and some knowledge of the regional geology. Magnetic anomaly maps are also useful to petroleum exploration in indicating the presence of igneous plugs, intrusives, or lava flows, areas to be avoided in the search for hydrocarbons.

In conclusion magnetic surveys are a quick and cost-effective way of defining broad-basin architecture, but can seldom be used to locate drillable petroleum prospects.

Gravity Surveying

Methodology

According to Newton's law of gravitation:

$$F = \frac{GM_1M_2}{d^2}$$

where F is the gravitational force between two point masses,
        M_1 and M_2
        d is the distance between M_1 and M_2
        G is the universal gravitational constant (usually taken as
            6.670×10^{-11} m^3/(kg)(s^2) or 6.670×10^{-8} in cgs units).

According to Newton's second law of motion, the acceleration a when a point, mass M_1, is attracted to another point, mass M_2, may be expressed as:

$$a = \frac{F}{M_1} = \frac{GM_2}{r^2}$$

If M_2 is considered to be the mass of the earth and r its radius, then a is the gravitational acceleration on the earth's surface. Were the earth a true sphere of uniform density, then a would be a constant anywhere

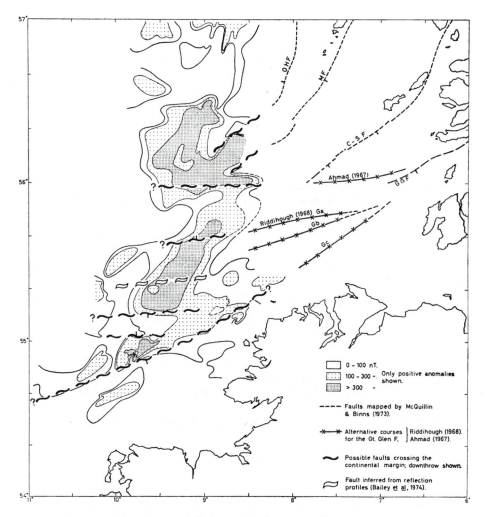

FIGURE 3.45 *Interpretation of the magnetic intensity map shown in Figure 3.44 indicating location of inferred faults. (From R. J. Bailey et al., 1975. Reprinted with permission.)*

on the earth's surface. However, the value of *a* varies from place to place across the earth. This variation is due to the effect of latitude, altitude, and topography, as well as geology. The first three of these variations must be removed before the last residual one can be detected.

The earth is not a true sphere, but is compressed at the poles such that the polar radius is 21 km less than the equatorial radius. Thus the force of gravity increases with increasing latitude. Furthermore, gravity is affected by the speed of the earth's rotation, which increases the effect of latitudinal variation.

The correction for latitude is generally calculated using the international gravity formula of 1967, which supersedes an earlier version of 1930:

$$G = 978031.8[1/(0.0053024 \sin^2 y - 0.0000058 \sin^2 2y)]$$

where y = the latitude

The acceleration due to gravity is measured in gals (from Galileo): 1 gal is an acceleration of 1 cm/s^2. For practical purposes the milligal (mgal), which is 1 thousandth of 1 gal, or the gravity unit (gu), which equals 0.00001 gal, is generally used.

When allowance has been made for variations in G due to latitude, the gravity value still varies from place to place. This variation occurs because the force of gravity is also affected by local variations in the altitude of the measuring station and topography. Gravity decreases with elevation, that is, distance from the center of the earth, at a rate of 0.3086 mgal/m. Thus a *free-air* correction must be made to compensate for this effect. A second correction must be made to compensate for the force due to the mass of rock between the survey station and a reference datum, generally taken as sea level. This difference, the *Bouguer anomaly*, is 0.04191d mgal/m, where d is the density of the rock.

The free-air and Bouguer effects (E) can be removed simultaneously:

$$E = (0.3086 - 0.04191d)h$$

where d = density of rock
h = height in meters above reference datum

In mountainous terrain a further correction is necessary to compensate for the gravitational pull exerted by adjacent cliffs or mountains. When the free-air, Bouguer, and terrain corrections have been made for the readings at each station in a gravity survey, they may be plotted on a map and contoured in milligals (Fig. 3.46).

Gravity surveys can be carried out on land and at sea, both on the sea floor and aboard ship. In the latter case, corrections must be made for the motion of the ship and for the density and depth of the seawater. Recently, airborne gravity surveys have become feasible (Hammer, 1982).

Interpretation

The interpretation of gravity maps presents many problems, the simplest of which are caused by different subsurface bodies producing the same

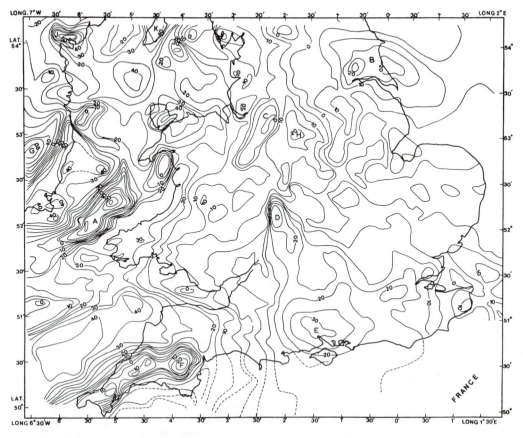

FIGURE 3.46 *Bouguer anomaly map of southern Great Britain. Contours in milligals. (From Maroof, 1974. Reprinted with permission.)*

anomaly on the surface. For example, distinguishing between a small sphere of large density and a large sphere of low density at similar depths is impossible.

The precise cause of an individual anomaly can be tested by a series of models, each model being posited on various depths, densities, or geometries for the body. Gravity maps are seldom used for such detailed interpretation, however, since seismic surveys are generally more useful for small areas. Like magnetic maps, gravity maps are more useful for showing the broad architecture of a sedimentary basin. In general terms depocenters of low-density sediment appear as negative anomalies, and ridges of dense basement rock show up as positive anomalies (Fig. 3.47).

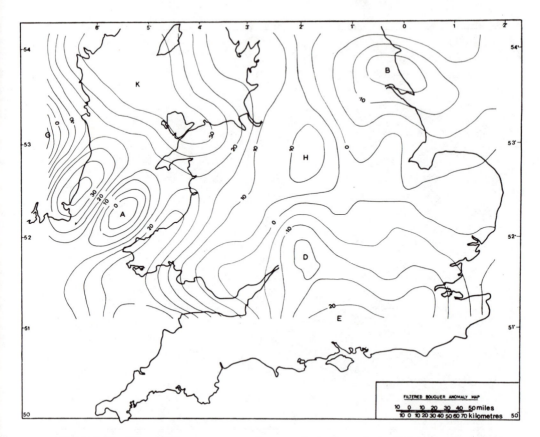

FIGURE 3.47   *Filtered Bouguer anomaly map of southern Great Britain computed
from the raw data shown in Figure 3.46. (From Maroof, 1974. Reprinted with
permission.)*

    In some circumstances gravity maps may indicate drillable prospects
by locating salt domes and reefs. Salt is markedly less dense than most sediments and, because of this lower density, often flows up in domes. These
salt domes can generate hydrocarbon traps in petroliferous provinces (see
p. 296). Because of their low density, salt domes may often be located from
gravity maps. Likewise, reefs may trap hydrocarbons (see p. 307), and they
may also show up as gravity anomalies because of the density contrast between the reef limestone and its adjacent sediments (Ferris, 1972).

Magnetic and Gravity Surveys: Summary

Both magnetic and gravity surveys are cost-effective methods of carrying
out reconnaissance surveys of areas before drilling. Their main use is in

defining the limits and scale of sedimentary basins and the internal distribution of structural highs and lows (Nettleton, 1972). When used in combination, a far more accurate basement map can be prepared than when they are interpreted separately (Fig. 3.48).

Neither method is sufficiently responsive to geological variations to be used to locate individual petroleum prospects, although gravity surveys can sometimes locate reef and salt dome traps (Fig. 3.49).

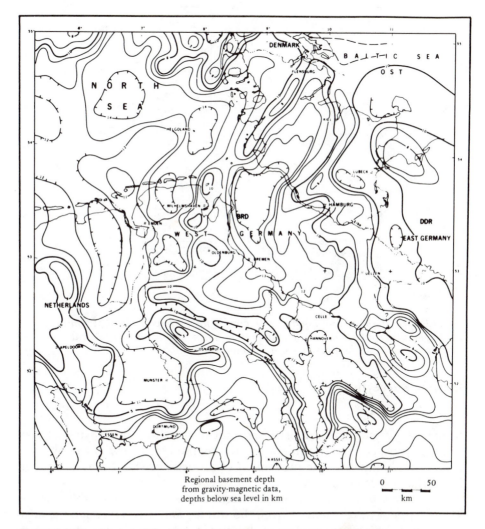

Regional basement depth
from gravity-magnetic data,
depths below sea level in km

0   50
km

FIGURE 3.48 *Regional depth-to-basement map (contours in kilometers) of northwest Germany based on a combination of magnetic and gravity surveys and geological data. (From Pratsch, 1979. Reprinted with permission.)*

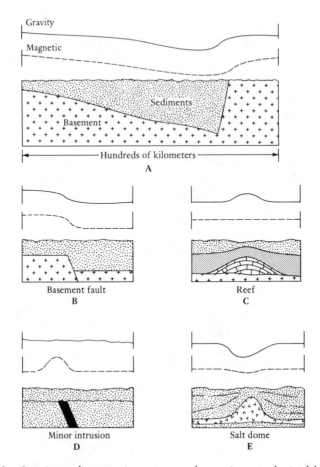

FIGURE 3.49 *Gravity and magnetic responses for various geological features.*
(A) Regional responses across a sedimentary basin. (B) through (E) Local
anomalies that may be present on a scale of a few kilometers.

Seismic Surveying

Basic Principles

Seismic surveying is the most important of the three main types of geophysical prospecting. To understand the method, a review of the physical
principles that govern the movement of acoustic or shock waves through
layered media is necessary.

Consider a source of acoustic energy at a point on the earth's surface.
Three types of wave emanate from the surface and travel through the adjacent layers, which have acoustic velocities and densities $v_1 p_1$ and $v_2 p_2$. *Sur-*
face, or *longitudinal, waves* move along the surface. The two other types of

wave, termed *body waves*, move radially from the energy source: *P (push) waves* impart a radial movement to the wave front; *S (shake) waves* impart a tangential movement. P waves move faster than S waves. Surface waves are of limited significance in seismic prospecting. They move along the ground at a slower velocity than body waves. The surface disturbance is termed *ground roll* and may include Rayleigh and other vertical and horizontal modes of propagation.

Seismic surveying is largely concerned with the primary P waves. When a wave emanating from the surface reaches a boundary between two media that have different acoustic impedance (the product of density and velocity), some of the energy is reflected back into the upper medium. Depending on the angle of incidence, some of the energy may be refracted along the interface between the two media or may be refracted into the lower medium.

The laws of refraction and reflection that govern the transmission of light also pertain to sound waves. Thus it is convenient to consider the movement of waves in terms of their ray paths. A ray path is a line that is perpendicular to successive wave fronts as an acoustic pulse moves outward from the source. Figure 3.50 shows the ray paths for the three types of wave just described.

The fundamental principle of seismic surveying is to initiate a seismic pulse at or near the earth's surface and record the amplitudes and travel times of waves returning to the surface after being reflected or refracted from the interface or interfaces of one or more layers of rock. Seismic surveying is more concerned with reflected ray paths than refracted ones.

If the average acoustic velocity of the rock is known, then it is possible to calculate the depth D to the interface:

FIGURE 3.50 *Cross-section illustrating various seismic wave paths.*

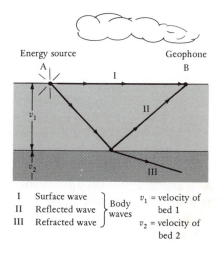

I	Surface wave	} Body waves
II	Reflected wave	
III	Refracted wave	

v_1 = velocity of bed 1

v_2 = velocity of bed 2

$$D = \frac{vt}{2}$$

where v is the acoustic velocity
        t is the two-way (i.e., there and back) travel time

The acoustic velocity of a rock varies according to its elastic constants and density. As sediments compact during burial, their density (and hence their acoustic velocities) generally increases with depth. Theoretically, the velocity of a P wave can be calculated as follows:

$$v = \sqrt{(k + \tfrac{4}{3}h)/p}$$

where k is the bulk modulus
        h is the shear modulus
        p is the density

The relationships between the velocity and density of common sedimentary rocks are shown in Figure 3.51. The product of the velocity and density of a rock is termed the *acoustic impedance*. The ratio of the amplitudes of the reflected and incident waves is termed the *reflectivity* or *reflection coefficient*:

$$\text{Reflection coefficient} = \frac{\text{amplitude of reflected wave}}{\text{amplitude of incident wave}}$$

$$= \frac{\text{change in acoustic impedance}}{2 \times \text{mean acoustic impedance}}$$

$$= \frac{p_2 v_2 - p_1 v_1}{p_2 v_2 + p_1 v_1}$$

where p_1 is the density of the upper rock
        p_2 is the density of the lower rock
        v_1 is the acoustic velocity of the upper rock
        v_2 is the acoustic velocity of the lower rock

Since velocity and density generally increase with depth, the reflection coefficient is usually positive. Negative coefficients occur where there is a downward decrease in velocity and density; for example, at the top of an overpressured shale diapir. A reflection coefficient that is positive for a downgoing wave is negative for an upgoing wave and, of course, vice versa.

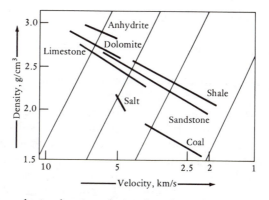

FIGURE 3.51   *The velocity-density relationships for sedimentary rocks. Note how velocity increases with density for most rocks except salt. Diagonals are lines of equal acoustic impedance. (After Gardner et al., 1974.)*

Velocity and density generally increase with depth for a homogenous lithology, but at the boundary between two units we can, and usually will, get a lithological contrast. Therefore the positive and negative reflection coefficients tend to appear with approximately equal frequency (taking the case of a downgoing wave only). Seismic processing is often based on the assumption that the earth's reflection coefficient series is random, which is empirically true. This assumption would not be the case if most reflection coefficients were positive. Negative coefficients occur where there is a decrease in acoustic impedance across the reflecting interface. Reflection coefficients have a maximum range limited by $+1$ and -1. Values of $+0.2$ or -0.2 would occur for very strong reflectors. In practice most reflection coefficients lie between -0.1 and $+0.1$. The boundary between a porous sand and a dense, tight limestone will have a high reflection coefficient and show up as a prominent reflecting surface, but the interface between two shale formations of similar impedance will have a negligible reflection coefficient and will reflect little energy.

It is now appropriate to see how these general principles are used in seismic surveying. Traditionally, seismic exploration involves three steps: data acquisition, data processing, and interpretation.

Data Acquisition

Seismic surveys are carried out on land and at sea in different ways. On land the energy source may be provided by detonating explosives buried in shot holes, by dropping a heavy weight off the back of a lorry (the thumper technique is actually a rather sophisticated procedure), or by vibrating a metal plate on the ground (Vibroseis). The returning acoustic waves are recorded on geophones arranged in groups. The signals are transmitted from the geophones along cables to the recording truck. Equipment in this truck controls the firing of the energy source and records the incoming signals from the geophones on magnetic tapes.

The shot points and the receiving geophones may be arranged in many ways. Many groups of geophones are commonly on line with shot points at the end or in the middle of the geophone spread. Today, common depth point, or CDP, coverage is widely used. In this method the shot points are gradually moved along a line of geophones. In this way up to 48 signals may be reflected at different angles for a common depth point.

The basic method of acquiring seismic data offshore is much the same as that of onshore, but it is simpler, faster, and, therefore, cheaper. A seismic boat replaces a truck as the controller and recorder of the survey. This boat trails an energy source and a cable of hydrophones (Fig. 3.52). In marine surveys dynamite is seldom used as an energy source. For shallow, high-resolution surveys, including sparker and transducer surveys, high-frequency waves are used. In sparker surveys an electric spark is generated between electrodes in a sonde towed behind the boat. Every time a spark is generated, it implodes after a few milliseconds. This implosion creates shock waves, which pass through the sea down into the strata. Transducers alternately transmit and receive sound waves. These high-resolution techniques generally only penetrate up to 1.0 s two-way time. They are useful for shallow geological surveys (e.g., to aid pipeline construction) and can sometimes indicate gas seepages (Fig. 3.53).

For deep exploration the *air gun* is a widely used energy source. In this method a bubble of compressed air is discharged into the sea; usu-

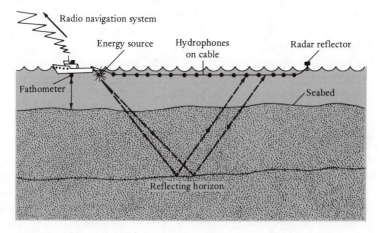

FIGURE 3.52 *Sketch showing how seismic data are acquired at sea.*

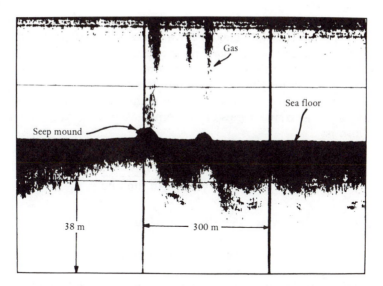

FIGURE 3.53 *Transducer record showing gas seeps and mounds in offshore Texas. (From Sieck and Self, 1977. Reprinted with permission.)*

ally a number of energy pulses are triggered simultaneously from several air guns. The air guns can emit energy sufficient to generate signals at between 5 and 6 s two-way travel time. Depending on interval velocities, these signals may penetrate to over 5 km.

The reflected signals are recorded by hydrophones on a cable towed behind the ship. The cable runs several meters below sea level and may be between 2 and 5 km in length. As with land surveys the CDP method is employed and up to 48 recorders may be used. The reflected signals are transmitted electronically from groups of hydrophones along the cable to the recording unit on the survey ship.

Other vital equipment on the ship includes a fathometer and position-fixing devices. The accurate location of shot points at sea is obviously far more difficult than it is on land. It is generally done either by radio positioning or by getting fixes on two or more navigation beacon transmitters from the shore, backed up by a satellite navigation system.

Data Processing

Once seismic data has been acquired, it must be processed into a format suitable for geological interpretation. This process involves the statistical manipulation of vast numbers of data using mathematical techniques far beyond the comprehension of geologists. Seismic data processors include mathematicians, physicists, electronic engineers, and computer programmers.

Consider the signals of three receivers arranged in a straight line from a shot point. The times taken for a wave to be reflected from a particular layer will appear as a wiggle on the receiver trace. The arrival times will increase with the distance of receivers from the shot point. Thus a time-distance graph may be constructed as shown in Figure 3.54. When time squared is plotted against distance squared, a straight line can be drawn through the arrival times of the signal. Velocity of the wave through the medium is:

$$\text{Velocity} = \sqrt{\frac{1}{\text{slope of the line}}}$$

The depth from the surface can then be calculated as follows:

$$\text{Depth} = \text{velocity} \times \text{time (one way)}$$

But since the travel time recorded is two-way (there and back):

$$\text{Depth} = \text{velocity} \times \frac{\text{time (two way)}}{2}$$

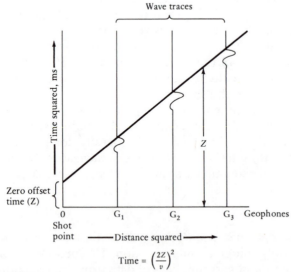

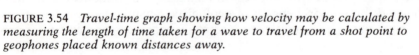

FIGURE 3.54 *Travel-time graph showing how velocity may be calculated by measuring the length of time taken for a wave to travel from a shot point to geophones placed known distances away.*

With common depth point shooting (discussed previously), reflections from the same subsurface points are recorded with a number of different combinations of surface source and receiver positions, and the signals are combined, or stacked. Thus wave-time graphs can be displayed in a continuous seismic section, and a single seismic reflecting horizon can be traced across it. Wave traces are displayed in several ways. They may be shown as a straightforward graph or *wiggle* trace. Reflectors show up better on a variable area display in which deviations in one direction (positive or negative polarity) are shown in black (Fig. 3.55). This type of display is the basis for the conventional seismic section used today.

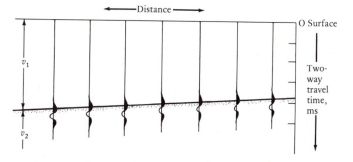

FIGURE 3.55 *Distance-time graph showing common depth point wave traces for a series of geophones recording a signal reflected from the boundary between rocks with velocities v_1 and v_2. The wave traces are shaded in the variable area format. This is the basis for the conventional seismic section.*

Four main steps are involved in the processing of raw seismic data before the production of the final seismic section (McQuillan and Ardus, 1977):

1. Conversion of field magnetic tape data into a state suitable for processing.

2. Analysis of data to select optimum processing parameters (e.g., weathering correction and seawater velocity).

3. Processing to remove multiple reflectors and enhance primary reflectors.

4. Conversion of data from digital to analogue form, and printout on graphic display (i.e., final seismic line).

During processing, many unwanted effects must be filtered out. All seismic records contain both the genuine signal from the rocks and the background "noise" due to many extraneous events, which may range from a microseism to a passing bus. The signal-noise ratio is a measure of the

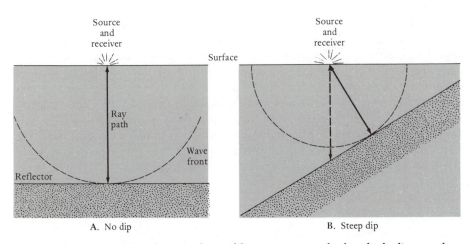

FIGURE 3.56   *Diagrams showing the problem encountered when beds dip steeply. In the left-hand section (A) the strata are horizontal. Thus the wave fronts first strike reflectors vertically beneath the energy source and receiver. In the right-hand section (B), however, the beds are steeply dipping. Thus signals return to the receiver from points up-dip earlier, and thus apparently shallower, than from the true vertical depth. This phenomenon may be corrected by using a process termed migration.*

quality of a particular survey. Noise should obviously be filtered out. The frequency and amplitude of the signal is the result of many variables, including the type of energy source used and its resultant wave amplitude, frequency, and shape, and the parameters of the various rocks passed through. Deconvolution is the reverse filtering that counteracts the effects of earlier undesirable filters.

One particularly important aspect of processing is wave migration. When beds dip steeply, the wave returns from the reflector from a point not immediately beneath the surface location midway between the shot point and each individual geophone but from a point up-dip from this position (Fig. 3.56). During processing, the data must be migrated to correct this effect. This migration causes several important modifications in the resultant seismic section. Anticlines become sharper, synclines gentler, and faults more conspicuous.

The preceding discussion briefly covered some aspects of seismic processing. The main point to note is that digitally recorded seismic data can be processed time and time again, either by using new, improved computer programs or by tuning the signal to bring out aspects of particular geological significance.

Interpretation

Geologists looking at seismic lines and maps inevitably tend to see them as representations of rock and forget that they do, in fact, represent time.

According to Tucker and Yorston (1973), three main groups of pitfalls in seismic interpretation are to be avoided:

1. Pitfalls due to processing
2. Pitfalls due to local velocity anomalies
3. Pitfalls due to rapid changes in geometry

Pitfalls due to processing are the most difficult for geologists to avoid. One of the most common processing pitfalls is multiple reflections—a series of parallel reflections caused by the reverberation between two reflectors. The most prevalent variety is the seabed multiple, which, as its name states, is caused by reverberation between the sea surface and the sea floor (Fig. 3.57A). Harder to detect are multiples caused by events within the sediments themselves. These events generally occur where there is a formation with high reflection coefficients above and below. Multiples can be removed by deconvolution and filtering during processing.

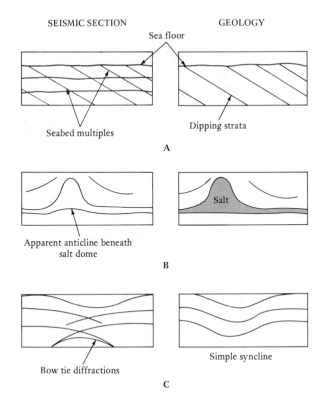

FIGURE 3.57 *Examples of each of the three major groups of pitfalls in seismic interpretation: (A) processing pitfall, (B) velocity pitfall, and (C) geometry pitfall.*

Rapid local variations in formation velocity cause many pitfalls. Two of the best-known examples of these variations are produced by salt domes and reefs. In both cases salt and limestone may have faster travel times than the adjacent sediments. Thus the pre-salt or pre-reef reflector may appear on the seismic section as an apparent anticline, when it is, in fact, a velocity pull-up (Fig. 3.57B). A number of salt domes were drilled in the quest for such fictitious anticlines until the method of undershooting them was developed as an effective aid to mapping pre-salt structure (Krey and Marshall, 1975). Less serious a pitfall is the sub-reef "anticline," since the reef itself is a valid petroleum prospect and the well will probably be scheduled to drill to pre-reef rocks anyway.

On a regional scale, units may sometimes apparently thin with increasing depth from shelf to basin floor. The formations may, in fact, have a constant thickness, but their velocities increase with increasing compaction. Thus the time taken for seismic waves to cross each interval decreases basinward.

The third group of pitfalls in seismic interpretation occur because of the departure of rock geometry from a simple layered model. This pitfall may cause reflectors to dip steeply, as in tight folds and diapirs, or even to terminate (adjacent to faults and diapirs). The former can give rise to distorted reflections, such as the *bow tie* effect produced by synclines (Fig. 3.57C). Reflector terminations cause diffractions, which cross-cut the reflectors on the seismic line.

When the seismic data have been correctly interpreted, thus avoiding the pitfalls just discussed, the various reflecting horizons can be mapped. Brightly colored pencils can be used to delineate laterally persistent reflecting horizons. Since seismic lines are generally shot on an intersecting grid, reflectors can be tied from line to line. For large surveys reflectors can be traced from line to line and tied in a loop to check that the reflector has been correctly picked and that the interpreter has not jumped a *leg* in the process.

The times for the reflectors for each shot point may then be tabulated. Contour maps may then be drawn for each horizon (red, blue, green, etc.). Note that these are time maps, not depth maps. They show isochrons (lines of equal two-way travel time), not structure contours. Figure 3.58 shows, in a very simple way, how reflectors may be picked on two intersecting seismic lines and used to construct isochron maps.

Geological Applications

Seismic lines and maps are applied to petroleum exploration in five main ways: regional mapping, prospect mapping, reservoir delineation, seismic modeling, and direct hydrocarbon detection. The first three of these applications form a continuous spectrum of endeavor, whereas the last two are quite separate.

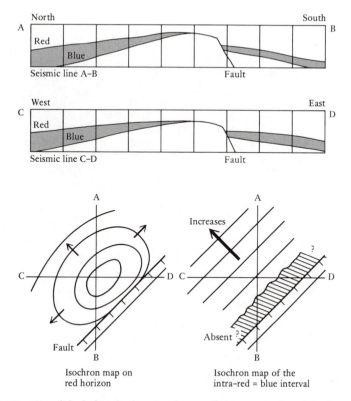

FIGURE 3.58 *Simplified sketch showing how reflectors may be picked on two intersecting seismic lines A–B and C–D and used to construct isochron maps of an individual reflector (bottom left) or the interval between two reflectors (bottom right).*

In the early stages of reconnaissance mapping of a new sedimentary basin, a broad seismic grid will be shot and integrated with magnetic and gravity surveys. Vail et al. (1977) have shown how seismic mapping can be used on a regional scale to delineate seismic sequences. Each sequence contains a series of concordant reflectors and is bounded above and below by discordances. The boundaries between sequences may be caused by onlap, top lap, downlap, or truncation. The age and lithology of a seismic sequence may not be directly apparent without well control, but they are, nonetheless, mappable units (Figs. 3.59 and 3.60). Studies by Vail and his colleagues (1977) showed the presence of global seismic stratigraphy in which a regular sequence of characteristic seismic sequences can be related to fluctuations in sea level. As discussed in Chapter 5, major transgressions appear to be related to episodes of source rock deposition (p. 181).

Seismic data may be mapped either in terms of seismic sequences, which are of global stratigraphic significance, or as seismic facies. Seismic facies are defined by the internal characteristics of seismic sequences, such

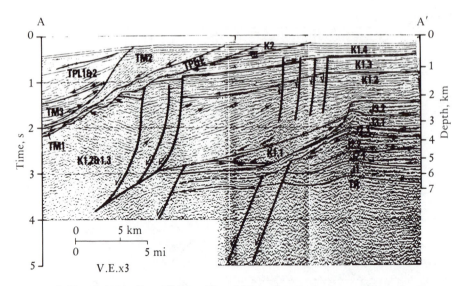

FIGURE 3.59   *Seismic line offshore West Africa showing sequence boundaries ranging in age from Tertiary (T) to Cretaceous (K), Jurassic (J), and Triassic (TR). For location see Figure 3.60. (From Vail et al., 1977. Reprinted with permission.)*

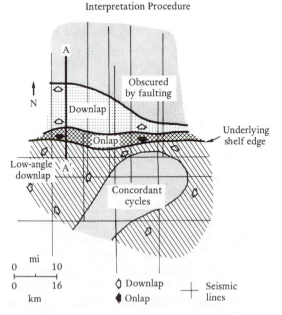

FIGURE 3.60   *Seismic facies map showing reflection patterns at lower surface of Lower Cretaceous sequences, offshore West Africa. A–A' is the seismic line shown in Figures 3.59 and 3.61. (After Vail et al., 1977.)*

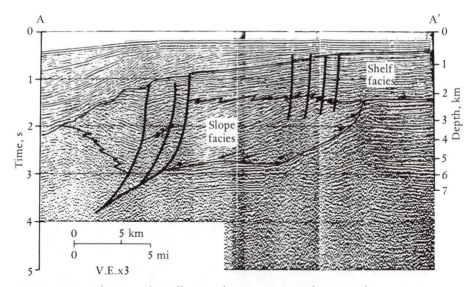

FIGURE 3.61   *The same line illustrated in Figure 3.59 (location shown in Figure 3.60) showing seismic facies boundaries. (From Vail et al., 1977. Reprinted with permission.)*

as the amplitude, continuity, and character of reflectors. These variables give an indication of the lithology and sedimentary environment of the facies (Fig. 3.61). Large-scale sedimentary features that may be recognized include prograding deltas, carbonate shelf margins, and submarine fans (Sheriff, 1976).

Moving from the regional to the local scale, once a closely spaced seismic grid has been shot, drillable prospects may be identified on sections and contour mapped in plan. Structural and occasionally even stratigraphic traps may be located. Chapter 7 deals with traps in much more detail. Suffice it to say for now, seismic maps may identify structural traps caused by folding and faulting (Figure 3.62). Because the quality of seismic data has improved in recent years, delineating small-scale sedimentary units, such as channels, sand bars, and reefs, has also become possible (Harms and Tackenberg, 1972; Lyons and Dobrin, 1972; Payton, 1977). Many of these features are stratigraphic traps (Fig. 3.63).

Once a trap has been drilled and found to contain hydrocarbons, it may be subjected to still more detailed seismic investigation. This further investigation will include careful mapping of a reflector on top of the reservoir (or as near to it as possible) as an aid to accurate reserve calculation. It may also include a three-dimensional seismic survey, in which a very close grid is shot. The abundance of data allows seismic lines to be prepared on any selected orientation, even horizontally, so that time slices may be displayed. Detailed study of the amplitude and character of individual seismic signals may also be instructive; it may give some clue as to the

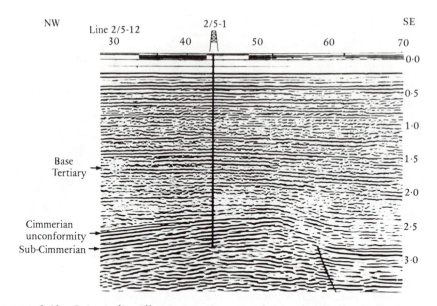

FIGURE 3.62   *Seismic line illustrating a structural trap: the Heather field of the northern North Sea. The reservoir is a Jurassic sand in a truncated fault block beneath the Cimmerian unconformity. (From Gray and Barnes, 1981. Reprinted with permission.)*

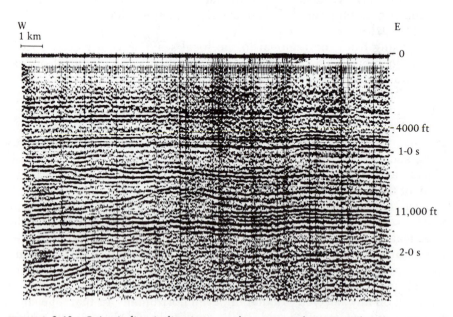

FIGURE 3.63   *Seismic line indicating a reef: a potential stratigraphic trap. (From Fitch, 1976. Reprinted with permission.)*

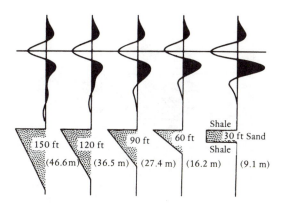

FIGURE 3.64 *Seismic responses for different types of sand-shale sequence. These signals may give some indication of the depositional environment of a sand body. Those signals to the left suggest progradational deposition. The right-hand example is more typical of some channels. (After Neidell and Poggiagliolmi, 1977.)*

thickness of a sand and the nature of its upper and lower boundaries (Fig. 3.64). This information can give an indication of depositional environment and, hence, reservoir geometry and continuity.

This last type of analysis of seismic data leads on logically to seismic modeling. The interpretation of seismic data becomes increasingly refined as wells are drilled. Not only the sonic log but velocity check shots give accurate information on the thickness and velocities of various formations. From this data, setting up a geological model in which the scale, shape, and velocities of rocks are specified is a small step (McDonald et al., 1982). A computer may then be programmed to construct simulated seismic sections for that model (Fig. 3.65).

The final application of seismic methods to hydrocarbon exploration is in direct detection. In certain situations oil-water or, more commonly, gas-water contacts may actually show up as a reflector on a seismic line (Stone, 1977). Thus, curved reflectors may define an anticline, within the core of which may be a horizontal reflector termed a *flat spot*. Sometimes, but unfortunately not always, this flat spot may indicate a hydrocarbon-water contact (Fig. 3.66). In addition a high-amplitude event may occur in the seismic trace where it goes from the cap rock into porous, gas-filled sand. This event is termed a *bright spot* or *dim spot* if the polarity is negative or positive, respectively.

The preceding review, although brief, was intended to show the ways in which a seismic survey can be used in petroleum exploration and to show some of the problems that lie in wait for the geologist who forgets that a seismic line displays time, not depth.

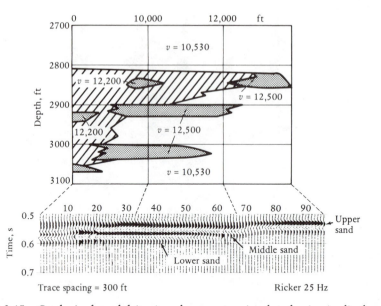

FIGURE 3.65 *Geological model (top) and computer-simulated seismic display (bottom) for regressing coastal barrier bar sands. (Modified from Meckel and Nath, 1977, with permission.)*

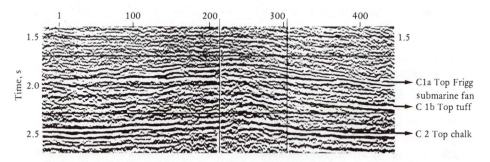

FIGURE 3.66 *Seismic line through the Frigg field of the northern North Sea. The reservoir is a submarine fan, whose paleotopography is clearly visible. Within the fan a flat reflector marks the gas-oil contact. (From Heritier et al., 1979. Reprinted with permission.)*

SUBSURFACE GEOLOGY

Successful petroleum exploration involves the integration of the wireline logs and geophysical surveys, described earlier in this chapter, with geological data and concepts. This section is concerned with the manipulation and presentation of subsurface geological information. It is written on the assumption that the reader is familiar with the basic principles of stratigraphy, correlation, and contour mapping. Many of the methods to be discussed are not, of course, peculiar to petroleum exploration, but have wider geological applications. Essentially, the two methods of representing geological data are cross-sections and maps. These methods are discussed in the following sections.

Geological Cross-Sections

Vertical cross-sections are extremely important in presenting geological data. This account begins with small-scale sections and well correlations and proceeds to regional sections.

Well Correlation

The starting point for detailed cross-section construction, as for example within a field, is consideration of well correlation. When a well has been drilled and logged, a composite log is prepared. This log correlates the geological data gathered from the well cuttings with that of the wireline logs. The formation tops then have to be picked, which is not always an easy task. The geologist, the paleontologist, and the geophysicist may all pick the top Cretaceous, for example, at a different depth. The geologist may pick the top at, say, the first sand bed; the paleontologist, at the first record of a particular microfossil (which may not have thrived in the environment of the sediments in question); and the geophysicist, at a velocity break in the shales, which is a prominent reflecting horizon.

When the formation tops have been selected, correlation with adjacent wells can be made. This correlation is an art in which basic principles are combined with experience. First, the most useful geophysical logs to use must be determined. In the early days of well correlation the IES array was the most widely used for composite logs. Now the gamma-sonic is often more popular, especially where the S.P. curve is ineffective. The gamma, sonic, and resistivity curves all tend to share the high amplitudes necessary for effective correlation. As a general guide, coal beds and thin limestones often make useful markers: they are both thin, but also give dramatic kicks on resistivity and porosity logs. Figure 3.67 illustrates a difficult series of logs to correlate. The correlation has been achieved by using the coals to the best advantage, but it allows them to be locally absent because of nondeposition or erosion by channels.

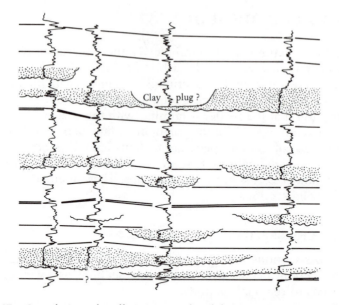

Clay ⚡ plug ?

FIGURE 3.67   *Correlation of wells in a complex deltaic reservoir. Note that the coals provide the best markers. They have been identified not just from the gamma logs shown here but also from the porosity logs. (After Rider and Laurier, 1979.)*

Sometimes when correlating wells, significant intervals of section may be missing. This phenomenon may be caused by depositional thinning, erosion, or normal faulting (Fig. 3.68). Examination of the appropriate seismic lines will generally reveal which of these three possibilities is the most likely. Repetition of sections may sometimes be noted. Repetition may be caused by reverse faulting, but this possibility should only be considered in regions known to have been subjected to compressional tectonics. An alternative explanation is that sedimentation is cyclic, causing the repetition of log motifs.

Constructing Cross-Sections

Once wells have had their formation tops picked and they have been correlated, they may be hung on a cross-section. This procedure is done with a datum, which may be sea level, a fluid contact, or a particular geological marker horizon. When sea level is used, the elevation of the log must be adjusted. Log depths are generally measured below the rotary table or kelly bushing. The elevation of whichever one of these was used is given on the log heading. The elevation above sea level is subtracted from each formation top to find its altitude or depth. This procedure is relatively simple

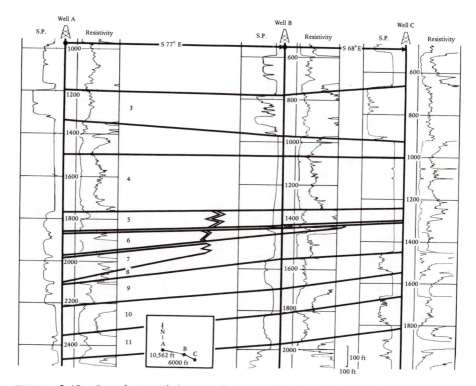

FIGURE 3.68   *Correlation of three wells in the Niger delta showing lateral thinning and facies change. (Courtesy of Schlumberger.)*

for vertical wells, but less so for deviated wells, such as those drilled from a marine production platform. For these types of wells the true vertical depth (TVD) must be determined, which requires a detailed and accurate knowledge of the path of the borehole.

When a cross-section is drawn to a horizontal datum, be it sea level or an oil-water contact, the result is a structural cross-section (Fig. 3.69). Alternatively, a cross-section can be constructed using a geological horizon as a datum (Fig. 3.70). The purpose of the cross-section should be considered before the datum is selected. In Figure 3.70 the purpose is to show the truncation of horizons below the Cimmerian unconformity.

Cross-sections of fields are generally based on well control. For regional studies a combination of seismic and well data is used. A development of the single cross-section is a series drawn using a sequence of different data horizons. Where these data horizons are selected to span a number of markers up to the present day, they can be used to document the evolution of a basin or an individual structural feature (Fig. 3.71).

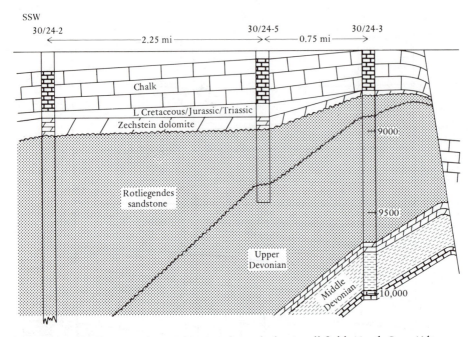

FIGURE 3.69 *Structural cross-section through the Argyll field, North Sea. (After Pennington, 1975.)*

Subsurface Geological Maps

Many different types of subsurface geological maps are used in oil exploration. Only a few of the more important kinds are reviewed in this section. The simplest, and probably most important, subsurface map is the structure contour map. This map shows the configuration of a particular horizon with respect to a particular datum, generally sea level. Structure contour maps may be regional or local, indicating the morphology of basins and traps, respectively. This information may be based on seismic data, well control, or, most effectively, a combination of both. Structure contour maps delineate traps and are essential for reserve calculations.

Next in importance are isopach maps, which record the thickness of formations. As with structural maps they may be regional or local and tend to be most reliable when seismic and well data are integrated. Care must be taken when interpreting isopach maps. There is a natural tendency to think that thickness increases with basin subsidence, which is not necessarily so. Both sand and carbonate sedimentation rates are often at a maximum a short way into a basin. Beds thin toward the shore because of erosion and basinward because of nondeposition or slow sedimentation rates (see also p. 334). Likewise, local or regional truncation may result in formation thickness being unrelated to syndepositional basin morphology.

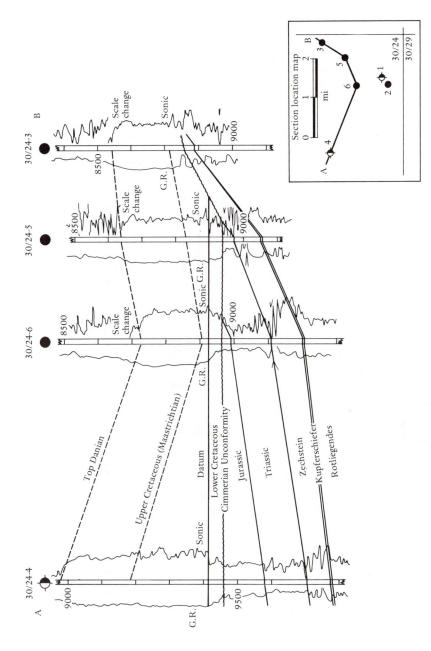

FIGURE 3.70 Cross-section of wells through the Argyll field of the North Sea, using the Cimmerian unconformity as a datum. This figure usefully demonstrates the pre-Lower Cretaceous truncation of strata. (After Pennington, 1975.)

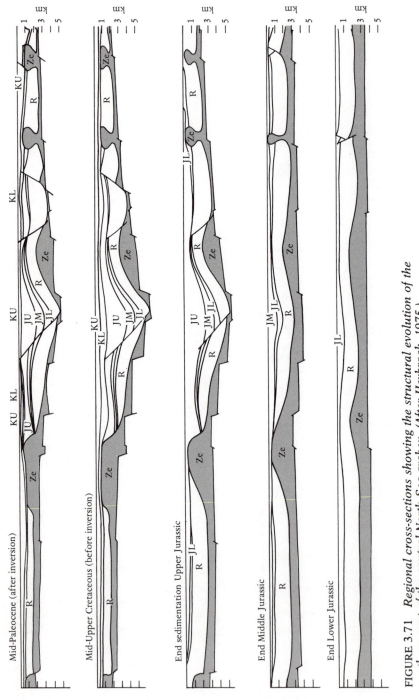

FIGURE 3.71 *Regional cross-sections showing the structural evolution of the southern part of the central North Sea graben. (After Heybroek, 1975.)*

116

An isochore map is a particular type of isopach map that indicates the thickness of the interval between the oil-water contact and the cap rock of a trap. A related type of subsurface geological map is the net pay map, which contours the ratio of gross pay to net pay within a reservoir. Sand-shale ratio maps can be drawn, which are very useful both on a local and a regional scale, because they may indicate the source of sand and, therefore, areas where good reservoirs may be found. Net sand and net pay maps are essential for detailed evaluation of oil and gas fields. Field development also requires maps that contour porosity and permeability across a field in *percent* and *millidarcies*, respectively. These maps are generally based on well information, but, as seismic data improve, net sand and porosity maps can be constructed without well control. Obviously, their accuracy is enhanced when they can be calibrated with log information. Figure 3.72 illustrates the various types of map just discussed for the Lower Permian (Rotliegendes) of the southern North Sea basin.

Net sand maps may often be combined with paleogeographic maps. Paleogeographic maps should be based on seismic and well data from which the depositional environment has been interpreted. This data may then be used to delineate the paleogeography of the area during the deposition of the sediments studied. Such maps can be used to predict the extent and quality of source rocks and reservoirs across a basin (Fig. 3.73).

Another useful type of map is the subcrop, or preunconformity map, which can be constructed from seismic and well data (Fig. 3.74). Regional subcrop maps are useful for showing where reservoirs are overlain by source rocks and vice versa. On a local scale subcrop maps may be combined with isopach maps to delineate truncated reservoirs. Subcrop maps also give some indication of the structural deformation imposed on the underlying strata.

Finally, maps of all the types previously discussed may be used in combination to delineate plays and prospects. A play map shows the probable geographic extent of oil or gas fields of a particular genetic type (i.e., a reef play, a rollover-anticline play, etc.). Figure 3.75 shows the extent of the Permian gas play of the southern North Sea. It selectively overlays several of the maps previously illustrated. Subject to a structure being present, the parameters that define this play include subcropping Carboniferous coal beds (the source rock), Lower Permian Rotliegendes of eolian type (the other facies lack adequate porosity and permeability), and capping Upper Permian Zechstein evaporites to prevent the gas from escaping. Prospect maps are also based on a judiciously selected combination of the previously discussed types of map, but will, of necessity, contain a structure contour map to define the prospect in question.

The preceding review shows some of the subsurface geological maps that may be constructed and some of the problems of making and interpreting them. Examples of these maps are used throughout the following chapters.

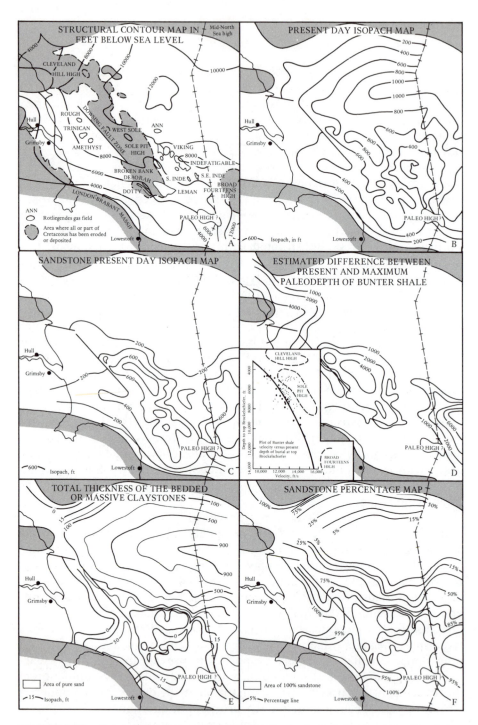

FIGURE 3.72 *Maps of the Lower Permian (Rotliegendes) of the southern North Sea basin showing the various types of map that can be constructed from the same basic data. (After Marie, 1975.)*

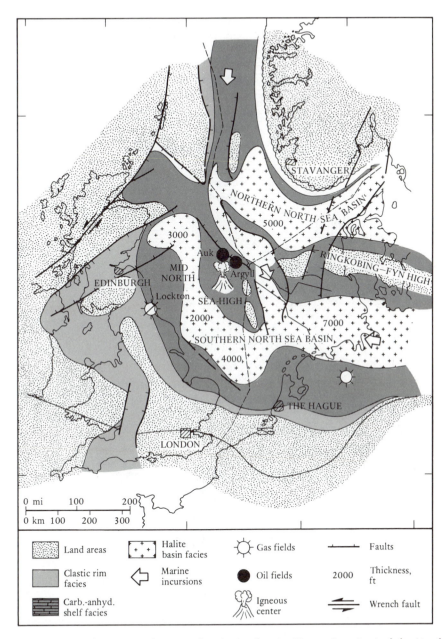

FIGURE 3.73 *Paleogeographic map for the Zechstein (Upper Permian) of the North Sea. (After Ziegler, 1975.)*

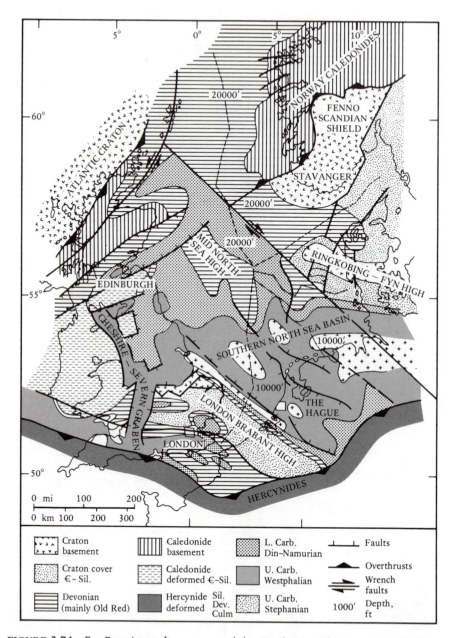

FIGURE 3.74 *Pre-Permian subcrop map of the North Sea. This map is useful for showing the nature and extent of pre-Permian structural deformation and erosion. It also indicates the extent of the Upper Carboniferous coal measures, which are the source for gas trapped in overlying Permian sands. (After Ziegler, 1975.)*

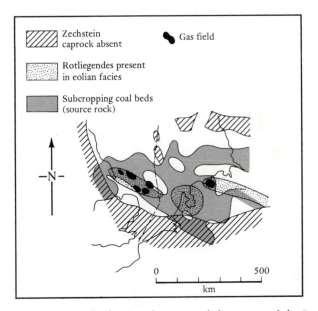

FIGURE 3.75 *Map showing the factors that control the extent of the Permian Rotliegende gas play of the southern North Sea.*

REMOTE SENSING

Remote sensing is the collection of data without the actual contact of the object being studied. Thus in terms of geology remote sensing includes aeromagnetic and gravity geophysical surveys, which have already been described (p. 85). This section is concerned only with remote sensing using electromagnetic waves.

The electromagnetic spectrum includes visible light, but ranges from cosmic radiation to radio waves (Fig. 3.76). Remote sensing can be carried out from various elevations, ranging from treetop height to satellites in the upper atmosphere. Thus the two parameters to remote sensing using electromagnetic waves are elevation and the wavelength analyzed. However, these techniques contain many common factors. An energy source is essential: it may be induced (as with microwave radiation used in radar) or natural (as with solar radiation). The sun's electromagnetic waves are filtered through the atmosphere and either absorbed by or reflected from the earth's surface. Reflected waves are modified by surface features of the earth. Some types of wave vary according to thermal variance of the surface, vegetation cover, geology, and so on. These wave variations can be measured photographically or numerically. They are then analyzed, either visually and subjectively or by more sophisticated computer processing. The results may then be interpreted (Fig. 3.77).

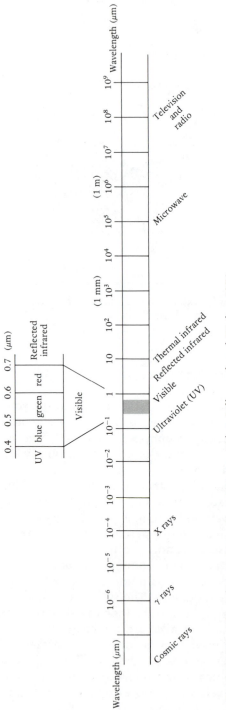

FIGURE 3.76 *The electromagnetic spectrum. (After Lillesand and Kiefer, 1979. Remote Sensing and Image Interpretation, © John Wiley & Sons, Inc.)*

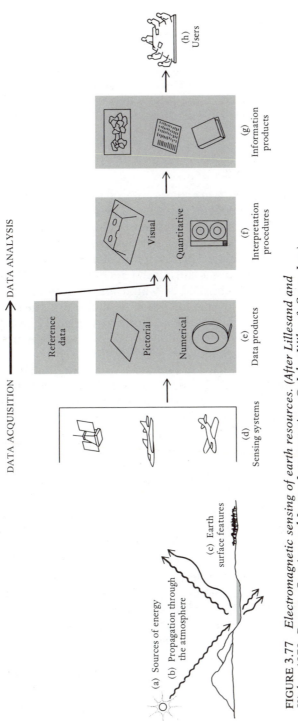

DATA ACQUISITION ⟶ DATA ANALYSIS

(a) Sources of energy
(b) Propagation through the atmosphere
(c) Earth surface features

(d) Sensing systems

Reference data

Pictorial
Numerical
(e) Data products

Visual
Quantitative
(f) Interpretation procedures

(g) Information products

(h) Users

FIGURE 3.77 *Electromagnetic sensing of earth resources. (After Lillesand and Kiefer, 1979. Remote Sensing and Image Interpretation, © John Wiley & Sons, Inc.)*

123

The field of remote sensing is large and currently expanding rapidly. Many new techniques are evolving and are being applied not only to petroleum exploration but to other geological surveys and the study of many other aspects of the earth's surface. The main remote sensing techniques applied to petroleum exploration are visual, radar, and multispectral. These techniques are described in the following sections. For further details see Lillesand and Kiefer (1979) and Siegal and Gillespie (1980).

Visual Remote Sensing

Conventional aerial photography is one of the oldest applications of remote sensing to petroleum exploration. The first aerial photographs were taken from balloons in the late nineteenth century. In the early part of the twentieth century, photos were taken from planes, mainly as isolated snapshots used for map making. The discipline of aerial photography advanced rapidly for military purposes during the Second World War. By this time a basic technique was established by which specially designed cameras took a sequence of photos at regular intervals along a plane's carefully controlled flight path, which must be accurately mapped. Photos are taken at a sufficiently close interval to produce a 50 to 60 percent overlap (endlap within a line and sidelap between adjacent lines).

The photos may be used in two ways. In one method they can be fitted together in a mosaic, with the unwanted overlapping photograph carefully removed. The resultant mosaic may then be rephotographed for ease of reproduction. Air photo mosaics may be used as a base for both topographic and geological surveys. Before a final map is prepared, however, the mosaic requires a number of corrections. The actual scale of the mosaic must be calibrated by reference to locations whose separation has been measured on the ground. Scale may not be constant over the mosaic if altitude has altered during the flight or if the elevation of the land surface varies. The terrain elevation is generally made with reference to sea level. The exact orientation and location of the mosaic on the earth's surface must also be determined. This task may be done by making astrofixes of ground locations, which can be clearly identified on the photograph.

A second technique widely employed in aerial photography is to use the overlap of adjacent photos. Two overlaps can be viewed side by side stereoscopically (Fig. 3.78). Thus the topography can actually be seen in three dimensions. In the early days of aerial photography the photos were viewed with a simple binocular stereoscope. Today far more sophisticated viewers are available. Accurately measuring ground elevation with respect to a datum is now possible. Thus spot heights may be digitized and located on a base map, or contours can be drawn straight off from the photographs. Therefore extremely accurate topographic surveys are possible.

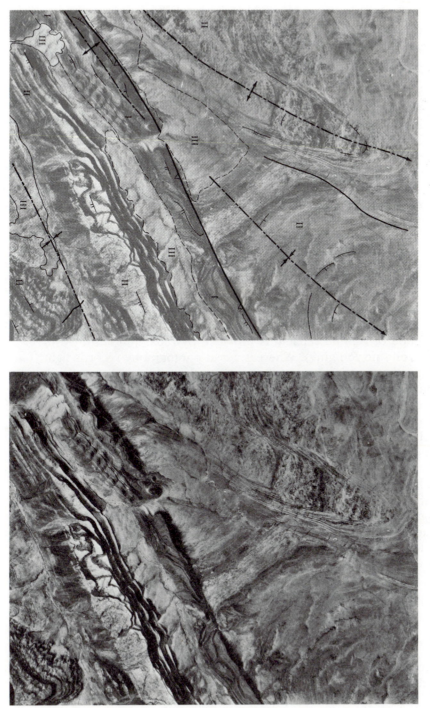

FIGURE 3.78   Stereo pair of air photos with interpretation map. The scale is approximately 1:30,000. I represents the oldest unit exposed and consists of well-bedded limestones; II consists of sands and shales; and III represents superficial deposits. Several folds and faults can be mapped. (Courtesy of Huntings Geology and Geophysics, Ltd.)

Since topography is closely related to geology, aerial photographs can also be used in geological mapping. The boundaries of various rock units may be delineated, but, more importantly, the actual dip and strike of strata can be measured from air photos, so regional structure can be mapped. Furthermore, linear features show up from the air far better than on the ground. Major structural lineaments, invisible on the ground, can be seen and traced for hundreds of kilometers using air photos. At the other extreme of scale, fracture systems may be detected from air photos and their orientation and frequency numerically analyzed.

The preceding account pertains to the interpretation of air photographs shot from planes at elevations of only a few kilometers. Satellite photographs are useful in a rather different way. Because of the great height, stereoscopic mapping is not feasible. On the other hand satellite photographs may show the presence of major lineations invisible from the ground or from the lower elevations of aircraft. Similarly, because of the large area covered, whole sedimentary basins may be seen in a single photograph, complete with concentric centripetally dipping strata.

Radar

Another method of remote sensing is provided by radar (an acronym of *RA*dio *De*tection *A*nd *R*anging). Whereas aerial photography records light reflected from the sun, radar uses an energy source on the plane or satellite that emits microwave radiation. The returning radiation is recorded and displayed in a format similar to an aerial photograph.

Radar has several advantages over light photographs. Microwaves penetrate cloud cover and haze and can be used at night as well as during the day. Thus radar surveys can be carried out with fewer constraints than can photographic surveys. A second advantage is that radar records data continuously, rather than in a series of separate photographs, so many of the problems of overlap and matching are eliminated.

The spatial resolution of radar imagery is largely related to the diameter of the antenna. The larger the antenna, the better the resolution for a given wavelength. Because of the difficulty in mounting a rotating antenna on a plane, it is generally mounted underneath and pointed out to the side. Thus most radar systems are side-looking airborne radar, referred to as SLAR or, more simply, SLR.

Multispectral Scanning

A conventional aerial photograph measures reflected light. Measuring the thermal radiation of the earth's surface using a thermal sensor is also possible. Multispectral scanning extends these concepts, measuring several different energy wavelengths simultaneously. Instead of recording light on film via an optic lens, the data are recorded electronically. This type of recording has the great advantage of eliminating the need to have the information transported to the ground in a film that has to be processed. The data can be transmitted from satellite (or spacecraft) by radio waves. The information can then be displayed visually in false-color photographs, which display variations not of light but of whatever wavelength was selected for recording and analysis. Whereas light photographs show wavelengths of 0.3 to 0.9 μm, multispectral scanning can extend this value to some 14 μm.

Because the information is recorded electronically rather than optically, it can be processed by computers in many ways. Three principle steps are used in processing before the data can be interpreted. First, the image must be restored as closely as possible to the original. Image defects and the effects of haze are corrected. The location of the image with respect to the earth's surface must be exactly fixed, and adjacent images fitted into a mosaic. In the second stage of processing, the image is improved by enhancing the edges of different types of surface by contrast stretching, density slicing, and selecting colors for various spectral bands. In the third and final stage of processing, the information is extracted and displayed in hard copy form (Plate 3) using false colors.

Multispectral scanning provides far more data than a conventional photo does. Because such a wide range of spectra can be recorded and because the data can be processed in so many ways, the information is open to a great diversity of interpretation. Multispectral scanning can be carried out from planes, but is also extensively employed in satellites, notably the LANDSAT and SEASAT systems. For further details see Sabins (1978).

Conclusions

Remote sensing is an invaluable technique in petroleum exploration. Indirectly, it is useful for topographic mapping; more directly, aerial photography has been widely used in geological mapping, especially in desert areas, where the effect of vegetation is minimal. The application of multispectral sensing from satellites is still in its infancy, and great care must be taken in its interpretation. Halbouty (1980) has shown how 15 giant oil and gas fields around the world are not revealed by LANDSAT imagery. As with all exploration tools, multispectral sensing is open to abuse and misinterpretation when used on its own. When used in conjunction with other techniques, such as gravity and magnetics, it may delineate anomalies that deserve further attention on the ground.

SELECTED BIBLIOGRAPHY

Oil well drilling and production:
MOORE, P. L. 1974.   *Drilling Practices Manual.* Tulsa: Pennwell, 448 pp.
NIND, T. E. W. 1981.   *Principles of Oil Well Production.* New York: McGraw-Hill, 391 pp.

Formation evaluation:
Books and chart books of the various wireline service companies are essential (and free), but some assume an understanding of the subject. Other useful books include:
ASQUITH, G. 1982.   *Basic Well Log Analysis for Geologists.* Tulsa: Am. Assoc. Petrol. Geol., 216 pp.
MERKEL, R. H. 1979.   *Well Log Formation Evaluation.* Am. Assoc. Petrol. Geol. Course Note Series, No. 4, 82 pp.
PIRSON, S. J. 1983.   *Geological Well Log Analysis,* Third Edition. Houston: Gulf Publishing Co., 424 pp.

Geophysics:
Books that cover gravity, magnetic, and seismic surveying:
DOBRIN, M. B. 1976.   *Introduction to Geophysical Prospecting.* New York: McGraw-Hill, 630 pp.
McQUILLAN, R. and ARDUS, D. A. 1977.   *Exploring the Geology of Shelf Seas.* London: Graham and Trotman, 234 pp.
NETTLETON, L. L. 1976.   *Gravity and Magnetics in Oil Prospecting.* New York: McGraw-Hill, 464 pp.
TELFORD, W. M., GELDART, L. P., SHERIFF, R. E., and KEYS, D. A. 1976.   *Applied Geophysics.* London: Cambridge University Press, 860 pp.

For seismic prospecting only, see:
ANSTEY, N. A. 1982.   *Simple Seismics.* Boston: International Human Resources Development Corporation, 168 pp.
COFFEEN, J. A. 1978.   *Seismic Exploration Fundamentals.* Tulsa: Petroleum Publishing Co., 277 pp.
FITCH, A. A. 1976.   *Seismic Reflection Interpretation.* Berlin: Gebruder Borntraeger, 148 pp.
KLEYN, A. H. 1982.   *Seismic Reflection Interpretation.* Barking: Applied Science Publishers, 269 pp.
PAYTON, C. E. 1977.   *Seismic Stratigraphy—Applications to Hydrocarbon Exploration.* Am. Assoc. Petrol. Geol., Mem. No. 26, 516 pp.
TUCKER, P. M. and YORSTON, H. J. 1973.   *Pitfalls in Seismic Interpretation.* Soc. Expl. Geophys., Monograph No. 2, 50 pp.
WATERS, K. H. 1978.   *Reflection Seismology.* New York: Wiley, 394 pp.

Remote sensing:
LILLESAND, T. M. and KIEFER, R. W. 1979.   *Remote Sensing and Image Interpretation.* New York: Wiley, 612 pp.

SABINS, F. F. 1978. *Remote Sensing: Principles and Interpretation.* San Francisco: Freeman, 175 pp.

SIEGAL, B. S. and GILLESPIE, A. R. 1980. Remote Sensing in Geology. London: Wiley, 720 pp.

REFERENCES

ARCHIE, G. E. 1942. The electrical resistivity log as an aid in determining some reservoir characteristics. *J. Petrol. Technol., 5,* Tech Paper No. 1422. Also published in *Trans. Am. Inst. Mech. Eng., 146,* 54–62.

BAILEY, R. J., BUCKLEY, J. S., and KIELMAS, M. M. 1975. Geomagnetic reconnaissance on the continental margin of the British Isles between 54° and 57° N. *J. Geol. Soc. Lond., 131,* 275–282.

BURKE, J. A., CAMPBELL, R. L., and SCHMIDT, A. T. 1969. The lithoporosity cross-plot. *The Log Analyst,* Nov.–Dec., 1–29.

CAMPBELL, R. L. 1968. Stratigraphic applications of dipmeter data in mid-continent. *Am. Assoc. Petrol. Geol. Bull., 52,* 1700–1719.

COLDRE, M. L. 1891. Les Salines et les Puits de Feu de la Province du Se-Tchouan. *Annales des Mines,* 8th Series, *19,* 441–528.

DICKEY, P. A. 1979. *Petroleum Development Geology.* Tulsa: PPC Books, 398 pp.

DOBSON, W. F. and SEELYE, D. R. 1981. Mining technology assists oil recovery from Wyoming field. *J. Petrol. Technol., 34,* 259–265.

FERRIS, C. 1972. Use of gravity meters in search for traps. In: *Stratigraphic Oil and Gas Fields.* R. E. King (ed.). Am. Assoc. Petrol. Geol., Mem. No. 16, 252–270.

GARDNER, G. H. F. L., GARDNER, L. W., and GREGORY, A. R. 1974. Formation velocity and density—the diagnostic basics of stratigraphic traps. *Geophys., 39,* 770–780.

GILREATH, J. A. and MARICELLI, J. J. 1964. Detailed stratigraphic control through dip computation. *Am. Assoc. Petrol. Geol. Bull., 48,* 1902–1910.

GOETZ, J. F., PRINS, W. J., and LOGAR, J. F. 1977. Reservoir delineation by wireline techniques. Jakarta: *Proc. 6th Cong. Indonesian Petrol. Assn.,* 40 pp.

GRAY, W. D. T. and BARNES, G. 1981. The Heather oil field. In: *Petroleum Geology of the Continental Shelf of Northwest Europe.* L. V. Illing and G. D. Hobson (eds.). London: Heyden Press, 335–341.

HALBOUTY, M. T. 1980. Geologic significance of Landsat data for 15 giant oil and gas fields. *Am. Assoc. Petrol. Geol., 64,* 8–36.

HAMMER, S. 1982. Airborne gravity is here! *Oil and Gas J.,* 11 Jan., 113–122.

HARMS, J. C. and TACKENBERG, P. 1972. Seismic signatures of sedimentation models. *Geophys., 37,* 45–58.

HASSAN, M., HOSSIN, A., and COMBAZ, A. 1976. Fundamentals of the differential gamma ray log. Trans. 4th European Symp. *Soc. Prof. Well Log Anal.,* F, 18 pp.

HERITIER, F. E., LOSSEL, P., and WATHNE, E. 1979. Frigg field—Large submarine-fan trap in Lower Eocene rocks of North Sea Viking Graben. *Am. Assoc. Petrol. Geol., 63,* 1999–2020.

HEYBROEK, P. 1975.   On the structure of the Dutch part of the Central North Sea Graben. In: *Petroleum and the Continental Shelf of North West Europe*. A. W. Woodland (ed.). London: Applied Science Publishers, 339–352.

IMBERT, P. 1828.   Letters describing brine wells and natural gas of Szechuan Province, China. *Ann. Assoc. Propag. Foi.*, *3*, 369–381.

JAGELER, A. H. and MATUSZAK, D. R. 1972.   Use of well logs and dipmeters in stratigraphic trap exploration. In: *Stratigraphic Oil and Gas Fields*. R. E. King (ed.). Am. Assoc. Petrol. Geol., Spec. Pub. No. 10, 107–135.

KREY, T. and MARSCHALL, R. 1975.   Undershooting salt-domes in the North Sea. In: *Petroleum and the Continental Shelf of North West Europe*. A. W. Woodland (ed.). London: Applied Science Publishers, 265–274.

LILLESAND, T. M. and KIEFER, R. W. 1979.   *Remote Sensing and Image Interpretation*. New York: Wiley, 612 pp.

LYONS, P. L. and DOBRIN, M. B. 1972.   Seismic exploration for stratigraphic traps. In: *Stratigraphic Oil and Gas Fields*. R. E. King (ed.). Am. Assoc. Petrol. Geol., Mem. No. 16, 225–243.

MARIE, J. P. P. 1975.   Rotliegendes stratigraphy and diagenesis, 1975. In: *Petroleum Geology and the Continental Shelf of North-West Europe*. A. W. Woodland (ed.). London: Applied Science Publishers, 205–210.

MAROOF, S. I. 1974.   A. Bouguer anomaly map of southern Great Britain and the Irish Sea. *J. Geol. Soc. Lond.*, *130*, 471–474.

McDANIEL, G. A. 1968.   Application of sedimentary directional features and scalar properties to hydrocarbon exploration. *Am. Assoc. Petrol. Geol. Bull.*, *52*, 1689–99.

McDONALD, J. A., GARDNER, G. H. F., and HILTERMAN, F. J. 1982.   Studies in 3-D seismics. Boston: *Physical Modelling*, I.H.R.D.C., 354 pp.

MECKEL, L. D. and NATH, A. K. 1977.   Geological consideration for stratigraphic modeling and interpretation. In: *Seismic Stratigraphy*. C. E. Payton (ed.). Am. Assoc. Petrol. Geol., Mem. No. 26, 417–438.

MERKEL, R. H. 1979.   *Well Log Formation Evaluation*. Continuing Education Course Note Series, No. 4. Am. Assoc. Petrol. Geol., 81 pp.

NEIDELL, N. S. and POGGIAGLIOLMI, E. 1977.   Stratigraphic models from seismic data. In: *Seismic Stratigraphy*. C. E. Payton (ed.). Am. Assoc. Petrol. Geol., Mem. No. 26, 389–416.

NETTLETON, L. L. 1972.   Use of gravity, magnetic and electrical methods in stratigraphic trap exploration. In: *Stratigraphic Oil and Gas Fields*. R. E. King (ed.). Am. Assoc. Petrol. Geol., Mem. No. 16, 244–251.

PAYTON, C. E. 1977.   *Seismic Stratigraphy—Applications to Hydrocarbon Exploration*. Am. Assoc. Petrol. Geol., Mem. No. 26, 516 pp.

PENNINGTON, J. J. 1975.   The geology of the Argyll field, 1975. In: *Petroleum and the Continental Shelf of North West Europe*. A. W. Woodland (ed.). London: Applied Science Publishers, 285–294.

PRATSCH, J. C. 1979.   Regional structural elements in Northwest Germany. *J. Petrol. Geol.*, *2.2*, 159–180.

RIDER, M. H. and LAURIER, D. 1979.   Sedimentology using a computer treatment of well logs. Trans. 6th European Symp. Soc. Prof. Well Log Anal., *J*, 12 pp.

SABINS, F. F. 1978.   *Remote Sensing: Principles & Interpretation*. San Francisco: Freeman, 175 pp.

SELLEY, R. C. 1976. Subsurface environmental analysis of North Sea sediments. *Am. Assoc. Petrol. Geol. Bull., 60*, 184–195.

SELLEY, R. C. 1978a. *Ancient Sedimentary Environments*, Second Edition. London: Chapman & Hall, 287 pp.

SELLEY, R. C. 1978b. *Concepts and Methods of Subsurface Facies Analysis.* Short Course Lecture Note Series, No. 6. Am. Assoc. Petrol. Geol., 80 pp.

SHERIFF, R. E. 1976. Inferring stratigraphy from seismic data. *Am. Assoc. Petrol. Geol. Bull., 60*, 528–542.

STONE, C. B. 1977. 'Bright spot' techniques. In: Developments in Petroleum Geology, vol. 1. G. D. Hobson (ed.). London: Applied Science Publishers, 275–292.

TUCKER, P. M. and YORSTON, H. J. 1973. *Pitfalls in Seismic Interpretation.* Soc. Explor. Geophys., Monograph No. 2, 50 pp.

VAIL, P. R., MITCHUM, R. M., TODD, R. G., WIDMEIR, J. M., THOMPSON, S., SANGREE, J. B., BUBB, J. N., and HATLEDID, W. G. 1977. Seismic stratigraphy and global changes of sea level. In: *Seismic Stratigraphy.* C. E. Payton (ed.). Am. Assoc. Petrol. Geol., Mem. No. 26, 49–212.

VERDIER, A. C., OKI, T., and ATIK, S. 1980. Geology of the Handil Field (East Kalimantan, Indonesia). In: *Giant Oil and Gas Fields of the Decade.* M. T. Halbouty (ed.). Am. Assoc. Petrol. Geol., 399–422.

WYLLIE, M. R. J. 1963. *The Fundamentals of Electric Log Interpretation*, Third Edition. New York: Academic Press, 238 pp.

WYLLIE, M. R. J., GREGORY, A. R., and GARDNER, G. H. F. 1956. Elastic wave velocities in heterogenous porous media. *Geophys., 21*, 41–70.

WYLLIE, M. R. J., GREGORY, A. R., and GARDNER, G. H. F. 1958. An experimental investigation of factors affecting elastic wave velocities in porous media. *Geophys., 23*, 459–493.

ZIEGLER, W. H. 1975. Outline of the geological history of the North Sea. In: *Petroleum and the Continental Shelf of North West Europe*, vol. 1. A. W. Woodland (ed.). London: Applied Science Publishers, 165–190.

The Subsurface Environment

4

Petroleum geology largely concerns the study of fluids, not just the oil and gas discussed in Chapter 2 but the waters with which they are associated and through which they move. Before examining the generation and migration of oil and gas in Chapter 5, the subsurface environment in which these processes operate should be considered.

This chapter begins with an account of subsurface waters, then considers pressure and temperature and their effects on the condensation and evaporation of gas and oil. The chapter concludes by putting these ingredients together and discussing the dynamics of fluids in basins.

SUBSURFACE WATERS

Two types of water can be recognized in the subsurface by their mode of occurrence:

1. Free water
2. Interstitial, or irreducible, water.

Free water is free to move in and out of pores in response to a pressure differential. *Interstitial water*, on the other hand, is bonded to mineral grains, both by attachment to atomic lattices as hydroxyl radicals and as a discrete film of water. Interstitial water is often referred to as the irreducible water because it cannot be removed during the production of oil or gas from a reservoir.

Analysis

Subsurface waters are analyzed for several specific reasons, apart from general curiosity. As discussed in Chapter 3 (p. 60), the measurement of the resistivity of formation water (R_w) is essential for the accurate assessment of S_w, and hence hydrocarbon saturation. R_w is, of course, closely related to salinity. Salinity varies both vertically and laterally across a basin. Salinity often increases with proximity to hydrocarbon reservoirs. Therefore regional salinity maps may be an important exploration tool. Similarly, subsurface waters contain traces of dissolved hydrocarbon gases, whose content increases with proximity to hydrocarbon accumulations.

Subsurface waters can be analyzed in two ways. The total concentration of solids, or salinity, can be calculated from R_w by using the S.P. log, as discussed on page 62. Alternatively, samples can be obtained from drill stem tests or during production. Care has to be taken when interpreting samples from drill stem tests because of the likelihood of contamination by mud filtrate.

Genesis

Traditionally, four types of subsurface water can be defined according to their genesis:

Meteoric waters occur near the earth's surface and are caused by the infiltration of rainwater. Their salinity, naturally, is negligible, and they tend to be oxidizing. Meteoric waters are often acidic because of dissolved humic, carbonic, and nitrous acid (from the atmosphere), although they may quickly become neutralized in the subsurface, especially when they flow through carbonate rocks.

Connate waters are harder to define. They were originally thought of as residual seawater that was trapped during sedimentation. Current definitions proposed for connate waters include "interstitial water existing in the reservoir rock prior to disturbance by drilling" (Case, 1956) and "waters which have been buried in a closed hydraulic system and have not formed part of the hydraulic cycle for a long time" (White, 1957). Connate waters differ markedly from seawater both in concentration and chemistry.

Juvenile waters are of primary magmatic origin. It may be difficult to prove that such hydrothermal waters are indeed primary and have received no contamination whatsoever from connate waters.

These three definitions lead naturally to the fourth class of subsurface waters—those of mixed origin. The mixed waters may be caused by the confluence of juvenile, connate or meteoric waters. In most basins a transition zone exists between the surface aquifer and the deeper connate zone. The effect of this transition zone on the S.P. curve was discussed previously (p. 55).

Chemistry of Subsurface Waters

The four characteristics of subsurface water to consider are Eh, pH, concentration, and composition.

Eh and pH

The current data on the Eh and pH of subsurface waters are summarized in Figure 4.1 (see Krumbein and Garrels, 1952; Pirson, 1983; and Friedman and Sanders, 1978). The data show that rainwater is oxidizing and acidic. It generally contains oxygen, nitrogen, and carbon dioxide in solution, together with ammonium nitrate after thunderstorms.

As rainwater percolates into the soil, it undergoes several changes as it becomes meteoric water. It tends to become reducing as it oxidizes organic matter. The pH of meteoric water may remain low in swampy environments because of the humic acids, but it increases in arid climates. If meteoric waters flow deep into the subsurface, they gradually dissolve salts and increase in pH. Deep connate waters show a wide range of Eh and pH values depending on their history and particularly on the extent to which they have mixed with meteoric waters or contain paleoaquifers trapped beneath unconformities. Oil field brines tend to be alkaline and strongly reducing. For further details on the Eh of subsurface fluids and its significance in petroleum exploration, see Pirson (1983).

The Eh and pH of pore fluids control the precipitation and solution of clays and other diagenetic mineral cements (see p. 243). Obviously, the study of their relationship with diagenesis and porosity evolution is important (see Selley, 1982, pp. 91–107 for further details).

Concentration

The significance of measuring the concentration of salts in subsurface waters has already been mentioned. Not only is it important for well evaluation, but the data may be plotted regionally as an exploration tool. Salinity

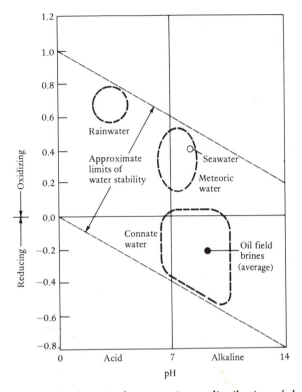

FIGURE 4.1 *Eh-pH graph showing the approximate distribution of the various types of subsurface fluid.*

or, more properly, the total dissolved solids is measured in parts per million, but is more conveniently expressed in milligrams per liter:

$$mg/1 = \frac{ppm}{density}$$

Average seawater has approximately 35,000 ppm (3.5 percent) of dissolved minerals. Values in subsurface waters range from about 0 ppm for fresh meteoric waters up to 642,798 ppm for a brine from the Salina dolomite of Michigan (Case, 1956). The latter value is extremely high because of solution from evaporites. In most connate waters the dissolved solids content seldom exceeds 350,000 ppm. In sands the salinity of connate waters generally increases with depth (Fig. 4.2) at rates ranging from 80 to 300 mg/(l)(m) (Dickey, 1979). For sources of data see Dickey (1966, 1969) and Russell (1951).

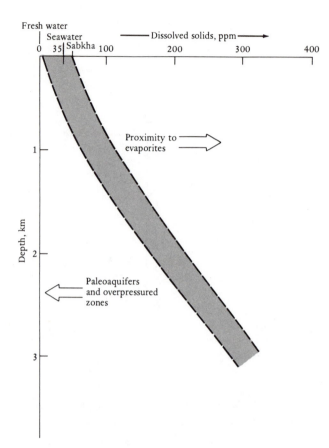

FIGURE 4.2 *Graph of salinity against depth for subsurface waters. (After Dickey,
1966, 1969, and Russell, 1951.)*

Local reversals of this general trend have been seen and are attributable to two causes. Meteoric water may sometimes be trapped beneath an
unconformity and is thus preserved as a paleoaquifer. A noted example
of this occurrence has been documented from the sub-Cretaceous unconformity of Israel (Bentor, 1969). Here the beds above the unconformity
have salinities of 60,000 ppm. The salinity drops to approximately 20,000
ppm beneath the unconformity before gradually increasing again to over
40,000 ppm.

Reversals of increasing salinity with depth are also noted in zones of
overpressure. This phenomenon is discussed in more detail later (p. 152),
but basically it reflects the fact that overpressured waters are trapped and
cannot move. In shales, however, the increase in salinity with depth is less
marked. The salinity of a sand is often about three times that of the shales

with which it may be interbedded (Dickey, 1979). This difference, together with the overall increase in salinity with depth, has been attributed to *salt sieving* (De Sitter, 1947). Shales may behave like semipermeable membranes. As water moves upward through compacting sediments, the shale prevents the salt ions from escaping from the sands, with the net result that salinity increases progressively with depth (see Magara, 1977, 1978, for further details).

Based on studies in the Gulf of Mexico and the Mackenzie delta, Overton (1973) and Van Elsberg (1978) have defined four major subsurface environments:

ZONE 1   Surface, depth of about 1 km: zone of circulating meteoric water. Salinity fairly uniform.

ZONE 2   Depth of about 1 to 3 km: salinity gradually increases with depth. Saline formation water is ionized (Fig. 4.2).

ZONE 3   Depth greater than 3 km: chemically reducing environment in which hydrocarbons form. Salinity uniform with increasing depth; may even decline if overpressured.

ZONE 4   Incipient metamorphism with recrystallization of clays to micas.

Having finished discussing vertical salinity variations, it is now apposite to consider lateral salinity changes. Salinity has long been known to tend to increase from the margins of a basin toward its center. Many instances of this salinity increase have been documented (Case, 1945; Youngs, 1975). This phenomenon is easy to explain: the basin margins are more susceptible to circulating meteoric water than is the basin center, where flow is negligible or coming from below (see p. 162). Regional isohaline maps can be a useful exploration tool, indicating areas of anomalously high salinity. These areas are presumably stagnant regions unaffected by meteoric flow, where oil and gas accumulations may have been preserved.

Composition

Subsurface waters contain varying concentrations of inorganic salts together with traces of organic compounds, including hydrocarbons. Table 4.1 presents analyses of many subsurface waters. For additional data see Krejci-Graf (1962, 1978) and Collins (1975).

Meteoric waters differ from connate waters not only in salinity but also in chemistry. Meteoric waters have higher concentrations of bicarbonate and sulfate ions and lower amounts of calcium and magnesium. These differences are the basis for the classification of subsurface waters proposed by Sulin (1946), which has been widely adopted by geologists (Table 4.2). For reviews of this classification and others, see Ostroff (1967).

TABLE 4.1 Representative oil field water analyses (ppm)

Pool	Reservoir rock, age	Cl⁻	SO₄⁻	CO₃⁻	HCO₃⁻	Na⁺ + K⁺	Ca²⁺	Mg²⁺	Total, ppm
Seawater, ppm		19,350	2690	150		11,000	420	1300	35,000
Seawater, %		55.3	7.7	0.2		31.7	1.2	3.8	
Lagunillas, western Venezuela	2000–3000 ft Miocene	89	—	120	5263	2003	10	63	7548
Conroe, Texas	Conroe sands Eocene	47,100	42	288		27,620	1865	553	77,468
East Texas	Woodbine sand U. Cretaceous	40,598	259	387		24,653	1432	335	68,964
Burgan, Kuwait	Sandstone Cretaceous	95,275	198	—	360	46,191	10,158	2206	154,388
Rodessa, Texas–La.	Oolitic limestone L. Cretaceous	140,063	284	—	73	61,538	20,917	2874	225,749
Davenport, Okla.	Prue sand Pennsylvanian	119,855	132	—	122	62,724	9977	1926	194,736
Bradford, Penn.	Bradford sand Devonian	77,340	730	—	—	32,600	13,260	1940	125,870
Oklahoma City, Okla.	Simpson sand Ordovician	184,387	286	—	18	91,603	18,753	3468	298,497
Garber, Okla.	Arbuckle limestone Ordovician	139,496	352	—	43	60,733	21,453	2791	224,868

From Levorsen, 1967. Reprinted with permission.

TABLE 4.2   Major classes of water by Sulin classification

Types of water (V. A. Sulin)	Ratios of concentrations, expressed as milliequivalent percent		
	$\dfrac{Na}{Cl}$	$\dfrac{Na–Cl}{SO_4}$	$\dfrac{Cl–Na}{Mg}$
Meteoric			
Sulfate-sodium	>1	<1	<0
Bicarbonate-sodium	>1	>1	<0
Connate			
Chloride-magnesium	<1	<0	<1
Chloride-calcium	<1	<0	>1

Connate waters differ from seawater not only because they contain more dissolved solids but also in their chemistry. Connate waters contain a lower percentage of sulfates, magnesium, and often calcium (possibly caused by the precipitation of anhydrite, dolomite and calcite) and a higher percentage of sodium, potassium, and chlorides than does seawater.

Because of the complex composition of subsurface waters, they can best be displayed and compared graphically. Schemes have been proposed by Tickell (1921) and Sulin (1946). The composition of seawater and typical connate waters are plotted according to these schemes in Figure 4.3.

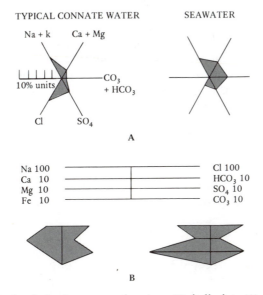

FIGURE 4.3   *Methods of plotting water chemistry. Tickell plots (A) and Sulin plots (B) for seawater (right) and typical connate waters (left).*

Connate waters also contain traces of dissolved hydrocarbons. The seminal work on this topic was published by Buckley et al. (1958). Basing their work on a study of hundreds of drill stem tests from the Gulf Coast, they found methane dissolved in subsurface waters at concentrations of up to 14 standard cubic feet per barrel. They also recorded ethane and propane and very minor concentrations of heavier hydrocarbons. The amount of dissolved gases correlated with salinity, increasing with depth and from basin margin to center. Halos of gas-enriched connate waters were recorded around oil fields.

These data are of great significance for two reasons. They raise the possibility of regionally mapping dissolved gas content as a key to locating new oil and gas fields. These data also have some bearing on the problems of the migration of oil and gas (for an example see Price, 1980). This topic is discussed in more detail in Chapter 5 (p. 198).

SUBSURFACE TEMPERATURES

Basic Principles

Temperature increases from the earth's surface toward its center. Bottom hole temperatures (BHTs) can be recorded from wells and are generally taken several times at each casing point. As each log is run, the BHT can be measured. Several readings at each depth is important because the mud at the bottom of the hole takes hours to warm up to the ambient temperature of the adjacent strata. Thus BHTs are recorded together with the number of hours since circulation. Figure 4.4 shows a BHT build-up curve. The true stabilized temperature can be determined from the Horner plot (Fertl and Wichman, 1977). In this method the recorded temperature is plotted against the following ratio:

$$\frac{\Delta t}{(t + \Delta t)}$$

where t = number of hours since circulation and logging
Δt = hours of circulation at that depth

An example of a Horner plot is shown in Figure 4.5. For a more detailed analysis of this topic, see Carstens and Finstad (1981).

Once several corrected BHTs have been obtained as a well is drilled, they can be plotted against depth to calculate the geothermal gradient (Fig. 4.6). These values range from approximately 1.8°C/100 m to 5.5°C/100 m. The global average is about 2.6°C/100 m. When several BHTs are plotted

FIGURE 4.4 *Graph showing how true bottom hole temperature can only be determined from several readings taken many hours apart.*

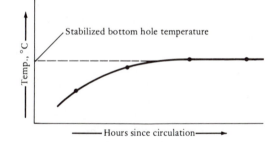

FIGURE 4.5 *Horner plot showing how true bottom hole temperature can be extrapolated from two readings.*

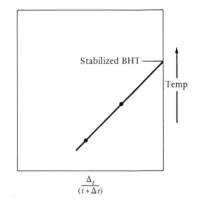

FIGURE 4.6 *Sketch showing how geothermal gradient may be determined from two or more BHTs taken at different log runs.*

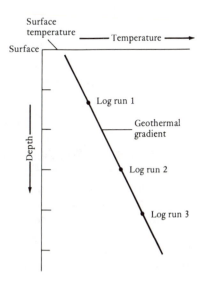

against depth for a well, they may show that the geothermal gradient is not constant with depth. This discrepancy is generally caused by variations in the thermal conductivity of the penetrated strata. This relationship may be expressed as follows:

Heat flow = geothermal gradient × thermal conductivity of rock

Table 4.3 shows the thermal conductivity of various sediments. Where sediments of different thermal conductivity are interbedded, the geothermal gradient will be different for each formation.

TABLE 4.3   Thermal conductivity of various rocks

Lithology	Thermal conductivity, $Wm^{-1}\,°C^{-1}$
Halite	5.5
Anhydrite	5.5
Dolomite	5.0
Limestone	2.8–3.5
Sandstone	2.6–4.0
Shale	1.5–2.9
Coal	0.3

From Evans, 1977; Oxburgh and Andrews-Speed, 1981.

A wide range of values can be expected for sands, shales, and limestones because of porosity variations. Because thermal conductivity is lower for water than it is for minerals, it increases with decreasing porosity and increasing depth of burial according to the following formula:

$$K_{pr} = K_w^{\phi}\, K_r^{1-\phi}$$

where K_{pr} = bulk saturated conductivity
K_w = conductivity of pore fluid
K_r = conductivity of the rock at zero porosity
ϕ = porosity

Figure 4.7 illustrates the vertical variations in conductivity, porosity, and geothermal gradient for a well in the North Sea. Note how conductivity increases and gradient decreases with depth and declining porosity. Local vertical variations are due to lithology, especially the high thermal conductivity of the Zechstein evaporites.

Once the geothermal gradient is established, isotherms can be drawn (Fig. 4.8). Note that the vertical spacing of isotherms decreases with conductivity and increasing geothermal gradient.

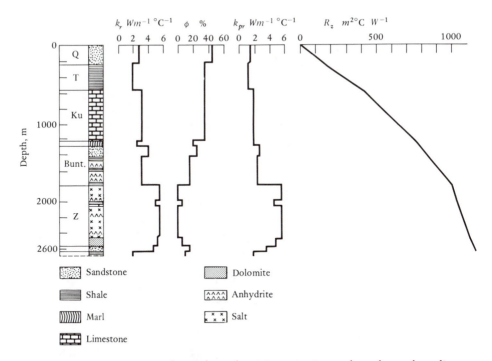

FIGURE 4.7 *Variations in thermal conductivity, porosity, and geothermal gradient for a well in the North Sea. Note how thermal conductivity and gradient gradually increase with depth and declining porosity. Note also the local fluctuations due to lithology, especially the high conductivity of the Zechstein evaporites. (After Oxburgh and Andrews-Speed, 1981.)*

Local Thermal Variations

Once the isotherms for a well have been calculated, extrapolating them across a basin is useful. The isotherms are seldom laterally horizontal for very far because of the following three factors:

1. Nonplanar geometry of sediments

2. Movement of fluids

3. Regional variations in heat flow

When strata are folded or where formations are markedly lenticular, anomalies are likely to occur in the geometry of isotherms. Two local variations are of considerable interest. Table 4.3 shows that salt has a far higher conductivity than most sediments have (in the order of 5.5 $Wm^{-1} °C^{-1}$).

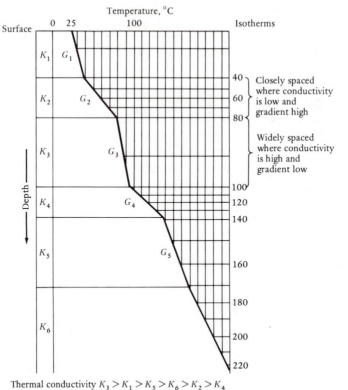

FIGURE 4.8 *Depth-temperature plot showing the effect of rocks of differing thermal conductivity (K) on geothermal gradient (G) and the vertical spacing of isotherms.*

Evans (1977), in a study of North Sea geothermal gradients, noted that isotherms would dome up over salt diapirs because of the high thermal conductivity of evaporties. Rashid and McAlary (1977) and Rashid (1978) studied the thermal maturation of kerogen in two wells on the Grand Banks of Newfoundland. The Adolphus 2–K–41 on the crest of a salt structure contained kerogen with a higher degree of maturation than the Adolphus D–50 some 3 km down flanks. This difference is attributable to the high thermal conductivity of the salt structure. Conversely, a mud diapir of highly porous clay has an anomalously low thermal conductivity. The isotherms within the clay will be closely spaced and depressed over the dome. Extensive overpressured clay formations act as an insulating blanket, trapping thermal energy and aiding source rock maturation (see p. 200).

Fluid flow is an additional cause of anomalous isotherms. Where waters are rapidly discharged to the surface, along open faults for example, isotherms are drawn upward. This phenomenon has been described for some of the Viking graben boundary faults of the North Sea (Cooper et al., 1975) and from growth faults of the Gulf Coast (Jones and Wallace, 1974).

Regional Thermal Variations

The heat flow of the earth's crust has previously been defined as the product of the geothermal gradient and the thermal conductivity. Data on global heat flow and discussions of its regional variation have been given by Lee (1965) and Sass (1971). The global average heat flow rate is in the order of 1.5 μcal/(cm^2)(s). Abnormally high heat flow occurs along mid-ocean ridges and intracratonic rifts, where magma is rising to the surface and the crust is thinning and separating. Conversely, heat flow is abnormally low at convergent plate boundaries, where relatively cool sediments are being subducted into the mantle (Klemme, 1975).

Regional variations in heat flow affect petroleum generation, as discussed in the next chapter. Data showing that oil generation occurs between temperatures of 60 and 120°C will be presented. In areas of high heat flow, and hence high geothermal gradient, the optimum temperature will be reached at shallower depths than in areas of low heat flow and geothermal gradient (Fig. 4.9A). Note also the effect of low conductive formations. With their high geothermal gradients they raise the depth at

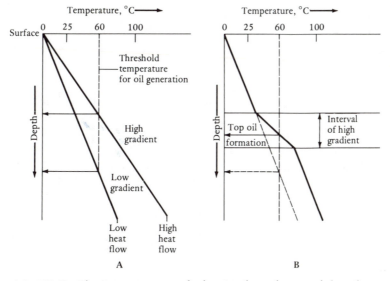

FIGURE 4.9   (A) Depth-temperature graph showing how the top of the oil generation zone rises with increasing geothermal gradient. (B) Depth-temperature graph showing how a formation with a low conductivity and high gradient raises the threshold depth of oil generation.

which the oil window is entered (Fig. 4.9B). Thus oil generation begins at greater depths in subductive troughs than in rift basins. Layers of low-conductivity rock may raise the depth at which oil generation begins.

SUBSURFACE PRESSURES

Measurement

Subsurface pressures can be measured in many ways. Some methods indicate pressure before a well is drilled; some operate during drilling; and others operate after drilling.

In wildcat areas with minimal well control the interval velocities calculated from seismic data may give a clue to pressure, or at least overpressure. Velocity generally increases with depth as sediments compact. A reversal of this trend may indicate undercompacted and hence overpressured shales (Fig. 4.10).

While a well is being drilled, a number of parameters may indicate abnormal pressure. These parameters include a rapid increase in the rate of penetration, a rapid increase in the temperature of the drilling mud, and

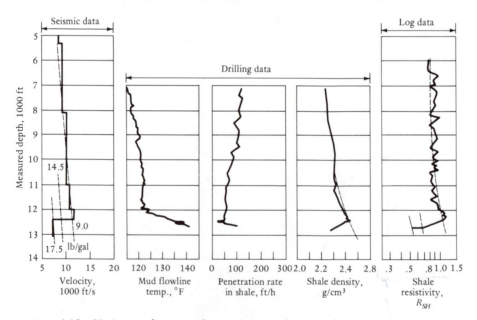

FIGURE 4.10 *Various indicators of overpressure detection. In this well a zone of overpressure is present below 12,400 ft. (After Fertl and Chilingarian, 1976 © SPE-AIME.)*

a decrease in the density of shale cuttings. A particularly useful method is the drilling (d) exponent (Jordan and Shirley, 1966). This method takes into account that the rate of penetration of the bit reflects not only the degree of compaction of the sediments but also the weight on the bit and the rotary speed:

$$d \text{ exponent} = \frac{\log (R/60N)}{\log (12W/10^6 D)}$$

where R = rate of penetration (ft/h)
N = rotary speed (rpm)
w = weight on bit (lb)
D = diameter of borehole (in)

The d exponent is plotted against depth as the well is drilled. It will decrease linearly until reaching the top of abnormal pressure, at which depth it will increase.

When these methods suggest that a zone of overpressure has been penetrated, it may be wise to stop drilling and run logs. Sonic, density, and neutron logs may all show a sudden increase in porosity, whereas resistivity may sharply decrease (Hottman and Johnson, 1965). Since logs respond to more than one variable, no single log is diagnostic. Lithological changes may give similar responses.

The problem with all of these methods of detecting overpressure is the difficulty of establishing at what point deviation from the normal is sufficiently clear for overpressure to be proven. By that time it may be too late, as the well may have kicked.

Once a well has been safely drilled, pressure can be measured by several methods. A pressure bomb, in which pressure is recorded against time, can be lowered into the hole. The drill stem test, in which the well is allowed to flow for several short periods while the pressure is monitored, is another method of measuring pressure. Drill stem tests may give good measurement results not only of the pressure but also of the flow rate (and hence permeability) of the formation. Fluids may also be allowed to flow to the surface for analysis. For further details see Bradley (1975) and Dickey (1979). Some wireline tools are designed to measure formation pressures and recover fluids from zones of interest. Sometimes they work.

Basic Principles

Pressure is the force per unit area acting on a surface. It is generally measured in kilograms per square centimeter (kg/cm²) or pounds per square inch (psi).

The several types of subsurface pressure can be classified as follows:

Overburden { 1. Lithostatic
pressure { 2. Fluid pressure { (a) Hydrostatic
                                { (b) Hydrodynamic

The lithostatic pressure is caused by the pressure of rock, which is transmitted through the subsurface by grain-to-grain contacts. The lithostatic pressure gradient varies according to depth, the density of the overburden, and the extent to which grain-to-grain contacts may be supported by water pressure. It often averages about 1 psi/ft.

The fluid pressure is caused by the fluids within the pore spaces. According to Terzaghi's Law (Terzaghi, 1936; Hubbert and Rubey, 1959),

$$s = p + o$$

where s = overburden pressure
p = lithostatic pressure
o = fluid pressure

As fluid pressure increases in a given situation, the forces acting at sediment grain contacts diminish and lithostatic pressure decreases. In extreme cases this effect may transform the sediment into an unstable plastic state.

In the oil industry fluid pressure is generally calculated as follows:

$$p = 0.052 \times wt \times D$$

where p = hydrostatic pressure (psi)
wt = mud weight (lb/gal)
D = depth (ft)

The two types of fluid pressure are hydrostatic and hydrodynamic. The hydrostatic pressure is imposed by a column of fluid at rest. For a column of fresh water (density 1.0) the hydrostatic gradient is 0.433 psi/ft, or 0.173 kg/(cm^2)(m). For water with 55,000 ppm of dissolved salts the gradient is 0.45 psi/ft; for 88,000 ppm of dissolved salts the gradient is about 0.465 psi/ft. These values are, of course, temperature dependent. Figure 4.11 shows the relationship between lithostatic and hydrostatic pressure.

The second type of fluid gradient is the hydrodynamic pressure gradient, or fluid potential gradient, which is caused by fluid flow. When a well is drilled, pore fluid has a natural tendency to flow into the well bore. Normally, this flow is inhibited by the density of the drilling mud. Nonethe-

FIGURE 4.11   *Depth-pressure graph illustrating hydrostatic and geostatic pressures and concepts.*

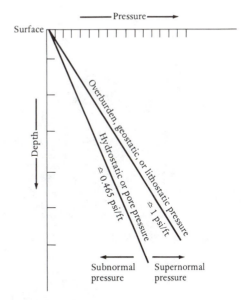

less, to be able to measure the level to which the fluid would rise if the well were open is important. This level is termed the *potentiometric* or *piezo-metric* level and is calculated as follows:

$$\text{Elevation to potentiometric level} = \frac{P}{W} - (D - E)$$

where P = bottom hole pressure (psi)
        W = weight of fluid (psi/ft)
        D = depth (ft)
        E = elevation of kelly bushing above sea level (ft)

The potentiometric level of adjacent wells may be contoured to de-fine the potentiometric surface. If this surface is horizontal, then no fluid flows across the region and the fluid pressure is purely hydrostatic. If the potentiometric surface is tilted, then fluid is moving across the basin in the direction of dip of the surface and the fluid pressure is caused by both hydrostatic and hydrodynamic forces (Fig. 4.12). Formation water salinity commonly increases in the direction of dip of the potentiometric surface (Fig. 4.13). Where the pressure gradient is hydrostatic ($\simeq$ 0.43 psi/ft), the pressure is termed *normal*. Abnormal gradients may be subnormal (less than hydrostatic) or supernormal (above hydrostatic). The level to which formation fluid would rise or fall to attain the potentiometric surface is ex-

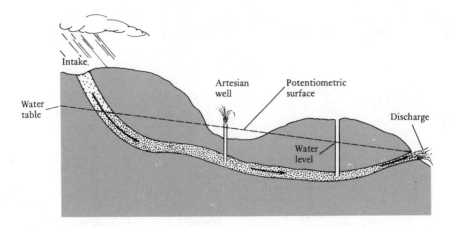

FIGURE 4.12   *Sketch illustrating the concept of the potentiometric, or piezometric, surface.*

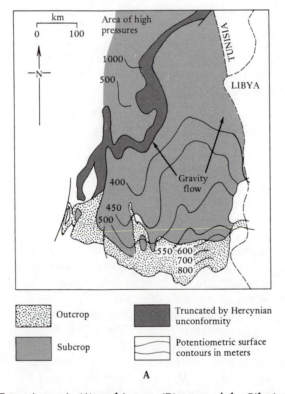

A

FIGURE 4.13   *Potentiometric (A) and isocon (B) maps of the Silurian–L. Devonian (Acacus and Tadrart) sandstones of eastern Algeria. Note how the potentiometric surface drops northward with increasing formation salinity. (After Chiarelli, 1978.)*

pressed as the *fluid potential.* Hubbert (1936) showed that this potential could be calculated as follows:

$$\text{Fluid potential} = Gz + \frac{p}{\rho}$$

where G = the acceleration due to gravity
z = datum elevation at the site of pressure measurement
p = static fluid pressure
ρ = density of the fluid

The fluid potential is also directly related to the head of water, h:

$$\text{Fluid potential} = hG$$

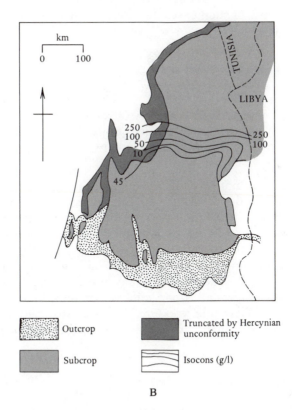

B

FIGURE 4.13   *(con'd)*

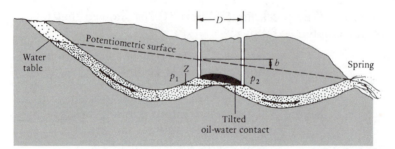

FIGURE 4.14 *Sketch showing how the slope of an oil-water contact may be due to hydrodynamic flow.*

Sometimes oil-water contacts are not horizontal, but tilted. Hydrodynamic flow is one of several causes of such contacts (see p. 278). Hubbert (1953) showed how in such cases the slope of the contact is related to the fluid potential (Fig. 4.14):

$$\frac{z}{d} = \frac{h}{d}\left(\frac{\rho_w}{\rho_w - \rho_o}\right)$$

where ρ_w = density of water
ρ_o = density of oil
d = distance between wells

Within any formation with an open-pore system, pressure will increase linearly with depth. When separate pressure gradients are encountered in a well, it indicates that permeability barriers separate formations (Fig. 4.15). This principle can be a useful aid to correlating reservoir formations between wells. In the situation shown in Figure 4.16, it is tempting to assume that the upper, middle, and lower sands are continuous from well to well. Pressure gradient plots may actually reveal a quite different correlation. This difference is particularly common in stacked regressive barrier bar sands. Borehole pressure data are also used to construct regional potentiometric maps, which may be used to locate hydrodynamic traps (p. 319).

Supernormal Pressures

Supernormal pressures are those pressures that are greater than hydrostatic. They are found in sediments ranging in age from Cambrian to Recent, but are especially common in Tertiary deltaic deposits, such as the

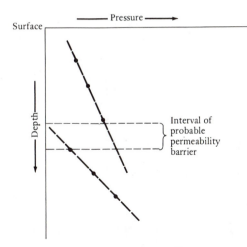

FIGURE 4.15 *Pressure-depth plot of a well showing different gradients. These gradients suggest the existence of a permeability barrier between the two intervals with uniform, but different, gradients.*

North Sea, Niger delta, and Gulf Coast of Texas. Study of the causes and distribution of overpressuring is very important because overpressuring presents a hazard to drilling and is closely related to the genesis and distribution of petroleum.

Overpressuring occurs in closed-pore fluid environments, where fluid pressure cannot be transmitted through permeable beds to the surface. Thus the two aspects to consider are the nature of the fluid barrier and the reason for the pressure build-up (Jacquin and Poulet, 1973; Bradley, 1975; Barker, 1979; Plumley, 1980).

The permeability barrier, which inhibits pressure release, may be lithological or structural. Common lithological barriers are evaporites and shales. Less common are the impermeable carbonates or sandstones, which may also act as seals. Structural permeability barriers may be provided by faults, although as discussed in detail later (p. 287), some faults seal and others do not.

Fertl and Chilingarian (1976) have listed 13 possible causes of overpressure. Only the more important causes will be discussed here:

1. Artesian
2. Structural
3. Compactional
4. Diagenetic

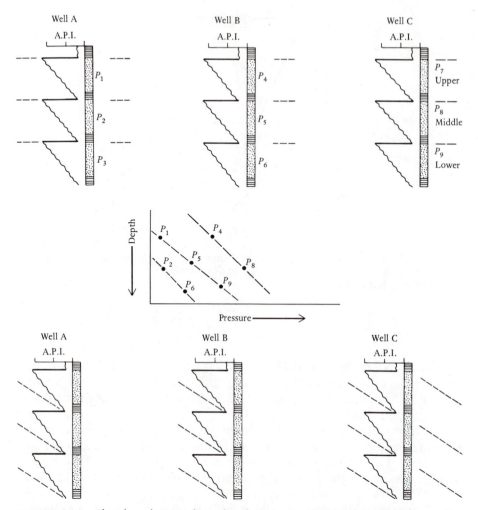

FIGURE 4.16   *Sketches showing how the obvious correlation of multistorey sands (top) may be shown to be incorrect from pressure data (center). The bottom correlation is common in laterally stacked barrier bar sands.*

The concept of the potentiometric surface and artesian pressure has just been dealt with. In cases such as the Silurian-Devonian sandstones of Algeria (Fig. 4.13) the potentiometric surface at the edge of the basin drops toward the center. In the central part of the basin, however, the surface rises, so very high pressures occur where the sandstones are sealed beneath the Hercynian unconformity.

Structural deformation can cause overpressure in several ways. At the simplest level a block of sediment can be raised between two sealing faults and, if fluids have no other egress, the pore pressure will be unable to adjust to the new lower hydrostatic pressure. In more complex settings compression of strata during folding can cause overpressure if excess fluid has no means of escape. This situation is most likely to happen when permeable strata are interbedded with evaporites, as in Iran. The evaporites can be involved in intense compression, deforming plastically and preventing excess fluid from bleeding off through fractures or faults.

The third and perhaps most common type of overpressure is caused by compaction, or lack of it. This situation is especially common in deltas where deposition is too fast for sediments to compact and dewater in the normal way. This topic has been studied in great detail (see, for example, Athy, 1930; Jones, 1969; Perry and Hower, 1970; Rieke and Chilingarian, 1973; and Fertl, 1977).

Figure 4.17 presents the basic effects of compaction on overpressure. On a delta plain, sands and clays are commonly interbedded. The sands are permeable and communicate with the surface. As these sediments are buried, the clays compact and lose porosity. Excess pore fluids move into the sands and escape to the surface. Thus although pore pressure increases with burial, it remains at hydrostatic level. On the seaward side of the delta, however, sands become discontinuous and finally die out. As the delta progrades basinward, the delta plain sediments overlie the pro-delta

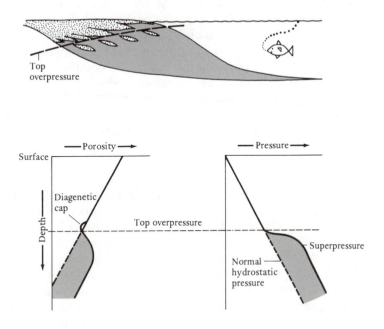

FIGURE 4.17 *(A) Cross-section through a delta and (B) depth-porosity and depth-pressure curves, indicating overpressuring due to undercompaction.*

muds. The overburden pressure increases on the pro-delta muds, but, lacking permeable conduits, they cannot dewater. Porosity remains high, and pore pressure exceeds hydrostatic. This type of overpressure has many implications for petroleum geology. The drilling problems have already been noted. More significantly, overpressured shales have lower densities and lower thermal conductivities than normally pressured shales have.

The lower density of overpressured shales means that they tend to flow upward and be replaced by denser, normally compacted sediments from above. This flowage causes growth faults and clay diapirs, which can form important hydrocarbon traps, as discussed and illustrated on pages 296 and 299. The significance of the low thermal conductivity of overpressured shales has been noted earlier (p. 144). By increasing the geothermal gradient, oil generation can occur at shallower depths.

A fourth cause of overpressure is mineralogical reactions during diagenesis. A number of such reactions have been noted, including the dehydration of gypsum to anhydrite and water and the alteration of volcanic ash to clays and carbon dioxide. Probably the most important reactions responsible for overpressure are those involving clays. In particular the dewatering of montmorillonitic clays may be significant. As montmorillonitic clays are compacted, they give off not only free-pore water but also water of crystallization. The temperatures and pressures at which these reactions occur are similar to those at which oil generation occurs. This phenomenon may be closely linked to the problem of the primary migration of oil, as discussed in the next chapter (p. 200). Other diagenetic reactions occur in shales, which may not only increase pore pressure but also decrease permeability. In particular a cap of carbonate-cemented shale often occurs immediately above an overpressured interval.

Finally, overpressuring can be caused during production either by fluid injection schemes or by faulty cementing jobs in which fluids move from an overpressured formation to a normally pressured one.

Subnormal Pressures

Subnormal pressures are those pressures that are less than the hydrostatic pressure. Examples and causes of subnormal pressures have been given by Fertl and Chilingarian (1976) and Dickey and Cox (1977). Subnormal pressures will only occur in a reservoir that is separated from circulating groundwater by a permeability barrier; were this not so, the reservoir would fill with water and rise to hydrostatic pressure (Fig. 4.18).

Subnormal pressures can be brought about by producing fluids from reservoirs, especially where they lack a water drive (see p. 264). Naturally occurring subnormal pressures are found around the world in both structural and stratigraphical traps. Fertl and Chilingarian (1976) cite pre-Pennsylvanian gas fields of the Appalachians and the Morrow sands of

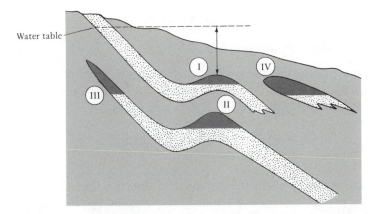

FIGURE 4.18 *Cross-section to illustrate the setting of a subnormally pressured trap. Trap I is in fluid communication with the surface and will be normally pressured. Traps II and III are in fluid communication with connate waters in the centre of the basin. They may be normally or overpressured. Trap IV is completely enclosed in impermeable shale. Virgin pressure may be sub- or supernormal, but will drop to subnormal during production because there is no aquifer to maintain reservoir pressure. (After Parke and Dickey, 1977).*

northwestern Oklahoma. Dickey and Cox (1977) cite subnormal pressures in the Cretaceous barrier bars from the Viking sands of Canada to the Gallup Sandstone of New Mexico.

In addition to fluid production, subnormal pressures may be caused by two main processes:

1. Increase in pore volume
    a. by decompression
    b. by fracturing
2. Decrease in reservoir temperature

In a confined system an increase of pore volume will cause pressure to drop as the fluids expand to fill the extra space. Pore volume may be expanded by decompression or tectonism. Erosion of rock above a normally pressured reservoir will cause a decrease in overburden pressure and hence an expansion of the reservoir rock, accompanied by a *pro rata* increase in pore volume. Alternatively, pore volume may increase when extensional fractures form over the crest of an anticline. This mechanism has been invoked to explain the subnormal pressure of the Kimmeridge Bay field of Dorset (Brunstrom, 1963).

The second main cause of subnormal pressure is a decrease in geothermal gradient, which will cause reservoir fluids to cool and shrink, and thus decrease in pressure.

SUBSURFACE FLUID DYNAMICS

Pressure-Temperature Relationships

Temperature and pressure in the subsurface have been reviewed separately; they are now examined together. From the laws of Boyle and Charles, the following relationship exists:

$$\frac{\text{Pressure} \times \text{volume}}{\text{temperature}} = \text{a constant}$$

This basic relationship governs the behavior of fluids in the subsurface, as elsewhere, and is particularly important in establishing the formation volume factor in a reservoir (p. 261).

For a given fluid at a constant pressure there is a particular temperature at which gas bubbles out of liquid and at which it condenses as temperature decreases. Similarly, at a uniform temperature there is a particular pressure at which liquid evaporates as pressure drops and gas condenses as pressure increases.

A pure fluid may exist in either the liquid or gaseous state, depending on the pressure and temperature (Fig. 4.19). Above the critical point (c), however, only one phase can exist. Subsurface fluids are mixtures of many compounds: connate water contains traces of hydrocarbons in solution; petroleum is a mixture of many different hydrocarbons in liquid and/or gaseous states.

FIGURE 4.19 *Pressure-temperature graph from a pure fluid. Above the critical point (c) only one phase can exist.*

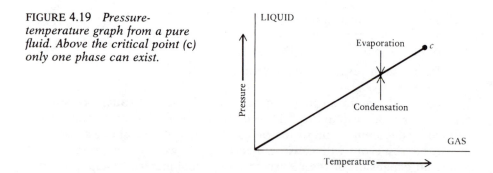

A pressure-temperature curve for a two-component system is shown in Figure 4.20. This figure shows three different phases: liquid, liquid and vapor, and gas. The bubble point line marks the boundary at which gas begins to bubble out of liquid. The dew point line marks the boundary at which gas condenses. These two lines join at the critical point (c).

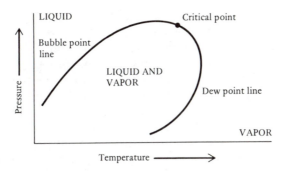

FIGURE 4.20 *Pressure-temperature graph for a two-phase fluid showing the position of the bubble point and dew point lines.*

Figure 4.21 shows pressure-temperature curves for various mixtures of hydrocarbons. Several general conclusions can be drawn from these graphs. The amount of gas that can be dissolved in oil increases with pressure and therefore depth. The lower the density (higher the API gravity) of an oil, the more dissolved gas it contains. Within a reservoir with separate oil and gas columns, the two zones are in thermodynamic equilibrium. As production begins and pressure drops, gas will come out of solution and the volume of crude oil will decrease.

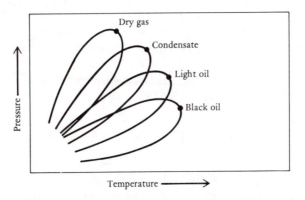

FIGURE 4.21 *Pressure-temperature graph showing phase behavior for hydrocarbons of different gravity.*

Secondary Migration of Petroleum

The effects of pressure and temperature on petroleum are important. They control both the way in which petroleum behaves in a producing reservoir and its migration from source rock to trap. Petroleum production from reservoirs is discussed in Chapter 6 (p. 263). Two types of migration occur.

Primary migration is the movement of hydrocarbons from the source rock into permeable carrier beds. Secondary migration is the movement of hydrocarbons through the carrier beds to the reservoir. Primary migration is still a mystery and is reviewed in Chapter 5 (p. 198).

Between leaving the source rocks and filling the pores of a trap, oil must exist as droplets. Therefore as long as the diameter of the droplets is less than that of the pore throats, buoyancy will move the droplets until they reach a throat whose radius is less than that of the droplet. Further movement can only occur when the displacement pressure of the oil exceeds the capillary pressure of the pore. Capillary pressure is discussed in more detail in Chapter 6 (p. 238), but basically the following relationship exists:

$$\text{Capillary pressure} = \frac{2i \cos \theta}{r}$$

where i = interfacial tension between the two fluids
θ = angle of contact
r = radius of pore

Neither buoyancy nor hydrodynamic pressure alone can exceed displacement pressure. As the oil droplets build up beneath the narrow throat, however, they increase pressure until a droplet is squeezed through the opening. Thus the process will continue until the oil reaches sediment whose pores are so small that the pressure of the column of oil beneath is insufficient to force further movement. The oil has thus become trapped beneath a cap rock.

The height of the oil column necessary to overcome displacement pressure has been shown by Berg (1975) to be:

$$Z_o = 2y \left(\frac{1}{r_t} - \frac{1}{r_p} \right) \bigg/ G(\rho_w - \rho_o)$$

where z_o = height of oil column
y = interfacial tension between oil and water
r_t = radius of pore throat in the cap rock
r_p = radius of pore throat in the reservoir rock
G = gravitational constant
ρ_w = density of water
ρ_o = density of oil

Figure 4.22 shows the relationship between height of oil column and grainsize for oil-water systems of various density contrasts (a difference of about 0.3 is usual). Oil is thus retained beneath a fine-grained cap rock by capillary pressure in what may be referred to as a *capillary seal*. A capillary

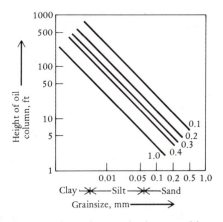

FIGURE 4.22 *Graph showing the column of oil trapped by various grades of sediment for oil-water systems of various density differences. Basically, the finer the grainsize (i.e., smaller the pore diameter), the thicker the oil column that it may trap. (After Berg, 1975.)*

seal is likely to be more effective for oil than for the more mobile gases. Gas entrapment may be due not so much to capillary seals but to pressure seals (Magara, 1977). A pressure seal occurs when the pressure due to the buoyancy of the hydrocarbon column is less than the excess pressure of the shales above hydrostatic. This seal commonly occurs where overpressured clays overlie normally pressured sands; that is, the pressure gradient is locally reversed from the usual downward increase to a downward decrease. Examples of this phenomenon are known from many overpressured basins, including the Gulf Coast (Schmidt, 1973). Once oil or gas has reached an impermeable seal, be it capillary or pressure, it will horizontally displace water from the pores.

The lateral distance to which petroleum can migrate has always been debated. Where oil is trapped in sand lenses surrounded by shale, the migration distance must have been short. Where oil occurs in traps with no obvious adjacent source rock, extensive lateral migration must have occurred. Table 4.4 cites some documented examples of long-distance lateral petroleum migration. Note that evidence has been produced for migration of up to 400 km (250 mi).

Summary: Fluid Dynamics of Young and Senile Basins

To understand the generation, migration, and entrapment of oil, familiarity with the subsurface environment is necessary. Basically, rock density, temperature, salinity, and pressure increase with depth, whereas porosity decreases. Local reversals of these trends occur in intervals of overpressure.

TABLE 4.4   Documented examples of long-distance lateral petroleum migration

Example	Distance km	(mi)	Reference
Phosphoria Shale, Wyoming and Idaho	400	(250)	Claypool et al. (1978)
Gulf Coast, U.S.A. Pleistocene	160	(100)	Hunt (1979)
Pennsylvanian, northern Oklahoma	120	(75)	Levorsen (1967)
Magellan basin, Argentina	100	(60)	Zielinski and Bruchhausen (1983)
Athabasca Tar Sands, Canada	100	(60)	Tissot and Welte (1978)

The dynamics of fluid flow in basins have been reviewed by Cousteau (1975), Neglia (1979), and Bonham (1980). A young basin has a very dynamic fluid system, but deep connate fluid movement declines with age. Immediately after deposition, overburden pressure causes extensive upward movement of water due to compaction. This movement is commonly associated with the migration and entrapment of hydrocarbons. Upward fluid movement may be locally interrupted by zones of overpressure.

As time passes and the basin matures, hydrocarbon generation ceases and overpressure slowly bleeds off to the surface. Pressures become universally hydrostatic or subnormal because of cooling or erosion and reservoir expansion. Meteoric circulation will continue in the shallow surficial sedimentary cover. Deep connate water may continue to move, albeit very slowly, in convection cells. This hypothesis is one explanation for anomalous heat flow data in the southern North Sea. Convection currents are known to exist within petroleum reservoirs (Combarnous and Aziz, 1970). Figure 4.23 summarizes and contrasts the characteristics of young and mature basins.

SELECTED BIBLIOGRAPHY

Chemistry of subsurface fluids:

COLLINS, A. G. 1975.   *Geochemistry of Oilfield Waters.* Amsterdam: Elsevier, 495 pp.
DREVER, J. I. 1982.   *The Geochemistry of Natural Waters.* Englewood Cliffs, N.J.: Prentice-Hall, 388 pp.

Subsurface temperatures:

KLEMME, H. D. 1975.   Geothermal gradients, heat flow and hydrocarbon recovery. In: *Petroleum and Global Tectonics.* A. G. Fisher and S. Judson (eds.). Princeton: Princeton Univ. Press, 251–304.

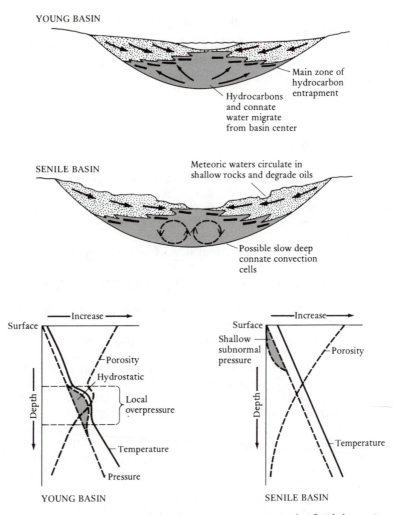

FIGURE 4.23 *Cross-sections and depth curves contrasting the fluid dynamics of young and senile sedimentary basins.*

Subsurface pressures:

VOCKROTH, G. B. (ed.). 1974. *Abnormal Subsurface Pressure.* Am. Assoc. Petrol. Geol., Reprint Series No. 11, 205 pp.

Subsurface fluid dynamics:

CHAPMAN, R. E. 1972. *Petroleum Geology, A Concise Study.* Amsterdam: Elsevier, 304 pp. Chapter 3 (pp. 48–82) gives an excellent exposé of fluid dynamics.

DAHLBERG, E. C. 1982. *Applied Hydrodynamics in Petroleum Exploration.* Heidelberg: Springer-Verlag, 161 pp.

NEGLIA, S. 1979.   Migration of fluids in sedimentary basins. *Am. Assoc. Petrol. Geol. Bull.*, *63*, 573–597.

YOUNG, A. and GALLEY, J. E. 1965.   *Fluids in Subsurface Environments*. Am. Assoc. Petrol. Geol., Mem. No. 4, 414 pp. Dated, but with a mass of data and several seminal papers.

REFERENCES

ATHY, L. F. 1930.   Density, porosity and compaction of sedimentary rocks. *Am. Assoc. Petrol. Geol. Bull.*, *14*, 1–24.

BARKER, C. 1979.   Role of temperature and burial depth in development of sub-normal and abnormal pressures in gas reservoirs. *Am. Assoc. Petrol. Geol. Bull.*, *63*, 414.

BENTOR, Y. K. 1969.   On the evolution of subsurface brines in Israel. *Chem. Geol.*, *4*, 83–110.

BERG, R. R. 1975.   Capillary pressures in stratigraphic traps. *Am. Assoc. Petrol. Geol. Bull.*, *59*, 939–56.

BONHAM, L. C. 1980.   Migration of hydrocarbons in compacting basins. *Am. Assoc. Petrol. Geol. Bull.*, *64*, 549–567.

BRADLEY, J. S. 1975.   Abnormal formation pressure. *Am. Assoc. Petrol. Geol. Bull.*, *59*, 957–973.

BRUNSTROM R. G. W. 1963.   Recently discovered oil fields in Britain. Frankfurt: *Proc. World Petrol. Cong.*, Section 1, Paper 49, 11–20.

BUCKLEY, S. E., HOCUTT, C. R., and TAGGART, M. S. 1958.   Distribution of dissolved hydrocarbons in subsurface waters. In: *The Habitat of Oil*. L. G. Weeks (ed.). Am. Assoc. Petrol. Geol., 850–882.

CARSTENS, H. and FINSTAD, K. 1981.   Geothermal gradients of the northern North Sea basin, 59–62°N. In: *Petroleum Geology of the Continental Shelf of North-West Europe*. L. V. Illing and G. D. Hobson (eds.). London: Heyden Press, 152–161.

CASE, L. C. 1945.   Exceptional Silurian brine near Bay City, Michigan. *Am. Assoc. Petrol. Geol. Bull.*, *29*, 567–570.

CASE, L. C. 1956.   The contrast in initial and present application of the term "connate water." *J. Petrol. Technol.*, *8*, 12.

CHIARELLI, A. 1978.   Hydrodynamic framework of eastern Algerian Sahara—influence on hydrocarbon occurrence. *Am. Assoc. Petrol. Geol. Bull.*, *62*, 667–685.

CLAYPOOL, G. E., LOVE, A. H., and MAUGHAM, E. K. 1978.   Organic geochemistry, incipient metamorphism, and oil generation in black shale members of phosphoria formation, western interior United States. *Am. Assoc. Petrol. Geol. Bull.*, *62*, 98–120.

COLLINS, A. G. 1975. *Geochemistry of Oilfield Waters*. Amsterdam: Elsevier, 495 pp.

COMBARNOUS, M. and AZIZ, K. 1970.   Influence de la convection naturelle dans les reservoirs d'huille ou de gas. *Revue Institute Francais de Petrol.*, *25*, 1335–1354.

COOPER, B. S., COLEMAN, S. H., BARNARD, P. C., and BUTTERWORTH, J. S. 1975. Palaeotemperatures in the northern North Sea basin. In: *Petroleum and the Continental Shelf of North West Europe*, vol. 1. A. W. Woodland (ed.). London: Applied Science Publishers, 487–492.

COUSTEAU, H. 1975. Classification hydrodynamiques des bassins sedimentaires. Tokyo: *Proc. 9th World Petrol. Cong.*, *2*, 105–119.

DE SITTER, L. V. 1947. Diagenesis of oilfield brines. *Am. Assoc. Petrol. Geol. Bull.*, *31*, 2030–2040.

DICKEY, P. A. 1966. Patterns of chemical composition in deep subsurface waters. *Am. Assoc. Petrol. Geol. Bull.*, *50*, 2472–2478.

DICKEY, P. A. 1969. Increasing concentration of subsurface brines with depth. *Chem. Geol.*, *4*, 361–370.

DICKEY, P. A. 1979. *Petroleum Development Geology*. Tulsa: Petroleum Publishing Co., 398 pp.

DICKEY, P. A. and COX., W. C. 1977. Oil and gas in reservoirs with subnormal pressures. *Am. Assoc. Petrol. Geol.*, *61*, 2134–2142.

EVANS, T. R. 1977. Thermal properties of North Sea rocks. *Log Analyst*, *18* (2), 3–12.

FERTL, W. H. 1977. Shale density studies and their application. In: *Developments in Petroleum Geology I*. G. D. Hobson (ed.). London: Applied Science Publishers, 293–328.

FERTL, W. H. and CHILINGARIAN, G. V. 1976. Importance of abnormal pressures to the oil industry. *Soc. Petrol. Eng.*, Paper 5946, 11 pp.

FERTL, W. H. and WICHMANN, P. A. 1977. How to determine static BHT from well log data. *World Oil*, *184* (*1*), 105–106.

FRIEDMAN, G. M. and SANDERS, J. E. 1978. *Principles of Sedimentology*. New York: Wiley, 792 pp.

HOTTMAN, C. E. and JOHNSON, R. K. 1965. Estimation of formation pressures from log-derived shale properties. *J. Petrol. Technol.*, June, 717–722.

HUBBERT, M. K. 1953. Entrapment of petroleum under hydrodynamic conditions. *Am. Assoc. Petrol. Geol. Bull.*, *37*, 1954–2026.

HUBBERT, M. K. and RUBEY, W. W. 1959. Role of fluid pressure in mechanics of overthrust faulting. *Am. Assoc. Petrol. Geol. Bull.*, *70*, 115–166.

HUNT, J. M. 1979. *Petroleum Geochemistry and Geology*. San Francisco: Freeman, 616 pp.

JACQUIN, C. and POULET, M. 1973. Essai de Restitution des conditions hydrodynamiques regnant dans un basin sedimentaire au cours de son evolution. *Rev. Institut Francais du Petrol.*, *28*, 269–298.

JONES, P. H. 1969. Hydrodynamics of geopressure in the northern Gulf of Mexico basin. *J. Petrol. Technol.*, *21*, 803–810.

JONES, P. H. and WALLACE, R. H. 1974. Hydrogeologic aspects of structural deformation in the northern Gulf of Mexico basin. *J. Res. U.S. Geol. Surv.*, *2*, 511–517.

JORDAN, J. R. and SHIRLEY, O. J. 1966. Application of drilling performance data to overpressure detection. *J. Petrol. Technol.*, *18*, 1387–1394.

KLEMME, H. D. 1975. Geothermal gradients, heat flow and hydrocarbon recovery. In: *Petroleum and Global Tectonics*. A. G. Fisher and S. Judson (eds.). Princeton: Princeton Univ. Press., 251–304.

KREJCI-GRAF, K. 1962.   Oilfield waters. *Erdol Kohle Petrochem., 15*, 102.

KREJCI-GRAF, K. 1978.   *Data on the Geochemistry of Oil Field Waters*. Frankfurt: K. Krejci, Graf. am Main, 174 pp.

KREY, T. and MARSCHALL, R. 1975.   Undershooting saltdomes in the North Sea. In: *Petroleum and the Continental Shelf of North West Europe*, vol. 1. A. W. Woodland (ed.). London: Applied Science Publishers, 265–274.

KRUMBEIN, W. C. and GARRELS, R. M. 1952.   Origin and classification of chemical sediments in terms of pH and oxidation-reduction potentials. *J. Geol., 60*, 1–33.

LEE, W. H. K. (ed.). 1965.   *Terrestrial Heat Flow*. Am. Geophys., Union Monograph 8.

LEVORSEN, A. I. 1967.   *Geology of Petroleum*. London: Freeman, 724 pp.

MAGARA, K. 1977.   Petroleum migration and accumulation. In: *Developments in Petroleum Geology*, vol. 1. G. D. Hobson (ed.). London: Applied Science Publishers, 83–126.

MAGARA, K. 1978.   *Compaction and Fluid Migration*. Amsterdam: Elsevier, 320 pp.

NEGLIA, S. 1979.   Migration of fluids in sedimentary basins. *Am. Assoc. Petrol. Geol. Bull., 63*, 573–597.

OSTROFF, A. G. 1967.   Comparison of some formation water classification systems. *Am. Assoc. Petrol. Geol. Bull., 51*, 404.

OVERTON, H. L. 1973.   Water chemistry analysis in sedimentary basins. 14th Ann. Logging Symp. *Soc. Prof. Well Log Analysts Assn.*, Paper L.

OXBURGH, E. H. and ANDREWS-SPEED, C. P. 1981.   Temperature, thermal gradients and heat flow in the southwestern North Sea. In: *Petroleum Geology of the Continental Shelf of Northwest Europe*. L. V. Illing and G. D. Hobson (eds.). London: Heyden Press, 141–151.

PERRY, E. D. and HOWER, J. 1970.   Burial diagenesis in Gulf Coast pelitic sediments. *Clays and Clay Miner., 18*, 165–177.

PIRSON, S. J. 1983.   *Geological Well Log Analysis*, Third Edition. Houston: Gulf Publishing Co., 370 pp.

PLUMLEY, W. J. 1980.   Abnormally high fluid pressure: survey of some basic principles. *Am. Assoc. Petrol. Geol. Bull., 64*, 414–430.

POTTER, P. E., MAYNARD, B., and PRYOR, W. A. 1980.   *Sedimentology of Shales*. New York: Springer-Verlag, 306 pp.

PRICE, L. C. 1980.   Aqueous solubility of crude oil at 400°C and 2,000 bars pressure in the presence of gas. *J. Petrol. Geol., 4*, 195–223.

RASHID, M. A. 1978.   The influence of a salt dome on the diagenesis of organic matter in the Jeanne d'Arc Subbasin of the northeast Grand Banks of Newfoundland. *Org. Geochem., 1*, 67–77.

RASHID, M. A. and McALARY, J. D. 1977.   Early maturation of organic matter and genesis of hydrocarbons as a result of heat from a shallow piercement salt dome. *J. Geochem. Explor., 8*, 549–569.

RIEKE, H. H. and CHILINGARIAN, G. V. 1973.   *Compaction of Argillaceous Sediments*. Amsterdam: Elsevier, 350 pp.

RUSSELL, W. L. 1951.   *Principles of Petroleum Geology*. New York: McGraw-Hill, 508 pp.

SASS, J. H. 1971.   The earth's heat and internal temperature. In: *Understanding the Earth*. I. G. Gass, P. J. Smith, and R. C. L. Wilson (eds.). Sussex: Artemis Press, 80–87.

SCHMIDT, G. W. 1973. Interstitial water composition and geochemistry of deep Gulf Coast shales and sandstones. *Am. Assoc. Petrol. Geol. Bull.*, 57, 321–37.

SELLEY, R. C. 1982. *Introduction to Sedimentology*, Second Edition. London: Academic Press, 417 pp.

SILVER, C. 1968. Principles of gas occurrence, San Juan basin. In: *Natural Gases of North America*. B. W. Beebe and B. F. Curtis (eds.). Am. Assoc. Petrol. Geol., Mem. 9, 1, 946–960.

SULIN, V. A. 1946. Vody neftyanykh mestorozhdenii v sisteme priroduikhvod (Waters of oil reservoirs in the system of natural waters). Moscow: Gostoptedkhizdat.

TERZAGHI, K. 1936. The shearing resistance of saturated soils. *Proc. 1st Int. Conf. Soil Mech.*, Harvard, *1*, 54–46.

TICKELL, F. G. 1921. Summary of Operations. *California Division of Mines*, 6, 7–16.

TISSOT, B. P. and WELTE, D. H. 1978. *Petroleum Formation and Occurrence*. Berlin: Springer-Verlag, 538 pp.

VAN ELSBERG, J. N. 1978. A new approach to sediment diagenesis. *Can. Petrol. Geol. Bull.*, *26*, 57–86.

WHITE, D. E. 1957. Magmatic, connate and metamorphic waters. *Geol. Soc. Am. Bull.*, *68*, 1659–1682.

WHITE, D. E. 1965. Saline waters in sedimentary rocks. In: *Fluids in Subsurface Environments*. Am. Assoc. Petrol. Geol., Mem. No. 4, 346–366.

YOUNGS, B. G. 1975. The hydrology of the Gidgealpa formation of the Western and Central Cooper Basin. *Rep. of Invest. 43. Geol. Surv. S. Aust.*, 35 pp.

ZIELINSKI, G. W. and BRUCHHAUSEN, P. M. 1983. Shallow temperature and thermal regime in the hydrocarbon province of Tierra del Fuego. *Am. Assoc. Petrol. Geol. Bull.*, *67*, 166–177.

Generation and Migration of Petroleum

The preceding chapters discussed the physical and chemical properties of crude oil and natural gas and the behavior of fluids in the subsurface. The object of this chapter is to discuss the genesis, migration, and maturation of petroleum. We now have a good understanding of what oil and gas are made of and plenty of empirically derived data that show how they occur. Thus we know how to find and use oil and gas. However, the precise details of petroleum generation and migration are still debatable. Recent advances in geochemistry, especially in analytical techniques, have resulted in rapid progress on this front, but many problems remain to be solved.

Any theory of petroleum generation must explain two sets of observations: geological and chemical.

Geological Facts That a Theory of Petroleum Genesis Must Explain

1. Major accumulations of hydrocarbon characteristically occur in sedimentary rocks (Table 5.1). In the words of Pratt (1942): "I believe that oil in the earth is far more abundant and far more widely distributed than is generally realized. Oil is characteristic of unmetamorphosed marine rocks of shallow water origin. In its native habitat in the veneer of marine sediments that came with time to be incorporated into the earth's crust, oil is a normal constituent of that crust; a creature of the direct reaction of common earth forces on common earth materials."

TABLE 5.1 Amount of organic matter in sedimentary rocks

		Mass, in 10^8 g
Organic sediments		
Petroleum	1	
Asphalts	0.5	
Coal . . . . ⎫		
Lignite . . . . ⎬	15	
Peat . . . . ⎭		
		16.5
Organic matter disseminated in sedimentary rock		
Hydrocarbons	200	
Asphalts	275	
Kerogen (nonextractable organic matter)	12,000	
		12,475
	Total	12,491.5

From Hunt, 1977. Reprinted with permission.

Nearly half a century later few geologists would disagree with this statement, except perhaps to note in passing that continental and deep marine sediments are also petroliferous.

2. Numerous examples of hydrocarbon accumulations in sandstone and limestone reservoirs are totally enclosed above, below, and laterally by impermeable rocks. The Devonian reefs of Alberta and the Cretaceous shoestring sands of the Rocky Mountain foothill basins are examples of this phenomenon (see p. 304).

3. Other geological occurrences to consider are:
   a. Commercial accumulations of hydrocarbons have been found in basement rocks (see p. 316), but a young geologist's career is not enhanced by recommending management to drill such prospects.
   b. Traces of indigenous hydrocarbons have been found in igneous and metamorphic rocks. Commercial accumulations in basement, however, are always in lateral fluid continuity with sedimentary rocks.
   c. Traces of hydrocarbons occur in chondritic meteorites.

Chemical Facts That a Theory of Petroleum Genesis Must Explain

1. Crude oils differ from Recent shallow hydrocarbons:
   a. They show a preference for even-numbered carbon chains (modern hydrocarbons tend to occur in odd-numbered chains).
   b. They contain over 50 percent light hydrocarbons, which are rare or absent in modern sediments.

2. Crude oils show the following affinities with modern organic hydro-
    carbons:
    a. Young oils show levorotation. This property of optical activity is
        characteristic of hydrocarbons produced biosynthetically.
    b. They contain certain complex molecules that occur either in mod-
        ern organic matter or as a product of their degradation (e.g., the
        porphyrins and steroids).

In discussing theories seeking to explain the preceding geological and
chemical facts, it is very important to make several distinctions. Whether
petroleum is of organic (biological) or of inorganic (cosmic or magmatic)
origin has been debated often. How petroleum is formed from its parent
material has also been debated. Origin and mode of formation are two related but quite distinct problems. Most geologists are clear in their own
minds that oil is organic in origin, although most have only the vaguest
ideas as to how it forms. Ignorance of oil's mode of formation need not invalidate the fact that its nature is organic. In an appropriate analogy, ignorance of the principles of brewing need not prevent one from knowing that
beer can be made from yeast, hops, and malted barley.

The next section discusses whether petroleum is organic or inorganic
in origin. This section is followed by a discussion of the mode of formation
of petroleum.

ORIGIN OF PETROLEUM:
ORGANIC OR INORGANIC

Early theories of petroleum generation postulated an inorganic origin (e.g.,
Berthelot, 1866; Mendele'ev, 1877, 1902). Jupiter and Saturn are known
to contain methane. A particular class of meteorites, the carbonaceous
chondrites, contain traces of various hydrocarbons, including complex
amino acids and the isoprenoids phytane and pristane (Mueller, 1963). The
presence of extraterrestrial hydrocarbons is taken by most authorities as
evidence of inorganic formation (Studier, Hayatsu, and Anders, 1965).
Link (1957) turned this argument around, suggesting that the presence of
cosmic hydrocarbons was evidence of extraterrestrial life. (The hydrocarbons in chondrites might be the remains of barbecued space monsters.)

Until fairly recent times some astronomers have thought that the
earth's oil was of cosmic origin. As recently as 1955 Professor Sir Frederic
Hoyle wrote, "The presence of hydrocarbons in the bodies out of which the
Earth is formed would certainly make the Earth's interior contain vastly
more oil than could ever be produced from decayed fish—a strange theory

that has been in vogue for many years. . . . If our prognostication that the oil deposits have also been squeezed out from the interior of the Earth is correct, then we must, I think, accept the view that the amount of oil still present at great depths vastly exceeds the comparatively tiny quantities that man has been able to recover" (Hoyle, 1955).

Let us return to earth to examine this theory. Until the turn of the century many distinguished scientists believed in the magmatic origin of hydrocarbons. Among the earliest and greatest of these scientists were the geographer Alexander Von Humboldt and the great chemist Gay-Lussac (see Becker, 1909). This theory was adopted by Mendele'ev, who suggested that the mantle contained iron carbide. This iron carbide could react with percolating water to form methane and other oil hydrocarbons (Mendele'ev, 1902), which is analogous to the reaction in which acetylene is produced by carbide and water:

$$CaC_2 + 2H_2O = C_2H_2 + Ca(OH)_2$$

There is little evidence for the existence of iron carbide in the mantle. Yet the belief in a deep, inorganic origin for hydrocarbons was widely held by many scientists (chemists and astronomers rather than geologists) throughout the nineteenth century (see Dott and Reynolds, 1969, for a historical review). Vestigial traces of this belief still survive today in Russia (Porfir'ev, 1974) and even in the United States. If this theory is true, one would expect hydrocarbons to be commonly associated with igneous rocks and areas of deep crustal disturbance and faulting. The evidence will now be examined for the exhalation of hydrocarbons from volcanoes, their occurrence in congealed magma, and their association with faults.

Gaseous hydrocarbons have been recorded emanating from volcanoes in many parts of the world. White and Waring (1963) have reviewed some of these occurrences. Methane is the most commonly found gaseous hydrocarbon, generally present as less than 1 percent, although readings of 15 percent or more are noted. In all these instances, however, the volcanoes erupt through a cover of sedimentary rocks, from which the methane may have been derived by the heating of organic matter.

Data have accumulated over the years describing the occurrence of hydrocarbons within igneous rocks. Three genetic types of igneous hydrocarbons may be classified:

1. Hydrocarbons within vesicles and microscopic inclusions in igneous rocks

2. Hydrocarbons trapped where igneous rocks intrude sediments

3. Hydrocarbons in weathered igneous basement trapped beneath unconformities

One of the most celebrated cases of the first type occurs in the Khibiny massif of the Kola peninsula in Russia (Petersil'ye, 1962; Ikorskiy, 1967). Here alkaline intrusives penetrate a series of volcanic and volcaniclastic sediments. Ninety percent of the intrusives are composed of nepheline syenite. Gaseous and liquid hydrocarbons occur within inclusions in the nepheline crystals. Some 85 to 90 percent of the hydrocarbons consist of methane, but there are traces of ethane, propane, and n- and isobutane, together with hydrogen and carbon dioxide. Traces of bitumens have also been recorded, with long-chain alkanes and aromatics. Even though the quantity of hydrocarbons present in the Khibiny intrusives is infinitesimal, their diversity is impressive. Further Russian occurrences of igneous oil are reviewed by Porfir'ev (1974).

A second celebrated case of igneous oil has been described by Evans, Morton, and Cooper (1964) from Dyvika in the Arendal area of Norway. Here oil occurs in vesicles in a dolerite dyke. The vesicles are rimmed with crystals of calcite, analcite, chalcedony, and quartz. Paraffinic hydrocarbons occur in the center of the vesicles. The dolerite dyke intruded Precambrian gneisses and schists.

The second type of occurrence of hydrocarbons in basement is far more common and contains many commercial accumulations of oil and gas. Oil sometimes occurs where sediments, especially organic shales, have been intruded by igneous material. Instances occur, for example, in the Carboniferous oil shales of the Midland Valley of Scotland. Commercial accumulations of this type occur in the Jatibonico field of Jatibonico, Cuba, and in the interior coastal plain of Texas (Collingwood and Rettger, 1926). In these instances oil occurs in various hydrothermally altered basic intrusions. Such fields lack water and, although they may show very high initial production rates because of oil or gas expansion drive, often deplete rapidly. The Texas intrusives, of which over 17 fields have been found, intrude Cretaceous limestones and shales (Fig. 5.1).

The third type of oil accumulation in igneous rock is frequently commercially significant. This type is where oil or gas occurs in fractures and solution pores within igneous and metamorphic rocks. Instances of this type of trap occur beneath unconformities. The porosity is obviously due to weathering. The basement rocks are overlain by sediments, and organic-rich clays are generally found adjacent to the accumulation. The Augila-Nafoora fields of Libya represent this type of accumulation (see p. 316).

Let us now consider the origins of these three genetic types of oil in igneous rock. For the first type, where hydrocarbons occur in vesicles or inclusions, the inorganic origin of the hydrocarbons is hard to dispute. Contamination of magma as it intruded organic sediments could be postulated, but in the two cases previously cited, sedimentary rocks are absent.

The second type of occurrence, where hydrocarbons occur at the margin of igneous intrusives, is genetically equivocal. It could be argued

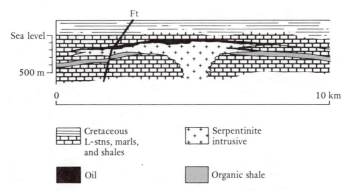

FIGURE 5.1 *Cross-section of the Lytton Springs oil pool, Caldwell County, Texas. Oil occurs in the crest of a circular serpentinite plug. There is no free water. Note adjacent organic shale. (After Collingwood and Rettger, 1926.)*

that either the oil was genetically related to the magma or it was formed by thermal metamorphism of kerogen in the intruded sediments.

The third type, where oil occurs in weathered basement beneath unconformities, can best be explained as being caused by emigration from the enveloping sediments. Genesis synchronous with intrusion is hard to accept, because oil emplacement must have postdated both the weathering of the basement and the deposition of the sediments that provide the seal. It has already been noted that where igneous degassing has been observed, it is noble gases, not hydrocarbons, that are exhaled (p. 17).

In concluding this review of hydrocarbons in igneous rocks, it must be stressed that the apparently unquestionable instances of indigenous magmatic oil are rare and not commercially important. Not only are the volumes of hydrocarbons trapped this way insignificant but the "reservoirs" are impermeable unless fractured. Commercial accumulations of hydrocarbons in igneous rocks only occur where the igneous rocks intrude or are unconformably overlain by sediments.

As discussed on page 287, faults play an important part in the migration and entrapment of petroleum. Some faults allow the upward movement of oil and gas; others are impermeable and act as a seal, thus causing fault traps. Some faults exhale flammable gases at the surface of the earth during earthquakes. Exhalation is preceded by bulging of the ground and accompanied by loud bangs, flashing lights, and visible ground waves. An ensuing evil smell is sometimes noted, due to the presence of hydrogen sulfide gas. These phenomena are all well documented and certainly demonstrate that some faults are conduits for flammable gas (Gold, 1979).

Detailed analyses of these gases reveal the presence of methane, carbon dioxide, hydrogen sulfide, and traces of noble gases (Sugisaki et al.,

1983). Peyve (1956) and Subbottin (1966) have argued that subcrustal abiogenic petroleum migrates up major faults to be trapped in sedimentary basins or dissipated at the earth's surface. Porfir'ev (1974) cites the flanking faults of the Suez, Rhine, Baikal, and Barguzin grabens as examples of such petroleum feeders.

This school of thought has recently been supported in the West (Gold, 1979; Gold and Soter, 1982). Gold argues that earthquake outgassing along faults allows methane to escape from the mantle. This process gives rise to deep gas reservoirs and, by polymerization, to oil at shallower horizons. To most geologists these ideas seem improbable and are rejected out of hand. Gold's earthquake outgassing theory implies that there is no energy crisis, and to find limitless quantities of oil and gas, all that needs to be done is to drill adjacent to faults deep enough and often enough.

The classic geologist's response has been cogently argued by North (1982). Since porosity and permeability decrease with depth, there can be no deep reservoirs to contain and disburse any gas that might exist. On a global scale gas reservoirs are closely related to coal-bearing formations of Upper Carboniferous (Pennsylvanian) and Cretaceous to Paleocene ages (the gas-generating potential of humic kerogen is discussed later).

Note that many of the most dramatic reports of earthquake outgassing occur in faulted basins with thick sequences of young sediments. These loosely packed sands are bound to be charged with biogenic methane and hydrogen sulfide. A seismic shock will cause the packing of the sand to tighten, resulting in a decrease in porosity and a violent expulsion of excess pore fluid.

It is possible, however, to differentiate biogenic from abiogenic methane. Carbon has two isotopes: C^{12} and C^{13}. The C^{12}-C^{13} ratio varies for different compounds, and this ratio is especially large for the different types of methane. Biogenic methane is enriched in C^{12}; abiogenic methane contains a higher proportion of C^{13} (Hoefs, 1980). C^{12}-C^{13} ratios of methane from the "hot spots" of the Red Sea, Lake Kivu (East Africa), and the East Pacific Rise suggest an abiogenic origin (McDonald, 1983).

Therefore some evidence exists to support the abiogenic origin of some methane and its movement by earthquake outgassing. Geologists will not fail to note, however, that commercial accumulations of oil are restricted to sedimentary basins. Petroleum seeps and accumulations are absent from the igneous and metamorphic rocks of continental shields, both far from and adjacent to faults. It is now routine to use gas chromatography "fingerprinting" to match the organic matter in shales with the petroleum in adjacent reservoirs (Flory et al., 1983). Geologists thus conclude that petroleum is formed by the thermal maturation of organic matter. This thesis will now be examined, beginning with an account of the production and preservation of organic matter on the earth's surface and then charting its evolution when buried with sediments.

MODERN ORGANIC PROCESSES ON THE EARTH'S SURFACE

The total amount of carbon in the earth's crust has been estimated to weigh 2.65×10^{20} g (Hunt, 1977). Some 82 percent of this carbon is locked up as CO_3 in limestones and dolomites. About 18 percent occurs as organic carbon in coal, oil, and gas (Schidlowski et al., 1974). The key reaction is the conversion of inorganic carbon into hydrocarbons by photosynthesis. In this reaction, water and atmospheric carbon dioxide are converted by algae and plants into water and glucose:

$$6CO_2 + 12H_2O = C_6H_{12}O_6 + 6H_2O + 6O_2$$

Glucose is the starting point for the organic manufacture of polysaccharides and other more complex carbon compounds. This polysaccharide production may happen within plants or within animals that eat the plants. In the natural course of events the plants and animals die and their organic matter is oxidized to carbon dioxide and water. Thus the cycle is completed (Fig. 5.2). In certain exceptional circumstances, however, the organic matter may be buried in sediments and preserved, albeit in a modified state, in coal, oil, or gas. In order to trace the evolution of living organic matter into crude oil and gas it is necessary to begin with an examination of the chemistry of organic matter.

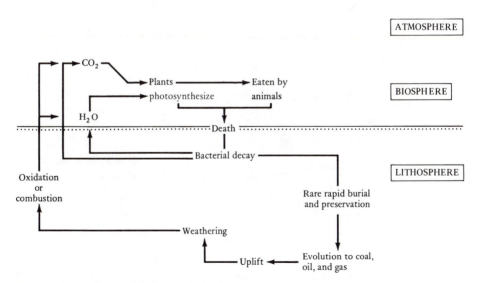

FIGURE 5.2 *The cycle of organic carbon.*

The major groups of chemicals that occur in organic matter are proteins, carbohydrates, lipids, and lignin. The proteins are found largely in animals and to a lesser extent in plants. They contain the elements hydrogen, carbon, oxygen, and nitrogen, with some sulfur and phosphorous. This combination of elements occurs in the form of amino acids. The carbohydrates are present in both animals and plants. They have the basic formula $C_n(H_2O)_n$ and include the sugars, such as glucose, and their polymers—cellulose, starch, and chitin. The lipids are also found in both animals and plants. They are basically recognized by their insolubility in water and include the fats, oils, and waxes. Chemically, the lipids contain carbon, hydrogen, and oxygen atoms. The basic molecule of the lipids is made of five carbon atoms (C_5H_8). From this base the steroids are built. The last of the four major groups of organic compounds is lignin, which is found only in the higher plants. Lignin is a polyphenol of high molecular weight, consisting of various types of aromatic carbon rings.

Figure 5.3 shows the distribution of these four groups of compounds in living organisms and in recent shallow sediments. Table 5.2 shows their abundance in different groups of animals and plants.

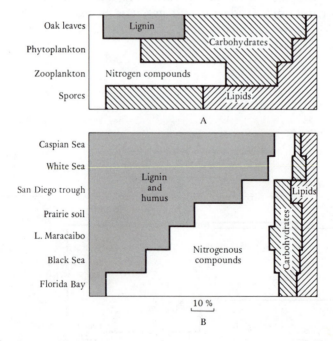

FIGURE 5.3 *Composition of organic matter in organisms (A) and shallow Recent sediments (B). (After Hunt, 1978.)*

TABLE 5.2   Chemical composition of various groups of animals and plants

| Substance | Weight percent of major constituents (ash-free) | | | |
	Proteins	Carbohydrates	Lignin	Lipids
Plants				
Spruce	1	66	29	4
Oak leaves	6	52	37	5
Pine needles	8	47	17	28
Phytoplankton	23	66	0	11
Diatoms	29	63	0	8
Lycopodium spores	8	42	0	50
Animals				
Zooplankton	60	22	0	18
Copepods	65	25	0	10
Higher invertebrates	70	20	0	10

From Hunt, 1979. Reprinted with permission.

Productivity and Preservation of Organic Matter

As previously discussed, the amount of organic matter buried in sediments is related to the ratio of organic productivity and destruction. Generally, organic matter is destroyed on the earth's surface, and only minor amounts are preserved. The deposition of an organic-rich sediment is favored by a high rate of production of organic matter and a high preservation potential. These two factors will now be discussed. Determining production and preservation for the present day is relatively easy; extrapolating back in time is harder. This problem is especially true for continental environments, whose production-destruction ratio of organic matter is largely related to the growth of land plants. Therefore, marine and continental environments should be considered separately.

Organic Productivity and Processes in Seas and Oceans

In the sea, as on the land, all organic matter is originally formed by photosynthesis. The photosynthesizers in the sea are pelagic phytoplankton and benthic algae. The biological productivity of these plants is related to both physical and chemical parameters. Of the former, temperature and light are of foremost significance. The amount of light is dependent on the depth, latitude, and turbidity of the water. The amount of organic productivity is highest in the shallow photic zone and decreases rapidly with increasing water depth and decreasing light and temperature.

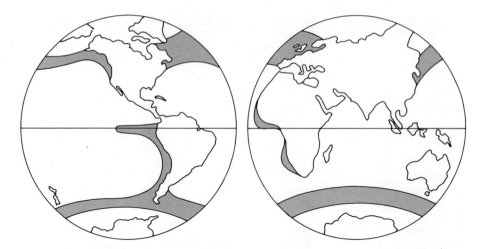

FIGURE 5.4 *Production of organic matter in the present-day world oceans. Note the low productivity in polar regions and the high productivity along eastward sides of the oceans. (After Debyser and Deroo, 1969, and Degens and Mopper, 1976.)*

Chemical conditions favoring organic productivity include the abundance of phosphates and nitrates. These chemicals are essential for the growth of plants and animals. Oxygenation is not important for the phytoplankton themselves, but is vital for the existence of animals that form later links in the food chain. Because oxygen is a by-product of photosynthesis, phytoplankton increases the oxygen content of the sea.

Figure 5.4 shows the biological productivity in the present-day oceans of the world. Three major areas of organic productivity can be recognized. Large zones, where less than 50 $g/m^2/year$ are produced, are found in the polar seas and in the centers of the large oceans. Two belts of higher productivity (200 to 400 $g/m^2/year$) encircle the globe along the boundaries between the polar and equatorial oceans. Three zones of above average organic productivity occur off the western coasts of North America, South America, and Africa. These zones are where deep, cold ocean currents well up from great depths (the California, Humboldt, and Benguela currents, respectively). The nutrients that these currents contain, principally nitrates and phosphates, are responsible for the high organic productivity of these zones.

It is important to stress that the areas of high oceanic productivity are not necessarily the areas where it is best preserved. The preservation of organic matter is favored by anaerobic bottom conditions and a rapid sedimentation rate. Johnson Ibach (1982) describes a detailed study of the relationship between rate of sedimentation and total organic carbon in cores recovered from the ocean floors by the Deep Sea Drilling Project.

The samples ranged in age from Jurassic to Recent. Figure 5.5 illustrates the results. Black shale, siliceous, and carbonate sediments are shown separately. For all three sediments, however, the total organic carbon content increases with sedimentation rate for slow rates of deposition, but decreases with sedimentation at high rates of deposition. The switchover occurs at slightly different sedimentation rates for the three lithologies.

Johnson Ibach interprets these results as showing that at slow sedimentation rates organic matter is oxidized on or near the seabed. Rapid deposition effectively dilutes the organic content of the sediment. For each lithology there appears to be an optimum rate for the preservation of organic matter; this rate is a balance between dilution and destruction of the organic fraction.

The second important condition favoring the preservation of organic matter is the presence of stratification within the waters from which sedi-

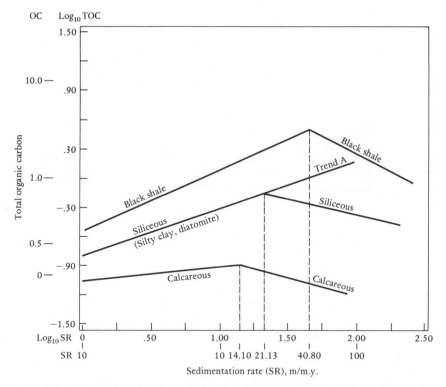

FIGURE 5.5 *Graph showing the relationship between total organic carbon and sedimentation rate for Jurassic to Recent cores recovered by the Deep Sea Drilling Project. The graph shows that each different type of sediment has an optimum sedimentation rate for the maximum preservation of organic matter. At too slow a rate, oxidation occurs; at too high a rate, organic carbon is diluted by sediment. (After Johnson Ibach, 1982.)*

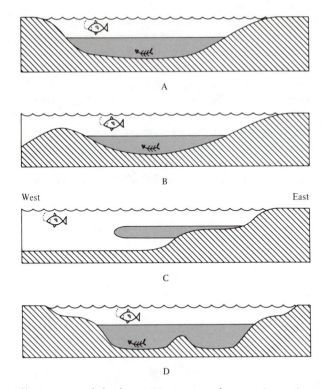

FIGURE 5.6 *Illustrations of the four main settings for anoxic environments favorable for the preservation of organic matter in sediments as classified by Demaison and Moore (1980): (A) A lake (thermally induced stratification); (B) a barred basin (stratification due to salinity contrast); (C) a continental shelf with upwelling; and (D) an ocean basin anoxic event.*

mentation occurs. Demaison and Moore (1979, 1980) recognize four major settings for the formation of organic-rich sediments in anoxic environments (Fig. 5.6).

The first of these settings is the freshwater lake, in which thermally induced stratification of water develops, with warm water near the surface and cooler, denser water deeper down. Life will thrive in the upper zone, with phytoplankton photosynthesizing and oxygenating the photic zone. In the deeper layer, oxygen, once used up, will not be replenished because photosynthesis does not occur in the dark and the density layering inhibits circulation. Sporadic storms may stimulate water circulation and cause mass mortality by asphyxiation of life throughout the lake, enhancing the amount of organic detritus deposited in the lake bed. This situation is well known in modern lakes, such as those of East Africa (Degens et al., 1971, 1973). Similar lakes in ancient times include the series of Tertiary lakes of Colorado, Utah, and Wyoming, in which the Green River Formation oil shales were deposited (Bradley, 1948; Robinson, 1976).

A second setting for the deposition of organic matter and its preservation in an anoxic environment occurs in barred basins (Fig. 5.6B). In this setting seawater enters semirestricted lagoons, gulfs, and seas. In arid climates evaporation causes the water level to drop and salinity to increase. Less dense water from the open sea enters across the bar and floats on the denser indigenous water. The same bar prevents the escape of the denser water to the sea. Thus a water stratification analogous to that of the freshwater lake develops. But in this case the density layering is due to differences not of temperature, but of salinity. Similar anoxic bottom conditions develop, however, and organic matter may be preserved. The Black Sea is a recent example of an anoxic barred basin (Deuser, 1971; Degens and Ross, 1974). The organic-rich Liassic shales of northwest Europe may have been deposited in this type of barred basin (Hallam, 1981).

A third situation favorable to the deposition and preservation of organic matter occurs today on the western sides of continents in low latitudes. The high productivity of these regions has already been described (Fig. 5.6C). In these settings the seawater is commonly oxygen-deficient between 200 and 1500 m (650 and 5000 ft) (Fig. 5.7).

The Phosphoria formation (Permian) and the Chattanooga-Woodford shales (Devonian-Mississippian) were probably deposited in analogous settings (Claypool et al., 1978; Duncan and Swanson, 1965). Both are relatively thin, but laterally extensive, formations that cover ancient continental shelves. The deposition of such organic-rich sediments is favored by a rise in sea level, probably caused by a global temperature rise, so that the

FIGURE 5.7 *Variation of oxygen content against depth in modern oceans. (After Emery, 1963, and Dietrich, 1963.)*

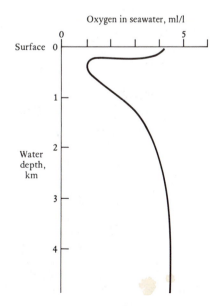

anoxic zone between 200 and 1500 m (650 and 5000 ft) extends across the continental shelves.

The fourth setting for the deposition of organic sediment is in an anoxic ocean basin. There is no known recent instance of this situation. Most recent ocean floors are reasonably well oxygenated. Deep oceanic circulation is caused by density currents, where cold polar waters flow beneath warmer low-latitude waters. It has been argued, however, that deep oceanic circulation may not have existed in past periods when the earth had a uniform, equable climate without the polar ice caps of today. At such times global "anoxic" events may have been responsible for the worldwide deposition of organic-rich sediment (Fig. 5.6D). This mechanism has been proposed to explain the extensive Upper Jurassic–Lower Cretaceous black shales that occur in so many parts of the world (Arthur and Schlanger, 1979; Schlanger and Cita, 1982). Tissot (1979) has shown that about 85 percent of the world's oil was sourced from these formations. Figure 5.8 shows that there is a correlation between organic sedimentation and sea level, thus supporting the thesis that transgressions favor anoxic environments.

Waples (1983), however, has disputed the existence of a global anoxic event in the early Cretaceous, pointing out that the evidence comes mainly

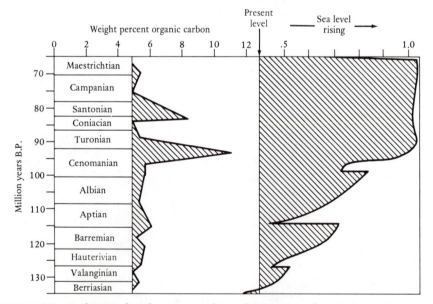

FIGURE 5.8 *(Left) Graphs of average values of organic carbon in Deep Sea Drilling Project cores in the North Atlantic. (After Arthur and Schlanger, 1979.) (Right) Global sea level curve. (After Vail, Mitchum, and Thompson, 1977.)*

from the Atlantic Ocean and its borderlands. During this time the Atlantic was but a narrow seaway in which local restriction of water circulation was to be expected.

This review shows that the preservation of organic matter in sediments is favored by an optimum sedimentation rate and anoxic conditions. These two conditions may occur in diverse settings, from localized lakes to whole oceans. Many geologists have documented the distribution of organic-rich sediments in space and time (North, 1979, 1980; Grunau, 1983). Parrish (1982) has prepared a series of global paleogeographic maps for the Paleozoic and has used these maps to predict paleoclimate and paleo-oceanic circulation patterns. These predictions show that many Paleozoic black shales were deposited in low latitudes on westerly facing continental shelves. Ager (1975) made a similar study for the Jurassic.

Organic Productivity and Preservation in Continental Environments

The productivity and preservation of organic matter in the continental realm is rather different from that in the oceans, seas, and lakes. Oxygen content is constant. Important variables are water and the number of growing days per year, which is a function of temperature and daylight hours. Thus the organic productivity of the polar regions is low. Given adequate humidity it increases with increasing temperature and daylight toward the equator.

Significant destruction of organic matter does not occur in upland areas below a mean temperature of approximately 10°C and, in swamps, of approximately 15°C (Gordon et al., 1958). Since any sediment with or without organic matter is unlikely to be preserved in upland environments, only swamp deposits need concern us. Gordon et al., (1958) also showed that the rate of production versus decay of organic matter increases from polar temperatures up to about 25°C. At this point organic productivity begins to decline, although the rate of decay continues to increase. Thus the preservation potential of organic detritus declines above 25°C. These data are of relatively little importance as far as the generation of fluid hydrocarbons are concerned, but are of considerable interest regarding the formation of coal. The same conclusion is reached for continental realms as for oceans, namely that the rate of decay of organic matter is more significant than its rate of formation.

Preservation of Organic Matter in Ancient Sediments

Extrapolating these data from the present day back in geological time is very difficult. Whereas the oceans appear to have contained diverse plant and animal life since way back in the Precambrian, plants only began to colonize the land toward the end of the Silurian period, and extensive coal formation did not begin until the Carboniferous. In addition to the possibility of fluctuation of the rate of production of organic matter in the past, the chemical composition of the various plants may have changed as well. In particular the amount of lignin may have increased in proportion to lipids, carbohydrates, and proteins.

Similarly, many paleontologists have traced the abundance of different groups of fossils back through geological time. Most studies use the number of species of a particular group as a criterion for their abundance. Although extrapolating from this information to actual numbers and biomass is hard, studies such as that by Tappan and Loeblich (1970) on phytoplankton may be particularly significant. They postulate a gradual increase in phytoplankton abundance from the Precambrian to the present day, with major blooms in the late Ordovician to early Silurian, and in the late Cretaceous.

FORMATION OF KEROGEN

Having now discussed the generation and preservation of organic matter at the earth's surface, it is appropriate to consider what happens to this organic matter when buried in a steadily subsiding sedimentary basin. As time passes, burial depth increases, exposing the sediment to increased temperature and pressure.

Tissot (1977) defines three major phases in the evolution of organic matter in response to burial:

1. *Diagenesis.* This phase occurs in the shallow subsurface at near-normal temperatures and pressures. It includes both biogenic decay, aided by bacteria, and abiogenic reactions. Methane, carbon dioxide, and water are given off by the organic matter, leaving a complex hydrocarbon termed *kerogen* (to be discussed in much greater detail shortly). The net result of the diagenesis of organic matter is the reduction of its oxygen content, leaving the hydrogen-carbon ratio largely unaltered.

2. *Catagenesis.* This phase occurs in the deeper subsurface as burial continues and temperature and pressure increase. Petroleum is released from kerogen during catagenesis—first oil and later gas. The hydrogen-carbon ratio declines, with no significant change in the oxygen-carbon ratio.

3. *Metagenesis.* This third phase occurs at high temperatures and pressures verging on metamorphism. The last hydrocarbons, generally only methane, are expelled. The hydrogen-carbon ratio declines until only carbon is left in the form of graphite. Porosity and permeability are now negligible.

The evolution of surface organic matter into kerogen and the ensuing generation of petroleum is discussed in detail in the following section.

Shallow Diagenesis of Organic Matter

The pH and Eh adjacent to the sediment-water interface is controlled by chemical reactions in which bacteria play an important role. In the ordinary course of events circulation continually mixes the oxygenated and oxygen-depleted waters. As previously discussed (p. 180), stratified water bodies inhibit this circulation. An oxygenated zone ($+$Eh) overlies a reducing zone ($-$Eh). Depending on circumstances, the interface between oxidizing and reducing conditions may occur within the water, at the sediment-water interface, or, if the sediments are permeable, below the sea bottom. In the lower reducing zone anaerobic sulfate-reducing bacteria of the genus *Desulfovibrio* remove oxygen from sulfate ions, releasing free sulfur. In the upper oxygenated zone the bacteria *Thiobacillus* oxidize the sulfur again:

$$SO_4 \underset{Thiobacillus}{\overset{Desulfovibrio}{\rightleftharpoons}} S + 2O_2$$

In the reducing zone the sulfur may combine with iron in ferrous hydroxide to form pyrite:

$$Fe(OH)_2 + 2S = FeS_2 + H_2O$$

Sulfate ions may also react with organic matter to form hydrogen sulfide:

$$SO_4 + 2CH_2O = 2HCO_3^- + H_2S$$
$$\underset{\substack{Organic \\ matter}}{}$$

This formula for organic matter is very simple, and it may be more appropriate to assume that the average content of marine organic matter has a carbon-nitrogen-phosphorous ratio of 106-16-1 (Goldhaber, 1978).

The first stage of biological decay is oxidation, which generates water, carbon dioxide, nitrates, and phosphates:

$$(CH_2O)_{106}(NH_3)_{16}H_3PO_4 + 138O_2$$
$$= 106CO_2 + 16NHO_3 + H_3PO_4 + 122H_2O$$

In the next stage, reduction of nitrates and nitrites occurs:

$$(CH_2O)_{106}(NH_3)_{16}H_3PO_4 + 94 \cdot 4NHO_3$$

$$= 106CO_2 + 55 \cdot 2N_2 + 177 \cdot 2H_2O + H_3PO_4$$

This process is followed in turn by sulfate reduction, which results in the generation of hydrogen sulfide and ammonia:

$$(CH_2O)_{106}(NH_3)_{16}H_3PO_4 + 52SO_4^{2-}$$

$$= 106HCO_3^- + 53H_2S + 16NH_3 + H_3PO_4$$

These equations are, of course, very simplified. As has been shown earlier, the organic compounds that are the starting point of this sequence of reactions are diverse and complex, consisting essentially of proteins, carbohydrates, lipids, and lignin; the relative rate of decay being in this order, with proteins being the least stable and woody lignin the most resistant. These compounds are all acted on by the enzymes of microbes to produce various biomonomers. For example, carbohydrates, such as starch and cellulose, are broken down to sugars. Cellulose is also converted into methane and carbon dioxide:

$$(C_6H_{10}O_5)_n \rightarrow CO_2 + CH_4$$

Methane is a major by-product of the bacterial decay of not only cellulose but many other organic compounds. As this biogenic methane moves upward from the decaying organic matter, it may cross the $-Eh:+Eh$ surface and be oxidized:

$$3CH_4 + 6O_2 = 3CO_2 + 6H_2O$$

In environments where the deposition of organic matter and its rate of decay are rapid, free methane may seep to the surface as bubbles of marsh gas. This gas occasionally ignites spontaneously, giving rise to the *will o' the wisp* phenomenon, which terrifies the uninformed in bogs and graveyards.

The proteins in decaying organic matter are degraded into amino acids and peptides. The lipid waxes and fats degrade to glycerol and fatty acids. The lignin of woody tissues breaks down to phenols, aromatic acids, and quinones. These changes all take place within the top few meters of sediment. With gradually increasing burial depth, the physical environment of the sediment changes, and chemical reactions alter in response to these changes. The increase in overburden pressure results in compaction of the sediment. Numerous studies of compaction have been made and are discussed elsewhere (p. 200). At this point it is sufficient to state that within the first 300 m of burial the porosity of a clay typically diminishes from some 80 percent down to only about 30 or 40 percent. During compaction, water escapes. Part of this water is primary pore water, but some of it is of biogenic origin. It contains in solution carbon dioxide, methane, hydrogen sulfide, and other organic compounds of decay, loosely termed *humic acids*.

At this time several important inorganic reactions may be taking place, causing the formation of early authigenic minerals. The generation of pyrite has already been discussed, but in carbonate-rich clays siderite ($FeCO_3$) is also common. Early calcium carbonate cementation may occur by direct precipitation or, in evaporitic environments, by the interreaction of calcium sulfate and organic matter (Friedman, 1972). Sulfate ions react with organic matter to produce hydrogen sulfide (as previously discussed) and HCO_3^-. The latter reacts with the calcium to form calcite, water, and carbon dioxide. Essentially:

$$CaSO_4 + 2CH_2O = CaCO_3 + H_2O + CO_2 + H_2S$$
Organic
matter

With the increasing depth of burial the ambient temperature increases and the role of bacteria in biogenic reactions gradually declines as they die out (although some thiophilic bacteria can occur deep in the subsurface). At the same time, however, increasing temperature allows inorganic reactions to accelerate. The generation of hydrocarbons now declines as the production of biogenic methane ceases. There are occasional reports of the shallow occurrence of biogenic hydrocarbons other than methane. One of the best known is the occurrence of up to 200 ppm of aromatics and paraffinic naphthenes in sand less than 10,000 years old and only 30 m deep off the Orinoco delta (Kidwell and Hunt, 1958). More recently, lipid-rich liquid hydrocarbons have been recorded from ocean floor samples. This *protopetroleum* is formed by the minor breakdown of organic oils. It is chemically distinct from crude oil, has not undergone thermal maturation, and does not appear to be present in commercial accumulations (Welte, 1972). Important changes continue to take place, however, within the preserved organic matter. Water and carbon dioxide continue to be expelled as the formation of kerogen begins.

Chemistry of Kerogen

The etymology and original definition of kerogen as recognized in oil shale is discussed on page 401. Today kerogen is the term applied to disseminated organic matter in sediments that is insoluble in normal petroleum solvents, such as carbon bisulfide. This insolubility distinguishes it from bitumen. Chemically, kerogen consists of carbon, hydrogen, and oxygen, with minor amounts of nitrogen and sulfur (Table 5.4).

TABLE 5.4   Chemistry of kerogens

	Weight percent					Ratios		Petroleum
	C	H	O	N	S	H-C	O-C	type
TYPE I Algal	75.9 (7 samples)	8.84	7.66	1.97	2.7	1.65	0.06	Oil
TYPE II Liptinic	77.8 (6 samples)	6.8	10.5	2.15	2.7	1.28	0.1	Oil and gas
TYPE III Humic	82.6 (3 samples)	4.6		2.1	0.1	0.84	0.13	Gas

Weight percent data from Tissot and Welte, 1978; ratios from Dow, 1977. Reprinted with permission.

Three basic types of kerogen are generally recognizable. The differences are chemical and are related to the nature of the original organic matter. Since these three kerogen types generate different hydrocarbons, their distinction and recognition are important (Tissot, 1977; Dow, 1977).

Type I kerogen is essentially algal in origin (Plate 2). It has a higher proportion of hydrogen relative to oxygen than the other types of kerogen have (H-O ratio is about 1.2-1.7). The H-C ratio is about 1.65 (Table 5.4). Lipids are the dominant compounds in this kerogen, with derivates of oils, fats, and waxes. This kerogen is particularly abundant in algae such as *Botryococcus*, which occurs in modern Coorongite and ancient oil shales (see p. 401). Similar algal kerogen is characteristic of many oil shales, source rocks, and the cannel or boghead coals. The basic chemical structure of type I algal kerogen is shown in Figure 5.9A.

Type II, or liptinic, kerogen is of intermediate composition (Plate 2). Like algal kerogen it is rich in aliphatic compounds, and it has an H-C ratio greater than one. The chemical structure is shown in Figure 5.9B. The original organic matter of type II kerogen consisted of algal detritus, but also contained material derived from zooplankton and phytoplankton. The Kimmeridge clay of the North Sea and the Tannezuft shale (Silurian) of Algeria are of this type.

Type III, or humic, kerogen has a much lower H-C ratio (<0.84). Chemically, it is low in aliphatic compounds, but rich in aromatic ones (Plate 2). The basic molecular structure is shown in Figure 5.10C. Humic kerogen is produced from the lignin of the higher woody plants, which grow on land. It is this humic material that, if buried as peat, undergoes diagenesis to coal. Type III kerogen tends to generate largely gas and little, if any, oil. Nonmarine basins were once thought to be gas prone because of an abundance of humic kerogen, whereas marine basins were thought to be oil provinces because of a higher proportion of algal kerogen. This type

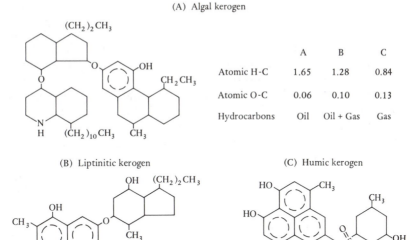

	A	B	C
Atomic H–C	1.65	1.28	0.84
Atomic O–C	0.06	0.10	0.13
Hydrocarbons	Oil	Oil + Gas	Gas

FIGURE 5.9   *The molecular structure of (A) type I, or algal, kerogen. (B) type II, or liptinitic, kerogen; and (C) type III, or humic, kerogen. (From Dow, 1977. Reprinted with permission.)*

of generalization is not valid. Many continental basins contain lacustrine shales rich in algal kerogen. The Green River Formation of Colorado, Utah, and Wyoming is a noted example (see p. 403); other examples occur in China (Ma Li et al., 1982). Further details on kerogen can be found in Tissot and Welte (1978), Hunt (1979), and Durand (1980).

This review of the three basic types of kerogen shows the importance of identifying the nature of the organic matter in a source rock so as to accurately assess its potential for generating hydrocarbons. A second important factor to consider is not just the quality of kerogen but also the quantity necessary to generate *significant* amounts of oil and gas suitable for commercial production. Several separate items are to be considered here, including the average amount of kerogen in the source bed, the bulk volume of the source bed, and the ratio of emigrated to residual hydrocarbons. The total organic matter in sediments varies from 0 percent in many Precambrian and continental shales to nearly 100 percent in certain coals. A figure of 1500 ppm total organic carbon is sometimes taken as the minimum requirement for further exploration of a source rock (Pusey, 1973a).

Maturation of Kerogen

During the phase of catagenesis, kerogen matures and gives off oil and gas. Establishing the level of maturation of kerogen in the source rocks of an area subject to petroleum exploration is vital. When kerogen is immature, no petroleum has been generated; with increasing maturity, first oil and then gas are expelled; when the kerogen is overmature, neither oil nor gas remains. Figure 5.10 shows the maturation paths for the different types of kerogen. The maturation of kerogen can be measured by several techniques to be described shortly. The rate of maturation may be dependent on temperature, time, and, possibly, pressure.

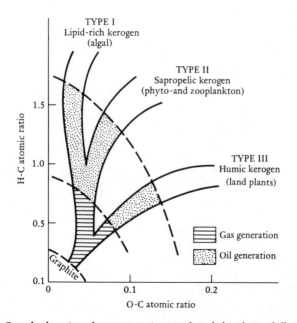

FIGURE 5.10   *Graph showing the maturation paths of the three different types of kerogen.*

A number of workers have documented an empirical correlation between temperature and petroleum generation (Pusey, 1973a, 1973b; Philippi, 1975; Hunt, 1979). Significant oil generation occurs between 60 and 120°C, and significant gas generation between 120 and 225°C. Above 225°C the kerogen is inert, having expelled all hydrocarbons; only carbon remains as graphite (Fig. 5.11). The temperatures just cited are only approximate boundaries of the oil and gas windows.

FIGURE 5.11   *Graph of depth vs thermal gradient showing how oil occurs in a liquid window of between 65 and 150°C, which extends from shallow depths with high thermal gradients to deep basins with lower gradients. (After Pusey, 1973a.)*

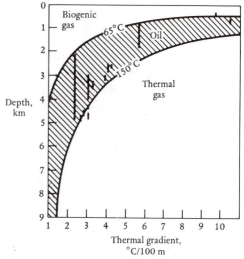

The chemical laws of kinetics, as expressed in the Arrhenius equation, state that the rate of a chemical reaction is related to temperature and time. Reaction rate generally doubles for each 10°C increase. Thus many geologists have considered that kerogen maturity is a function of temperature and time (Fig. 5.12). Petroleum may therefore have been generated from old, cool source rocks as well as from young, hot ones (e.g., Connan, 1974; Erdman, 1975; and Cornelius, 1975). Several techniques have been

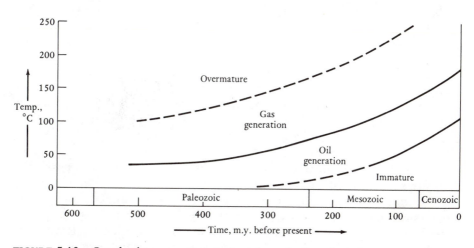

FIGURE 5.12   *Graph of temperature against time of oil and gas formation. For the problems of interpreting this figure, see text. (After Connan, 1974; Erdman, 1975; and Cornelius, 1975.)*

developed that try to quantify the relationship of temperature and time to kerogen maturity. These techniques are generally based on a burial history curve (Van Hinte, 1978), which is a graph on which burial depth is plotted against geological time for a particular region (see Fig. 5.13).

Two commonly used maturation indices are the thermal maturity index (TTI) and the level of organic maturation (LOM). The TTI was first proposed by Lopatin (1971) and developed by Waples (1980, 1981). The thermal maturity index is calculated from a formula that integrates temperature with the time spent in each temperature (in increments of 10°C) as a source rock is buried. The length of time spent in each temperature increment is established from the burial curve. The LOM was first proposed by Hood et al., (1975) and developed by Royden et al., (1980). The LOM is based on the assumption that reaction rate doubles for each 10°C increment of temperature. Oil generation occurs between LOM values of 7 and 13, and gas generation occurs between values of 13 and 18.

Both the TTI and LOM must be used with caution. They assume that geothermal gradient has been constant through time, which is seldom likely to be true. Burial curves must also be used carefully, not just considering the present day depth of the formations but also allowing for compaction, uplift, and for the fact that in few basins do subsidence and sedimentation balance one another. Price (1983) has cast still more doubts on these methods. His careful analysis of his own data and previously published data led him to conclude that only temperature affects organic metamorphism; time is not important.

Whatever may be the case regarding the effects of temperature and time, measuring the maturity of kerogen is obviously necessary. Many techniques are available, some of which will be described in the following sections.

Paleothermometers

Because of the important relationship between temperature and petroleum generation, measuring the maturity of the kerogen is important. It is not enough to be able to answer the questions: Are there organic-rich source rocks in the basin? Are they present in large enough volumes? Are they oil prone or gas prone? It is also necessary to know whether or not the source rocks have matured sufficiently to generate petroleum or whether they are supermature and barren.

To measure only the bottom hole temperature from boreholes does not answer the question of kerogen maturity. This measurement only indicates the present day temperature, which may be considerably lower than that of the past. It is necessary to have a paleothermometer, which can measure the maximum temperature to which the source rock was ever subjected. Several such tools are available, each varying in efficiency and

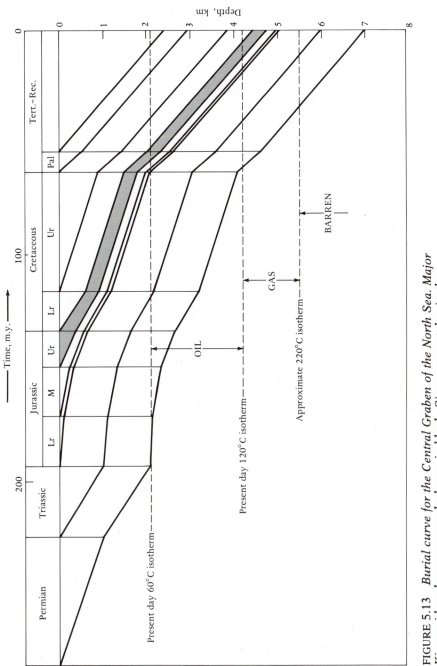

FIGURE 5.13 *Burial curve for the Central Graben of the North Sea. Major Kimmeridge clay source rock shown in black. Since present-day isotherms are used, the graph only gives a rough idea of when the source rock entered the oil and gas windows. (After Selley, 1980.)*

each requiring expensive laboratory equipment and a considerable degree
of technical expertise. Broadly speaking, two major groups of techniques
are used for measuring the maximum paleotemperature to which a rock
has been heated (Cooper, 1977):

1. Chemical paleothermometers
    a. Organic
        i.   Carbon ratio
        ii. Electron spin resonance
        iii. Pyrolysis
    b. Inorganic
        i.   Clay mineral diagenesis
        ii. Fluid inclusions
2. Biological paleothermometers
    a. Pollen coloration
    b. Vitrinite reflectance

Chemical Paleothermometers

One of the oldest and most fundamental of maturation indices is the carbon ratio method. Obviously, as organic matter matures, organic carbon is
lost: first as carbon dioxide during decarboxylation and later during petroleum generation. The net result of kerogen maturation is thus a gradual
decrease in the proportion of movable organic carbon and a relative increase in the proportion of fixed carbon.

The carbon ratio method compares the total organic carbon (C_T) in
the sample with the residual carbon (C_R) after pyrolysis to 900°C. The ratio
C_R-C_T increases with maturation (Gransch and Eisma, 1970). This paleothermometer is not easy to calibrate, however, because of the variations in
the original carbon content in the different types of kerogen and the risky
assumption that their carbon ratios change at the same rate.

A second chemical thermometer is the electron spin resonance (ESR)
technique developed by Conoco (Pusey, 1973b). The ESR method is based
on the fact that the atomic fraction of kerogen contains free electrons,
whose number and distribution are related to the number and distribution
of the benzene rings. These electron characteristics change progressively
as the kerogen matures (Marchand et al., 1969). Kerogen can be extracted
from the source rock and placed in a magnetic field whose strength can be
controlled. If the sample is then exposed to microwaves, the free electrons
resonate (hence the term ESR.) and, in so doing, alter the frequency of the
microwave. With an electrometer certain properties of the electrons can
be measured, including their number, signal width, and the location of the
resonance point. As already stated, these parameters have been shown to
vary with kerogen maturation.

The ESR technique was originally developed as a paleothermometer in the Gulf Coast of the United States (Pusey, 1973b) and later in the North Sea (Cooper et al., 1975). The method has the advantage of being quick (once a laboratory is established), and it can be used on minute and impure samples of kerogen. As with most paleothermometers, great care must be taken that the shale samples gathered during drilling are not caved; ideally, core or sidewall core material should be used. Again, as with the carbon ratio method, the ESR has certain problems caused by original variations in kerogen composition and the presence of recycled organic detritus (Durand et al., 1970).

Pyrolysis, the heating of kerogen or source rock, was the process developed to produce petroleum from oil shales. Today the pyrolysis of kerogen samples is an important laboratory technique in source rock analysis (see Tissot and Welte, 1978; Hunt, 1979). Pyrolysis is now generally carried out with a flame ionization detector. A sample of rock or extracted kerogen is heated in a stream of helium. The temperature is gradually raised at a carefully measured rate, and the expelled hydrocarbon gases recorded with a hydrogen flame ionization detector. At relatively low temperatures (200 to 300°C) any free hydrocarbons in the sample are volatilized. These hydrocarbons are referred to as S_1. With increasing temperature hydrocarbons are expelled from the kerogen. These hydrocarbons are termed S_2. Carbon dioxide and water, which are also expelled, are grouped as S_3. Pyrolysis generally continues up to 800 to 900°C. The three readings can be used to determine the maturation level of the source rock (Espitalié, 1977). Where migration has not occurred, the ratio S_1-$(S_1 + S_2)$ shows the amount of petroleum generated compared with the total amount capable of being generated. This ratio is referred to as the *transformation ratio*. Plotted against depth, the transformation ratio will generally show a gradual increase as the source rock matures.

Clay mineral analysis is one of the oldest of the inorganic paleothermometers, having been studied quite independently of kerogen maturation. Newly deposited clay minerals include smectites (montmorillonite), illite, and kaolin. The varying abundance of different mineral types depends partly on the source area and partly on the depositional environment (p. 243). As clays are buried, they undergo various diagenetic changes. Illite and kaolin authigenesis is largely related to the pH of the pore fluid. With increasing depth of burial, higher temperatures recrystallize the clays. Of particular interest is the behavior of the smectites, which dewater and change to illite within the optimum temperature for oil generation (80 to 120°C). This phenomenon is considerably important to oil migration (Power, 1967) and is discussed later in this chapter (p. 200).

As temperatures increase to the upper limits at which oil and gas form, another major change in clay takes place. Kaolin and illite recrystallize into micas, whereas in ferromagnesian-rich environments they form chlorites. A number of studies have documented these changes (e.g., Fuchtbauer, 1967). These calibration points are rather imprecise to pick on the kerogen maturation scale. A greater degree of sophistication was achieved, however, by Gill et al., (1979), who showed how the crystallinity of kaolin and illite could be mapped across the South Wales coal basin and how those variations correlated with coal rank. This technique can undoubtedly be applied to other regions.

Another inorganic paleothermometer is provided by the study of fluid inclusions. This method has been used for many years by hard rock geologists interested in finding the temperature of formation of crystalline minerals (Roedder, 1967). Attempts have been made to extend this technique to the fluid inclusions of quartz and calcite in fractures (Currie and Nwachukwu, 1974). At present this method is unlikely to become widespread because of the relatively low temperatures involved in hydrocarbon generation.

Biological Paleothermometers

In addition to chemical methods for paleotemperature measurement, several biological paleothermometers are used. One of these techniques measures the color of the organic matter in the source rock. Kerogen has many colors and shades, which are dependent on both maturation and composition. Spores and pollen, however, begin life essentially colorless. As they are gradually heated, they change to yellow, orange, brown (light to dark), and then to black. Palynologists have used this color change pattern for years to differentiate recycled spores and pollen from indigenous ones, since the former will be darker than average for a particular sample.

More recently, it has been realized that these color changes can be related to the degree of maturation of the associated kerogen. Several color codes are used by palynologists: some developed by palynologists themselves, others borrowed from brewers. One popular system is the 10-point Spore Color Index of Barnard et al., (1978); another is the 5-point Thermal Alteration Index of Staplin (1969). Using the simple brewer's color code, the following relationship between pollen color and hydrocarbon generation has been noted: water: immature; lager: oil generation; bitter: condensate generation; Guinness: dry gas. It has been argued that pollen color measurement is a subjective assessment, but with reference standards and adequate operator experience the results seem to be valid. The technique is relatively fast and economical, since the palynologists will review their samples as a routine check for contamination and recycling anyway.

The last paleothermometer to consider is vitrinite reflectance (Dow, 1977; Dow and O'Connor, 1979; Cooper, 1977), a very well established technique used by coal petrographers to assess the rank of coal samples. Basically, the shininess of coal increases with rank from peat to anthracite. This shininess, or reflectance, can be measured optically. Vitrain, the coal maceral used for measurement, occurs widely, if sparsely, throughout sedimentary rocks. Kerogen, which includes vitrain, is separated from the sample by solution in hydrofluoric and hydrochloric acids. The residue is mounted on a slide, or in a resin block with a low-temperature epoxy resin and then polished. A reflecting-light microscope is then used to measure the degree of reflectivity, termed R_o. This step requires great care and can only be done by a well-trained operator. The prepared sample may contain not only indigenous kerogen but also caved and recycled material, as well as organic material from drilling mud (lignosulfonate). Reflectivity may also vary from the presence of pyrite or asphalt. A number of reflectivity readings are taken for each sample and plotted on a histogram, reference being made to standard samples of known R_o values. R_o may be measured for a series of samples down a borehole, and a graph of reflectivity versus depth may be plotted.

An empirical relationship has been noted between vitrinite reflectance and hydrocarbon generation. Crude oil generation occurs for R_o values between 0.6 and 1.5. Gas generation takes place between 1.5 and 3.0; at values above 3.0 the rocks are essentially graphitic and devoid of hydrocarbons. These values are only approximate because, as already discussed, the oil- or gas-generating potential of kerogen varies according to its original composition.

Reflectivity is generally assumed to increase with temperature and time. McTavish (1978) has shown, however, the retarding effect of abnormal pressure, and, as already noted, Price (1983) considers temperature to be the only important variable.

The measurement of vitrinite reflectivity is one of the most widely used techniques for establishing the maturity of a potential source rock in an area. Among other uses, borehole logs that plot R_o against depth indicate the intervals in which oil or gas may have been generated. Abrupt shifts in the value of R_o with depth may indicate faults or unconformities. An abrupt increase in R_o with depth followed by a return to the previous gradient may be caused by igneous intrusives.

This review of paleothermometers shows that a wide range of techniques are available. All the methods require elaborate laboratory equipment, carefully trained analytical operators, and skilled interpretation of the resultant data. There are many pitfalls for the inexperienced interpreter. Probably, pollen coloration and vitrinite reflectivity are the most reliable and widely used maturation indices at the present time. Figure 5.14 shows the relation between R_o, spore coloration, and hydrocarbon generation.

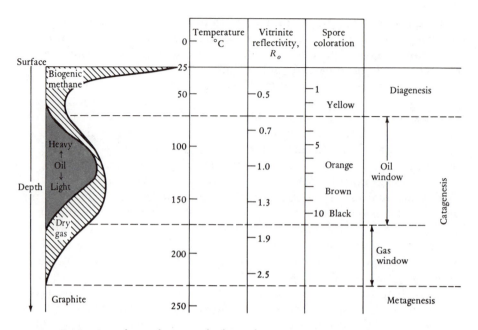

FIGURE 5.14   *Correlation between hydrocarbon generation, temperature, and some paleothermometers.*

PETROLEUM MIGRATION

The preceding sections of this chapter showed that petroleum is of organic origin; discussed the production of organic matter and its preservation in sediments; and traced the diagenesis of organic matter into kerogen and the maturation of kerogen as petroleum is expelled. The object of this section is to discuss how fluid hydrocarbons may emigrate from the source rock to the reservoir.

A number of lines of observational evidence show that oil and gas do not generally originate in the rock in which they are found, but that they must have migrated into it from elsewhere. This theory is proved by the following observations:

1. As previously discussed, organic matter is easily destroyed by oxidization in porous, permeable sediments at the earth's surface. It must therefore have invaded the reservoir rock after considerable burial and raised temperature.

2. Oil and gas often occur in solution pores and fractures that must have formed after the burial and lithification of the host rock.

3. Oil and gas are trapped in the highest point (structural culmination, or stratigraphic pinchout) of a permeable rock unit, which implies upward and lateral migration.

4. Oil, gas, and water occur in porous, permeable reservoir rock stratified according to their relative densities. This stratification implies that they were, and are, free to migrate vertically and laterally within the reservoir.

These observations all point to the conclusion that hydrocarbons migrate into reservoir rocks at a considerable depth below the surface and some time after burial.

An important distinction is made between primary and secondary migration. Primary migration is understood as the emigration of hydrocarbons from the source rock (clay or shale) into permeable carrier beds (generally sands or limestones). Secondary migration refers to subsequent movement of oil and gas within permeable carrier beds and reservoirs. There is consensus that secondary migration occurs when petroleum is clearly identifiable as crude oil and gas, and, although gas may be dissolved in oil, their solubility in connate water is negligible. Secondary migration occurs by buoyancy due to the different densities of the respective fluids and in response to differential pressures, as discussed in Chapter 4 (p. 160).

Primary migration, the emigration of hydrocarbons from source bed to carrier, is still a matter for debate. The object of this section is to examine the problem of primary hydrocarbon migration which is to many people the last great mystery of petroleum geology. Seminal works on this topic are by Magara (1977), Tissot and Welte (1978), Barker et al., (1978), McAuliffe (1979), Hunt (1979), and Roberts and Cordell (1980). A number of basic questions need to be answered before the problem of primary migration can be fully understood. These questions include: When did migration occur? What was the physical environment (temperature, pressure, permeability, porosity) of the source bed at that moment? What was the chemical composition of the source bed at that moment (clay mineralogy may be one critical factor; water content another)? What was the nature of the hydrocarbons? Were they in the form of *protopetroleum*, some nebulous substance transitional between kerogen and crude oil or gas? Did oil and gas emigrate in discrete phases, in solution (one in the other), in pore water, or in complex organic compounds in the pore fluid? These questions are numerous and complex. They require much geological data that are difficult to acquire, and they are hard to attack experimentally because of the time, temperatures, and pressures involved. In addition, a knowledge of physics and chemistry beyond that of the average geologist is required.

The study of primary migration contains a major paradox. Oil and gas are trapped in porous, permeable reservoirs. The source rocks from which they emigrated can be identified (Welte et al., 1975; Hunt, 1979; Flory et al., 1983). Yet these same source rocks are impermeable shales. How then did the fluids emigrate? It would be nice to believe that oil and gas were squeezed from the source clay during early burial before compaction destroyed permeability. This process cannot be so, however, because the temperatures necessary for hydrocarbon generation are not reached until compaction has greatly diminished permeability and water saturation.

At this point a review of the relationship between clay porosity, permeability, compaction, water loss, and hydrocarbon migration would be appropriate. Numerous compaction curves for argillaceous sediments have been published (Fig. 5.15). These curves show that most water expulsion by compaction occurs in the upper 2 km of burial. Pore water expelled by compaction is minimal below this depth. Note that for an average geothermal gradient (25°C/km), oil generation begins below the depth at which most of the compactional pore water has been expelled. The migration of oil by the straightforward flushing of pore water is not therefore a viable proposition. However, other factors are relevant to this problem, including consideration of the role of supernormal temperatures and pressures and the diverse kinds of water present in clays. Many major hydrocarbon provinces have been widely noted to be areas of supernormal temperatures and pressures. The Tertiary clastic prism of the Gulf of Mexico in the United States, which has undergone intense detailed studies, is a case in point.

Powers (1967) pointed out that there are two types of water in clays: normal pore water and structured water that is bonded to the layers of montmorillonitic clays (smectites), as shown in Fig. 5.16. When illitic or kaolinitic clays are buried, a single phase of water emission occurs because of compaction in the first 2 km of burial. When montmorillonitic-rich muds are buried, however, two periods of water emission occur: an early phase and a second, quite distinct phase when the structured water is expelled during the collapse of the montmorillonite lattice as it changes to illite (Fig. 5.17). Further work by Burst (1969) detailed the transformation of montmorillonite to illite and showed that this change occurred at an average temperature of some 100 to 110°C, right in the middle of the oil generation window (Fig. 5.18). The actual depth at which this point is reached varies with the geothermal gradient, but Burst (1969) was able to show a normal distribution of productive depth at some 600 m above the clay dehydration level (Fig. 5.19). By integrating geothermal gradient, depth, and the clay change point, it was possible to produce a fluid redistribution model for the Gulf Coast area. Similar studies have been reported from other regions (Fascolos and Powell, 1978).

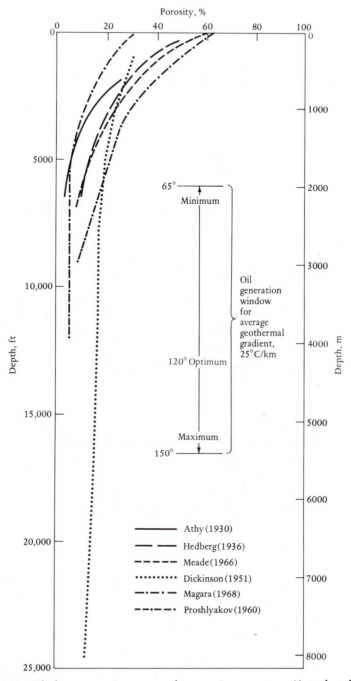

FIGURE 5.15 *Shale compaction curves from various sources. Note that there is minimal water loss through compaction over the depth range of the oil window.*

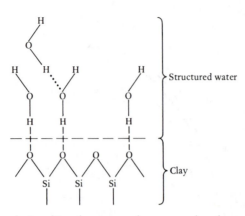

FIGURE 5.16 *The relationship of structured water molecules to clay mineral lattice. (After Barker, 1978.)*

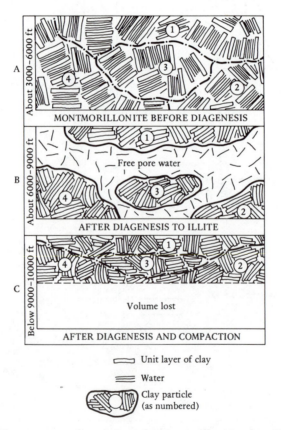

FIGURE 5.17 *The two-stage dewatering of montmorillonitic clay. (After Powers, 1967.)*

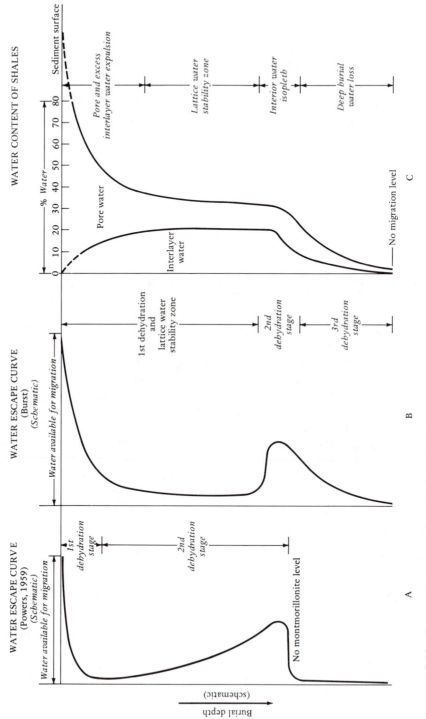

FIGURE 5.18 Dewatering curves for clay burial. (After Burst, 1969.)

FIGURE 5.19 *Depth-frequency curve of the top of 5368 oil reservoirs in the Gulf Coast of the United States. This curve shows that the peak production depth is about 600 m above the average depth at which the second phase of clay dehydration occurs. (After Burst, 1969.)*

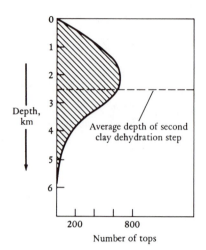

Barker et al., (1978) have pursued this idea, showing that not only water but also hydrocarbons may be attached to the clay lattice (Fig. 5.20). Obviously, the hydrocarbons will be detached from the clay surface when dehydration occurs. The exact physical and chemical process whereby oil is expelled from the source rock is still not clear, but Figure 5.18 demonstrates an empirical relationship between clay dehydration and hydrocarbon accumulation. Regional mapping of the surface at which this change occurs is thus a valid exploration tool, although the processes responsible for the relationship may not be fully understood.

Several qualifications must be placed on this technique. These data pertain to the Tertiary Gulf Coast of the United States. In many other hydrocarbon provinces in the world, smectitic clays are largely absent. The dewatering of clay cannot therefore be advocated as the dominant process of emigrating hydrocarbons from source rocks. Overpressure is obviously

FIGURE 5.20 *Sketch showing how hydrocarbon molecules (in this case an R-C-C-COH alcohol) may be attached to a clay mineral lattice together with water molecules. (From Barker, 1978. Reprinted with permission.)*

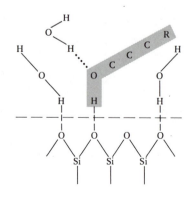

a factor that may aid petroleum generation by maintaining porosity and permeability and inhibiting the formation of a rigid framework to the rock. Several geologists have suggested that fluid emigration from clays is aided by the development of microfractures (e.g., Palciauskus and Domenico, 1980). These microfractures would cause a marked increase in permeability and thus allow fluid to escape. The microfractures would then close as pore pressure dropped. It has been suggested that petroleum globules could migrate by shouldering aside the unfixed clay grains. But clay dehydration is only one of several causes of supernormal pressure. Inhibition of normal compaction due to rapid sedimentation, and the formation of pore-filling cements, can also cause high pore pressures (see also p. 154). Furthermore, some major hydrocarbon provinces do not have supernormal pressures. Nevertheless, the late expulsion of water, of whatever origin, must play an important role in the primary migration of petroleum.

The various theories for primary hydrocarbon migration may be grouped as follows:

1. Expulsion as protopetroleum
2. Expulsion as petroleum
   a. In solution
      i. Dissolved in water $\begin{cases} \text{Derived from compaction,} \\ \text{expelled from clays,} \\ \text{or dissolved from meteoric flushing} \end{cases}$
      ii. Within micelles
      iii. Solution of oil in gas
   b. Globules of oil in water
   c. Continuous phase

The merits and limitations of these various mechanisms are described and discussed in the following sections.

Expulsion of Hydrocarbons as Protopetroleum

To actually study hydrocarbons in the act of primary migration is, of course, extremely difficult. The evolutionary sequence from kerogen to crude oil and/or gas is very complex. Assessing whether this transformation is completed before, during, or after migration from source rock to carrier bed is very difficult.

One of the major problems in understanding hydrocarbon migration is their low solubility in water. Hunt (1968) suggested that emigration occurs before the hydrocarbons are recognizable crude oil, that is, while they are in the form of ketones, acids, and esters, which are soluble in water. This transitional phase is termed *protopetroleum*.

This mechanism contains several problems (Cordell, 1972). The observed concentrations of ketones, acids, and esters in source rocks are low, and it is difficult to see how they can actually migrate to the carrier bed and, once there, separate out from the water. These compounds are likely to be adsorbed on the surface of clay minerals and to resist expulsion from the source rock. If, however, they do emigrate to a carrier bed, it is difficult to envisage how they evolve into immiscible crude oil, since they are soluble in water.

Expulsion of Hydrocarbons in Aqueous Solution

One obvious possibility to consider is that the hydrocarbons emigrate from the source bed fully formed, yet dissolved in water. The solubility of hydrocarbons is negligible at the earth's surface, but may be enhanced by temperature or the presence of micelles. These two mechanisms are examined in the following sections.

The Hot Oil Theory

Figure 5.21 shows the solubility of various crude oils plotted against temperature. This graph shows that solubilities are negligible below about 75°C and do not become significant until about 150°C. It is worth remembering that paleotemperature analysis shows that optimum oil generation occurs at about 120°C (p. 190); at this temperature experimental data suggest solubilities in the order of 10 to 20 ppm.

This information does not seem very helpful. Hydrocarbon solubility is worth looking at in more detail. Figure 5.22 shows solubility of hydrocar-

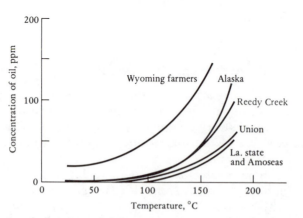

FIGURE 5.21   *Graph showing solubilities of various crude oils plotted against temperature. (After Price, 1976.)*

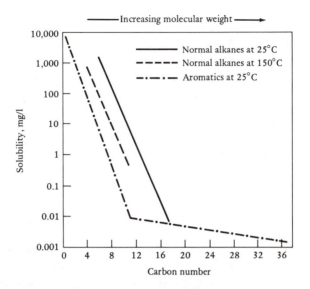

FIGURE 5.22 *Solubility of hydrocarbons with different carbon numbers. Solubility decreases with increasing molecular weight. (After McAuliffe, 1979.)*

bons plotted according to hydrocarbon numbers. This graph shows that hydrocarbon solubility increases with decreasing carbon number for both the normal alkanes (paraffins) and the aromatic series. Note particularly that solubilities are very high, several hundred to several thousand mg/1, for the paraffin gases (C_1–C_5). Perhaps this information has solved part of the problem of understanding primary migration. It seems quite easy for the gaseous hydrocarbons to emigrate from the source bed dissolved in pore water. Exsolution may then occur when they reach the lower temperatures and pressures of the carrier bed.

Only about 25 percent of a typical crude oil is composed of the moderately soluble alkanes, and another 25 percent is composed of the heavy naphtheno-aromatics and resins, which are virtually insoluble in water. Nonetheless, there are advocates for primary oil emigration by solution in water (e.g., Price, 1976, 1977, 1978, 1980, 1981a, 1981b). Experimental data show that solubilities of as much as 10 weight percent occur for hydrocarbons in pressured salty water. Solubility is enhanced by the presence of gas. Even so, Price advocates temperatures of between 300 and 350°C for oil generation. As already seen, these temperatures are well above those that paleothermometers indicate as optimal for oil generation. Furthermore, at these temperatures diagenesis will probably have destroyed porosity and permeability of source and carrier beds alike. The stability of hydrocarbons at these temperatures has also been queried (Jones, 1978).

Micelle Theory

Another way in which the solubility of hydrocarbons in water may be enhanced is by the presence of micelles (Baker, 1962; Cordell, 1973). Micelles are colloidal organic acid soaps whose molecules have hydrophobic (water insoluble) and hydrophylic (water soluble) ends. Their presence may thus enhance the solubility of hydrocarbons in water by acting as a link between OH radicals on their hydrophylic ends and hydrocarbon molecules on their hydrophobic ends. Baker (1962) showed that the particle size of micelles in crude oils have a bimodal lognormal distribution. This distribution is related to two basic micelle types—the small *ionic* and large *neutral* micelles. The principles by which soaps may be used to enhance the solubility of hydrocarbons are familiar to petroleum production engineers. The process of micellar flooding of a reservoir is frequently used to enhance recovery (p. 269).

If this micelle theory is correct, then the proportions of different hydrocarbons in a crude oil should be related to their micellar solubility. Experimental data show this to be the case. The naphthenic and aromatic fractions are present in monomodal lognormal frequency distributions. By contrast the paraffins, both normal and branched, show a frequency distribution suggesting that they are solubilized by both ionic and neutral micelles. Hydrocarbons of low atomic weight (C_5 or less) show a normal frequency distribution. As already shown, this distribution is probably related to the fact that they are naturally soluble in water and do not require the assistance of micelles.

Numerous writers have reviewed the micelle theory. Major objections listed by McAuliffe (1979) include the requirement of a very high ratio of micelles to hydrocarbons to provide an effective process. Micelles are commonly only present in traces. In addition, the size of micelle molecules is greater than the diameter of pore throats in clay source beds. The processes whereby the micelles lose their hydrocarbon molecules and presumably break down are not clear.

Expulsion of Oil in Gaseous Solution

Other theories of hydrocarbon migration have considered the role of gases in acting as catalysts or transporting media. Momper (1978), in particular, has discussed the role of carbon dioxide, which is known to be driven off during kerogen maturation (p. 247). Directly and indirectly, carbon dioxide may have considerable influence on hydrocarbon migration. Indirectly, by combining with calcium ions it precipitates calcite cement, which diminishes pore volume and thus increases pore pressure. The presence of carbon dioxide gas in solution lowers the viscosity of oil, thus increasing its mobility. It causes the precipitation of the N-S-O heavy ends of oil, thus making the residual oil lighter and increasing the gas-oil ratio.

On the debit side, however, the precipitation of calcite cement may diminish permeability. Furthermore, the main phase of decarboxylation of kerogen is known to occur before hydrocarbon generation (p. 247). Concentrations of carbon dioxide at the time of petroleum generation may thus be too low to assist migration in the ways previously outlined.

Gases may play a direct role in oil migration in another way. Hydrocarbon gases in deep wells are well known to carry oil in gaseous solution. The oil condenses when pressure and temperature drop as the hydrocarbon gases are brought to the surface. Therefore, oil may possibly migrate in this manner from the source bed. This mechanism quite satisfactorily explains the secondary migration of *condensates* through carrier and reservoir beds. As McAuliffe (1979) points out, however, it does not quite satisfactorily explain the secondary migration of normal crudes or the primary migration of any oil. The solubility of the heavier naphtheno-aromatics and resins appears to be negligible. The expulsion of a gas bubble through a shale pore throat is as difficult as it is for an oil droplet. As shall be shown shortly, this presents considerable problems for theories of primary oil phase migration.

Primary Migration of Free Oil

A whole spectrum of theories postulate that oil emigrates from the source bed not in any kind of solution but as a discrete oil phase. There are two major types of such migration: the expulsion of discrete droplets associated with pore water and the expulsion of a three-dimensional continuous phase of oil.

Hobson and Tiratsoo (1981) have shown mathematically how in water-wet pores it is impossible for a globule of oil to be squeezed through a pore throat in a clay source rock, either by virtue of its own properties (e.g., buoyancy or immiscibility) or by the assistance of flowing water. This mathematical hypothesis is based on the assumption that the diameter of the oil globule is greater than that of the pore throat. Hobson and Tiratsoo (1981) cite pore diameters in the order of 50 to 100 Å for shales at 2000 m. Individual molecules of asphaltenes and complex ring-structured hydrocarbons have diameters of 50 to 100 Å and 10 to 30 Å, respectively (Tissot and Welte, 1978). Therefore it is hard to see how discrete oil globules could have diameters less than those of pore throats.

It is possible, however, that organic-rich source rocks are not water-wet, but oil-wet. In this situation petroleum would not migrate as discrete globules of oil in water, but as a continuous three-dimensional phase. This mechanism is the so-called *greasy wick* theory, pointing to the analogy of molten wax moving through the fiber of a candle wick. Such a mechanism may work for rich source rock, but is less likely in leaner source rocks, which are probably water-wet.

Estimation of Volumes of Petroleum Generated

It is useful to be able to assess the amount of petroleum that has been generated in a sedimentary basin. Such an assessment is obviously very difficult in a virgin area with no data. In a mature petroleum province where a large quantity of data are available, it is considerably easier. Knowledge of the quantity of reserves yet to be discovered is important for deciding whether continuing exploration is worth the expense if only small reserves remain to be found.

The volume of oil generated in an area may be calculated using the geochemical material balance method (White and Gehman, 1979). The basic equation may be expressed as follows:

$$\begin{array}{c} \text{Volume of oil} \\ \text{generated} \end{array} = \begin{array}{c} \text{volume of} \\ \text{source rock} \end{array} \times \begin{array}{c} \text{volume of} \\ \text{organic matter} \end{array}$$

$$\times \begin{array}{c} \text{genetic} \\ \text{potential} \end{array} \times \begin{array}{c} \text{transformation} \\ \text{ratio} \end{array}$$

$$\times \begin{array}{c} \text{fraction of oil} \\ \text{in hydrocarbon yield} \end{array} \times \begin{array}{c} \text{volume increase} \\ \text{on oil yield} \end{array}$$

The volume of source rock can be calculated from isopach maps. The average amount of organic matter must be estimated from the geochemical analysis of cores and cuttings, extrapolating from wireline logs where possible. The *genetic potential* of a formation is the amount of petroleum that the kerogen can generate (Tissot and Welte, 1978). The *transformation ratio* is the ratio of petroleum actually formed to the genetic potential, and, as described on page 195, these values are determined from the pyrolysis of source rock samples.

Estimates of the volumes of petroleum generated in various basins have been published by Hunt and Jamieson (1956), Hunt (1961), Conybeare (1965), Pusey (1973a), Fuller (1975), and Goff (1983). These studies show that the transformation ratio needs to exceed 0.1 for significant oil generation and is usually in the range of 0.3 to 0.7 in major petroleum provinces.

Hydrocarbon Generation and Migration: Summary

Understanding the primary migration of hydrocarbons is one of the last problems of petroleum geology. Research in this field is currently very active, so any review is bound to date quickly; nonetheless, a summary of this complex topic is appropriate:

1. Commercial quantities of oil and gas form from the metamorphism of organic matter.

2. Kerogen, a solid hydrocarbon disseminated in many shales, is formed from buried organic detritus and is capable of generating oil and gas.

3. Three types of kerogen are identifiable: Type I (algal), type II (liptinic), and type III (humic). Type I tends to generate oil; type III, gas.

4. The maturation of kerogen is a function of temperature and, to a lesser extent, time. Oil generation occurs between 60 and 120°C, and gas generation between 120 and 225°C.

5. Source rocks generally contain over 1500 ppm organic carbon, but yield only a small percentage of their contained hydrocarbon.

6. Several techniques may be used to measure the maturity of a source rock.

7. The exact process of primary migration, whereby oil and gas migrate from source beds, is unclear. Solubility of hydrocarbons in water is low, but, for the lighter hydrocarbons, is enhanced by high temperature and the presence of soapy micelles. An empirical relationship between oil occurrence and clay dehydration suggests that the flushing of water from compacting clays plays an important role in primary migration.

SELECTED BIBLIOGRAPHY

For concise reviews of the geochemistry of source rocks, the problems of migration, and paleothermometers, see:

COOPER, B. S. 1977.   Estimation of the maximum temperatures attained in sedimentary rocks. In: *Developments in Petroleum Geology*, vol. 1. G. D. Hobson (ed.). London: Applied Science Publishers, 127–146.

MAGARA, K. 1977.   Petroleum migration and accumulation. In: *Developments in Petroleum Geology*, vol. 1. G. D. Hobson (ed.). London: Applied Science Publishers, 83–126.

TISSOT, B. 1977.   The application of the results of organic geochemical studies in oil and gas exploration. In: *Developments in Petroleum Geology*, vol. 1. G. D. Hobson (ed.). London: Applied Science Publishers, 53–82.

For papers expounding the various theories of hydrocarbon migration, see:

BARKER, C. et al. 1978.   *Physical and Chemical Constraints on Petroleum Migration.* Tulsa: Am. Assoc. Petrol. Geol., 283 pp.

For detailed accounts of the generation and migration of oil, see:

HUNT, J. M. 1979.   *Petroleum Geochemistry and Geology.* San Francisco: Freeman, 617 pp.

KINGHORN, R. R. F. 1983.   *Introduction to the Physics and Chemistry of Petroleum.* Chichester: Wiley, 436 pp.

ROBERTS, W. H. and CORDELL, R. J. 1980. *Problems of Petroleum Migration.* AAPG Studies in Geology, No. 10, 273 pp.

TISSOT, B. P. and WELTE, D. H. 1978. *Petroleum Formation and Occurrence.* Berlin: Springer-Verlag, 530 pp.

WAPLES, D. W. 1981. *Organic Geochemistry for Exploration Geologists.* Minneapolis: CEPCO, 144 pp.

REFERENCES

AGER, D. V. 1975. The Jurassic world ocean. In: *Jurassic Northern North Sea Symposium.* K. Finstad and R. C. Selley (eds.). Oslo: Norwegian Petrol. Soc. Paper No. 1, 43 pp.

ARTHUR, M. A. and SCHLANGER, S. O. 1979. Cretaceous 'Oceanic Anoxic Events' as Causal factors in Development of Reef-Reservoired Giant Oil Fields. *Am. Assoc. Petrol. Geol. Bull., 63,* 870–885.

ATHY, L. F. 1930. Density, porosity and compaction of sedimentary rocks. *Am. Assoc. Petrol. Geol. Bull., 14,* 1–35.

BAKER, E. G. 1962. Distribution of hydrocarbons in petroleum. *Am. Assoc. Petrol. Geol. Bull., 46,* 76–84.

BARNARD, P. C. and COOPER, B. S. 1981. Oils and source rocks of the North Sea area. In: *Petroleum Geology of the Continental Shelf of North-West Europe.* L. V. Illing and G. D. Hobson (eds.). London: Inst. Petrol., 169–175.

BARNARD, P. C.; COOPER, B. S.; and FISHER, M. J. 1978. Organic maturation and hydrocarbon generation in the Mesozoic sediments of the Sverdrup Basin, Arctic Canada. Lucknow: *Proc. 4th Internat. Palynol. Cong.,* 163–175.

BECKER, G. F. 1909. Relations between local magnetic disturbances and the genesis of petroleum. *Bull. U.S. Geol. Surv., 401,* 23 pp.

BERTHELOT, M. 1860. Sur l'Origine de Carbures et des Combustibles Mineraux. *Compt. Rend., 63,* 949–951.

BRADLEY, W. H. 1948. Limnology and the Eocene lakes at the Rocky Mtn. region. *Bull. Geol. Soc. Am., 59,* 635–648.

BURST, J. F. 1969. Diagenesis of Gulf Coast clay sediments and its possible relation to petroleum migration. *Am. Assoc. Petrol. Geol. Bull., 53,* 73–93.

CLAYPOOL, G. E.; LOVE, A. H.; and MAUGHAN, E. K. 1978. Organic geochemistry, incipient metamorphism, and oil generation in black shale members of Phosphoria formation, western interior, United States. *Am. Assoc. Petrol. Geol. Bull., 62,* 98–120.

COHEN, C. R. 1981. Time and temperature in petroleum formation: application of Lopatin's method to petroleum exploration: Discussion. *Am. Assoc. Petrol. Geol. Bull., 65,* 1647–1648.

COLLINGWOOD, D. M. and RETTGER, R. E. 1926. The Lytton Springs oil field, Caldwell County, Texas. *Am. Assoc. Petrol. Geol. Bull., 10,* 1199–1211.

CONNAN, J. 1974. Time-temperature relation in oil genesis. *Am. Assoc. Petrol. Geol. Bull., 58,* 2516–2521.

CONYBEARE, C. E. B. 1965. Hydrocarbon-generation potential and hydrocarbon-yield capacity of sedimentary basins. *Can. Petrol. Geol. Bull., 13,* 509–528.

COOPER, B. S. 1977. Estimation of the maximum temperatures attained in sedimentary rocks. In: *Developments in Petroleum Geology I*. G. D. Hobson (ed.). London: Applied Science Publishers, 127–146.

COOPER, B. S.; COLEMAN, S. H.; BARNARD, P. C.; and BUTTERWORTH, J. S. 1975. Palaeo-temperatures in the northern North Sea basin. In: *Petroleum and the Continental Shelf of Northwest Europe*. A. W. Woodland (ed.). London: Applied Science Publishers, 487–492.

CORDELL, R. J. 1972. Depths of oil origin and primary migration: a review and critique. *Am. Assoc. Petrol. Geol. Bull.*, 56, 2029–2067.

CORDELL, R. J. 1973. Colloidal soap as a proposed primary migration medium for hydrocarbons. *Am. Assoc. Petrol. Geol. Bull.*, 57, 1618–1643.

CORNELIUS, C. D. G. 1975. Geothermal aspects of hydrocarbon exploration in the North Sea area. In: Proc. Bergen Conf. Hydrocarbon Geol. *Norsk. Geol. Undersok.*, 26–67.

CURRIE, J. B. and NWACHUKWU, S. O. 1974. Evidence of incipient fracture porosity in reservoir rocks at depth. *Can. Petrol. Geol. Bull.*, 22, 42–57.

DEBYSER, J. and DEROO, G. 1969. Faits d'observation sur la genese du petrole. I. Facteurs controlant la repartition de la matiere organiques dans les sediments. *Rev. Inst. Fr. Petr.*, 24, 21–48.

DEGENS, E. T. and MOPPER, K. 1976. Factors controlling the distribution and early diagenesis of organic material in marine sediments. *Chem. Oceanog.*, 6, 59–113.

DEGENS, E. T. and ROSS, D. A. (eds.). 1974. *The Black Sea—Geology, Chemistry and Biology*. Am. Assoc. Petrol. Geol., Mem. No 20, 633 pp.

DEGENS, E. T.; Von HERTZEN, R. P.; and WONG, H. K. 1971. Lake Tanganika: water chemistry, sediments, geological structure. *Naturwissenschaften*, 58, 224–291.

DEGENS, E. T.; Von HERTZEN, R. P.; and WONG, H. K. 1973. Lake Kivu: structure, chemistry and biology of an East African rift lake. *Geol. Rundsch.*, 62, 245–277.

DEMAISON, G. J. and MOORE, G. T. 1979. Anoxic environments and oil source bed genesis. *Org. Geochem.*, 2, 9–13.

DEMAISON, G. J. and MOORE, G. T. 1980. Anoxic environments and oil source bed genesis. *Am. Assoc. Petrol. Geol. Bull.*, 64, 1179–1209.

DEUSER, W. G. 1971. Organic-carbon budget of the Black Sea. *Deep Sea Res.*, 18, 995–1004.

DICKINSON, G. 1951. Geological Aspects of Abnormal Reservoir Pressures in the Gulf Coast Region of Louisiana, U.S.A. *3rd Wld. Petrol. Cong. Proc.* Section 1, 1–16, and *Am. Assoc. Petrol. Geol. Bull.*, 10, 40–432.

DIETRICH, G. 1963. *General Oceanography, an Introduction*. New York: Wiley, 588 pp.

DOTT, R. H. and REYNOLDS, M. J. 1969. *Sourcebook for Petroleum Geology*. Tulsa: Am. Assoc. Petrol. Geol. Bull., Mem. No. 5, 471 pp.

DOW, W. G. 1977. Kerogen studies and geological interpretation. *J. Geochem. Explor.*, 7, 79–100.

DOW, W. G. and O'CONNOR, D. I. 1979. Vitrinite reflectance—what, how and why. *Am. Assoc. Petrol. Geol. Bull.*, 63, 441.

DUNCAN, D. C. and SWANSON, V. E. 1965. Organic-rich shale of the United States and World Land Areas. *U.S. Geol. Surv. Circ.*, 523, 30 pp.

DURAND, B. 1980. *Kerogen—Insoluble Organic Matter from Sedimentary Rocks.* London: Graham and Trotman, 550 pp.

DURAND, B.; MARCHAND, A.; AMIELL, J.; and COMBAZ, A. 1970. Etude de Kerogenes par RPE. In: *Advances in Organic Geochemistry.* R. Campos and J. Goni (eds.). London: Pergamon, 154–195.

EMERY, K. O. 1963. Oceanic factors in accumulation of petroleum. Frankfurt: *Proc. 6th World Petrol. Cong.*, Section I, Paper 42, PD2, 483–491.

ERDMAN, J. G. 1975. Geochemical formation of oil. In: *Petroleum and Global Tectonics.* A. G. Fischer and S. Judson (eds.). Princeton Univ. Press, 225–250.

ESPITALIÉ, J.; LAPORTE, J. L.; MADEC, M.; MARQUIS, F.; LEPLAT, P.; PAULET, J.; and BOUTEFOU, A. 1977. Méthode rapide de caracterisation des roches meres, de leur potentiel pétrolier et de leur degré d'évolution. *Rev. Inst. Fr. Petrol.*, 3, 23–42.

EVANS, W. D.; MORTON, R. D.; and COOPER, B. S. 1964. Primary investigations of the oleiferous dolerite of Dyvika. In: *Advances in Geochemistry.* U. Colombo and G. D. Hobson (eds.). Oxford: Pergamon, 264–277.

FLORY, D. A.; LICHTENSTEIN, H. A.; BIEMANN, K.; BILLER, J. E.; and BARKER, C. 1983. Computer process uses entire GC–MS data. *Oil & Gas J.*, 17 Jan., 91–98.

FOSCOLOS, A. E. and POWELL, T. G. 1978. Mineralogical and geochemical transformation of clays during burial-diagenesis: relation to oil generation. In: *International Clay Conference.* M. M. Mortland and V. C. Farner (eds.). Amsterdam: Elsevier, 261–270.

FRIEDMAN, G. M. 1972. Significance of Red Sea in problem of evaporites and basinal limestones. *Am. Assoc. Petrol. Geol. Bull.*, 56, 1072–1086.

FUCHTBAUER, H. 1967. Influence of different types of diagenesis on sandstone porosity. Mexico: *Proc. 7th World Petrol. Cong.*, 353–369.

FULLER, J. G. C. 1975. Jurassic source-rock potential and hydrocarbon correlation, North Sea. In: *Jurassic Northern North Sea Symposium.* K. Finstad and R. C. Selley (eds.). Oslo: Nor. Petrol. Soc., 18 pp.

GILL, W. D.; KHALAF, F. I.; and MASSOUD, M. S. 1979. Organic matter as indicator of the degree of metamorphism of the Carboniferous rocks in the South Wales coalfields. *J. Petrol. Geol.*, 1, 39–62.

GOFF, J. C. 1983. Hydrocarbon generation and migration from Jurassic source rocks in the E. Shetland Basin and Viking Graben of the northern North Sea. *J. Geol. Soc. London*, 140, 445–474.

GOLD, T. 1979. Terrestrial sources of carbon and earthquake outgassing. *J. Petrol. Geol.*, 1, 3–19.

GOLD, T. and SOTER, S. 1982. Abiogenic methane and the origin of petroleum. *Ener. Explor. and Exploit.*, 1, 89–104.

GOLDHABER, M. 1978. Euxinic facies. In: *Encyclopedia of Sedimentology.* R. W. Fairbridge and J. Bourgeois (eds.). Stroudsberg, Pa: Dowdon, Hutchinson & Ross, 296–300.

GORDON, M.; TRACEY, R. I.; and ELLIS, M. W. 1958. Geology of the Arkansas Bauxite Region. *U.S. Geol. Surv.*, Prof. Paper No. 299, 268 pp.

GRANSCH, J. A. and EISMA, E. 1970. Characterization of the insoluble organic matter of sediments by pyrolysis. In: *Advances in Organic Geochemistry.* G. D. Hobson and G. Speers (eds.). Oxford: Pergamon, 407–426.

GRUNAU, H. R. 1983. Abundance of source rocks for oil and gas worldwide. *J. Petrol. Geol.*, *6*, 39–54.

GRUNAU, H. R. and GRUNER, U. 1978. Source rock and origin of natural gas in the Far East. *J. Petrol. Geol.*, *1*, 3–56.

GUANGMING, Z. and QUANHENG, Z. 1982. Buried-hill oil and gas pools in the North China basin. In: *The Deliberate Search for the Subtle Trap*. M. T. Halbouty (ed.). Am. Assoc. Petrol. Geol., Mem. No. 32, 317–336.

HALLAM, A. 1981. *Facies Interpretation and the Stratigraphic Record*. San Francisco: Freeman, 291 pp.

HEDBERG, H. 1936. Gravitational compaction of clays and shales. *Am. J. Sci.*, *31*, 241–287.

HOBSON, G. D. and TIRATSOO, E. N. 1981. *Introduction to Petroleum Geology*. Beaconsfield: Scientific Press, 352 pp.

HOEFS, J. 1980. *Stable Isotope Geochemistry*. New York: Springer-Verlag, 352 pp.

HOOD, A.; GUTJAHR, C. C. M.; and HEACOCK, R. L. 1975. Organic metamorphism and the generation of petroleum. *Am. Assoc. Petrol. Geol. Bull.*, *59*, 986–996.

HOYLE, F. 1955. *Frontiers of Astronomy*. London: Heineman, 360 pp.

HUNT, J. M. 1961. Distribution of hydrocarbons in sedimentary rocks. *Geochem. Cosmochem. Acta*, *22*, 37–49.

HUNT, J. M. 1968. How gas and oil form and migrate. *World Oil*, *167*, No. 4, 140–150.

HUNT, J. M. 1977. Distribution of carbon as hydrocarbons and asphaltic compounds in sedimentary rocks. *Am. Assoc. Petrol. Geol. Bull.*, *61*, 100–116.

HUNT, J. M. 1979. *Petroleum Geochemistry and Geology*. San Francisco: Freeman, 617 pp.

HUNT, J. M. and JAMIESON, G. W. 1956. Oil and organic matter in source rocks of petroleum. *Am. Assoc. Petrol. Geol. Bull.*, *40*, 477–488.

IKORSKIY, S. V. 1967. Organic substances in minerals of Igneous rocks of the Khibina massif. (Translated from Russian by L. Shapino and I. A. Breger, 1976.) McLean, Va: Clark Co., 155 pp.

JOHNSON IBACH, L. E. 1982. Relationship between sedimentation rate and total organic carbon content in ancient marine sediments. *Am. Assoc. Petrol. Geol. Bull.*, *66*, 170–188.

JONES, R. W. 1978. Some mass balance and geologic constraints on migration mechanisms. In: *Physical and Chemical Constraints on Petroleum Migration*. W. H. Roberts and R. J. Cordell (eds.). Am. Assoc. Petrol. Geol. Course Notes, No. 8, A1–A43.

KIDWELL, A. L. and HUNT, J. M. 1958. Migration of oil in Recent sediments of Pedernales, Venezuela. In: *The Habitat of Oil*. L. G. Weeks (ed.). Am. Assoc. Petrol. Geol., 790–817.

LEVORSEN, A. I. 1967. *Geology of Petroleum*. San Francisco: Freeman, 724 pp.

LINK, T. A. 1957. Whence came the hydrocarbons? *Am. Assoc. Petrol. Geol. Bull.*, *41*, 1387–1402.

LOPATIN, N. V. 1971. Temperature and geologic time as factors in coalification. *Akad. Nauk SSR Izv. Serv. Geol.*, *3*, 95–106.

MA LI, GE TAISHENG, ZHAO XUEPING, ZIE TAIJUN, GE RONG, and DANG ZHENRONG. 1982. Oil basins and subtle traps in the eastern part of China. In: *The Deliberate Search for the Subtle Trap*. M. T. Halbouty (ed.). Tulsa: *Am. Assoc. Petrol. Geol.*, 287–316.

MacDONALD, G. J. 1983. The many origins of natural gas. *J. Petrol. Geol.*, *5*, 341–362.

McAULIFFE, C. D. 1979. Oil and gas migration—chemical and physical constraints. *Am. Assoc. Petrol. Geol. Bull.*, *63*, 761–781.

McTAVISH, R. 1978. Pressure retardation of vitrinite diagenesis, offshore northwest Europe. London: *Nature*, *271*, 648–650.

MAGARA, K. 1968. Compaction and migration of fluids in Miocene mudstone, Nagaoka plain, Japan. *Am. Assoc. Petrol. Geol. Bull.*, *52*, 2466–2501.

MAGARA, K. 1977. Petroleum migration and accumulation. In: *Developments in Petroleum Geology.* G. D. Hobson (ed.)., 83–126.

MAGARA, K. 1979. Structured water and its significance in primary oil migration. *Can. Petrol. Geol. Bull.*, *27*, 87–93.

MAGARA, K. 1981. Possible primary migration of oil globules. *J. Petrol Geol.*, *3*, 325–331.

MARCHAND, A.; LIBERT, P. A.; and COMBAZ, Z. 1969. Essai de caracterisation physico-chimique de la diagenesis de quelques roches organique, biologiquement homogenes. *Rev. Inst. Petrol.*, *24*, 3–20.

MEADE, R. H. 1966. Factors influencing the early stages of the compaction of clays and sands—review. *J. Sedim. Petrol.*, *36*, 1085–1101.

MENDELEEF, D. 1877. Entstehung und Vorkommen des Mineralols. *Deut. Chem. Ges. Ber.*, *10*, 229.

MENDELE'EV, D. 1902. *The Principles of Chemistry*, vol. 1. Second English edition translated from the sixth Russian edition. New York: Collier, 552 pp.

MOMPER, J. A. 1978. Oil migration limitations suggested by geological and geochemical considerations. In: *Physical and Chemical Constraints on Petroleum Migration.* W. H. Roberts and R. J. Cordell (eds.). Am. Assoc. Petrol. Geol. Course Notes, No. 8, B1–B60.

MUELLER, G. 1963. Properties of extraterrestrial hydrocarbons and theory of their genesis. Frankfurt: Proc. *6th World Petrol. Cong.*, Section 1, Paper 29, 14 pp.

NEEV, D. and EMERY, K. O. 1967. The Dead Sea—depositional processes and environments of evaporites. *Israel Geol. Surv. Bull.*, *41*, 147 pp.

NORTH, F. K. 1979. Episodes of source-sediment deposition, Part 1. *J. Petrol. Geol.*, *2*, 199–218.

NORTH, F. K. 1980. Episodes of source-sediment deposition, Part 2. The episodes in individual close up. *J. Petrol. Geol.*, *2*, 323–338.

NORTH, F. K. 1982. Review of Thomas Gold's Deep-Earth-Gas hypothesis. *Ener. Explor. & Exploit.*, *1*, 105–110.

OUDIN, J. L. 1976. Etude geochimique du bassin de la mer du Nord. *Bull. Du Centre de Recherches de Pau.*, *10*, 339–358.

PALCIAUSKUS, V. V. and DOMENICO, P. A. 1980. Microfracture development of compacting sediments: relation to hydrocarbon-maturation kinetics. *Am. Assoc. Petrol. Geol. Bull.*, *64*, 927–937.

PARRISH, J. T. 1982. Upwelling and petroleum source beds, with reference to Paleozoic. *Am. Assoc. Petrol. Geol. Bull.*, *66*, 750–774.

PETERSIL'YE, Ya. A. 1962. Origin of hydrocarbon gases and dispersed bitumens of the Khibina alkalic massif. (Translated from Russian.) *Geochem.*, *1*, 14–30.

PEYVE, A. V. 1956. General characteristics and spatial characteristics of deep faults. *Akad. Nauk. SSSR. Izv.*, *1*, 90–106.

PHILIPPI, H. W. 1975.   On the depth, time and mechanism of petroleum migration. *Geochem. Cosmochem. Acta, 29,* 1021–1049.

PORFIR'EV, V. B. 1974.   Inorganic origin of petroleum. *Am. Assoc. Petrol. Geol. Bull., 58,* 3–33.

POWERS, M. C. 1967.   Fluid-release mechanisms in compacting marine mudrocks and their importance in oil exploration. *Am. Assoc. Petrol. Geol. Bull., 51,* 1240–1254.

PRATT, W. E. 1942.   *Oil in the Earth.* Lawrence: University of Kansas Press, 105 pp.

PRICE, L. C. 1976.   Aqueous solubility of petroleum as applied to its origin and primary migration. *Am. Assoc. Petrol. Geol. Bull., 60,* 213–244.

PRICE, L. C. 1978.   New evidence for hot, deep origin and migration of petroleum. In: *Physical and Chemical Constraints on Petroleum Migration.* W. H. Roberts and R. J. Cordell (eds.). Am. Assoc. Petrol. Geol. Course Notes, No. 9, F9–F10.

PRICE, L. C. 1980.   Shelf and shallow basin oil as related to hot-deep origin of petroleum. *J. Petrol. Geol., 3,* 91–115.

PRICE, L. C. 1981a.   Primary petroleum migration by molecular solution: consideration of new data. *J. Petrol. Geol., 4,* 89–101.

PRICE, L. C. 1981b.   Aqueous solubility of crude oil to 400°C and 2,000 bars pressure in the presence of gas. *J. Petrol. Geol., 4,* 195–223.

PRICE, L. C. 1983.   Geologic time as a parameter in organic metamorphism and vitrinite reflectance as an absolute paleogeothermometer. *J. Petrol. Geol., 6,* 5–38.

PROSHLYAKOV, B. K. 1960.   Reservoir properties of rocks as a function of their depth and lithology. *Geol. Neft. i Gaza, 4,* 24–29. Assoc. Tech. Serv. Translation R. J. 3421.

PUSEY, W. C. 1973a.   How to evaluate potential gas and oil source rocks. *World Oil, 176* (5), 71–75.

PUSEY, W. C. 1973b.   Paleotemperatures in the Gulf Coast using the ESR-kerogen method. *Gulf Coast Assoc. Geol. Soc. Trans., 23,* 195–202.

ROBERTS, W. H. and CORDELL, R. J. 1980.   *Problems of Petroleum Migration.* Studies in Geology No. 10. Am. Assoc. Petrol. Geol., 273 pp.

ROBINSON, W. E. 1976.   Origin and characteristics of Green River oil shale. In: *Oil Shale.* G. V. Chilingarian and T. F. Yen (eds.). Amsterdam: Elsevier, 61–80.

ROEDDER, E. 1967.   Fluid inclusions in samples of ore deposits. In: *Geochemistry of Hydrothermal Deposits.* H. L. Barnes (ed.). New York: Holt, Rinehart & Winston, 151–172.

SCHIDLOWSKI, M.; EICHMANN, R.; and JUNGE, C. E. 1974.   Evolution des iridischen Sauerstof-Budgets und Entwicklung de Erdatmosphare. *Umschau., 22,* 703–707.

SCHLANGER, S. O. and CITA, M. B. 1982.   *Nature and Origin of Cretaceous Carbon-Rich Facies.* London: Academic Press, 250 pp.

SELLEY, R. C. 1980.   Economic basement in the northern North Sea. In: *Symposium on the Reservoir Rocks of the Northern North Sea.* Oslo: Norweg. Petrol. Soc., xx, 22 pp.

STAPLIN, F. L. 1969.   Sedimentary organic matter, organic metamorphism, and oil and gas occurrence. *Can. Petrol. Geol. Bull., 17,* 47–66.

STUDIER, M. H.; HAYATSU, R.; and ANDERS, E. 1965.   Organic compounds in Carbonaceous chondrites. *Science, 149,* 1455–1459.

SUBBOTIN, S. I. 1966.   Upper mantle and inorganic oil. In: *Problems in the Origin of Oil.* Kiev. Izd. Nauk. Dumka., 52–62.

SUGISAKI, R.; IDO, M.; TAKEDA, H.; and ISOBE, Y. 1983.   Origin of hydrogen and carbon dioxide in fault gases and its relation to fault activity. *J. Geol., 91,* 239–258.

TAPPAN, H. and LOEBLICH, A. R. 1970.   Geobiologic implications of fossil phytoplankton evolution and time/space distribution. In: *Symp. Palynology of Late Cretaceous and Early Tertiary.* R. M. Kosanke and S. T. Cross (eds.). Michigan State Univ. Geol. Soc., 247–340.

TISSOT, B. P. 1977.   The application of the results of organic chemical studies in oil and gas exploration. In: *Developments in Petroleum Geology,* vol. 1. G. D. Hobson (ed.). London: Applied Science Publishers, 53–82.

TISSOT, B. P. 1979.   Effects on prolific petroleum source rocks and major coal deposits caused by sea level changes. London: *Nature, 277,* 377–380.

TISSOT, B. P. and WELTE, D. H. 1978.   *Petroleum Formation and Occurrence: A New Approach to Oil and Gas Exploration.* Berlin: Springer-Verlag, 538 pp.

VAIL, P. R.; MITCHUM, R. M.; and THOMPSON, S. 1977.   Seismic stratigraphy and global changes of sea level, Part 3. In: *Seismic Stratigraphy—Applications to Hydrocarbon Accumulations.* C. Payton (ed.). Am. Assoc. Petrol. Geol., Mem. *26,* 63–82.

VAN HINTE, J. E. 1978.   Geohistory analysis—application of micropaleontology in exploration geology. *Am. Assoc. Petrol. Geol. Bull., 62,* 201–202.

WAPLES, D. W. 1980.   Time and temperature in petroleum exploration. *Am. Assoc. Petrol. Geol., 64,* 916–926.

WAPLES, D. W. 1981.   Organic Geochemistry for Exploration Geologists. Minneapolis: CEPCO, 144 pp.

WAPLES, D. W. 1983.   Reappraisal of anoxia and organic richness with emphasis on Cretaceous of North Atlantic. *Am. Assoc. Petrol. Geol. Bull., 67,* 963–978.

WELTE, D. H. 1972.   Petroleum exploration and organic geochemistry. *J. Geochem. Explor., 1,* 117–136.

WELTE, D. H.; HAGEMANN, H. W.; HOLLERBACH, A.; LEYTHAEUSER, D.; and STAHT, W. 1975.   Correlation between petroleum and source rock. Tokyo: *Proc. 9th World Petrol. Cong.,* Geology, vol. 2. London: Applied Science Publishers, 179–191.

WHITE, D. A. and GEHMAN, H. M. 1979.   Methods of estimating oil and gas resources. *Am. Assoc. Petrol. Geol. Bull., 63,* 2183–2203.

WHITE, D. E. and WARING, G. A. 1963.   Volcanic emanations. In: *Data of Geochemistry,* Second Edition. U.S.G.S. Prof. Paper 440-K, 29 pp.

6

The Reservoir

Chapter one showed that one of the five essential prerequisites for a commercial accumulation of hydrocarbons is the existence of a reservoir. Theoretically, any rock may act as a reservoir for oil or gas. In practice the sandstones and carbonates contain the major known reserves, although fields do occur in shales and diverse igneous and metamorphic rocks. For a rock to act as a reservoir it must possess two essential properties: It must have pores to contain the oil or gas, and the pores must be connected to allow the movement of fluids; in other words, the rock must have permeability.

This chapter is concerned with the nature of hydrocarbon reservoirs, that is, with the internal properties of a trap. It begins by describing porosity and permeability and discussing their relationship with sediment texture. This discussion is followed by an account of how postdepositional (diagenetic) changes may diminish or enhance reservoir performance. Following this discussion is a section on the vertical and lateral continuity of reservoirs and the calculation of oil and gas reserves. The chapter concludes with an account of the various ways of producing hydrocarbons from a reservoir.

POROSITY

Definition and Classification

Porosity is the first of the two essential attributes of a reservoir. The pore spaces, or voids, within a rock are generally filled with connate water, but contain oil or gas within a field. Porosity is either expressed as the void ratio, which is the ratio of voids to solid rock, or, more frequently, as a percentage:

$$\text{Porosity (\%)} = \frac{\text{volume of voids}}{\text{total volume of rock}} \times 100$$

Porosity is conventionally symbolized by the Greek letter phi (ϕ).

Pores are of three morphological types: catenary, cul-de-sac, and closed (Fig. 6.1). Catenary pores are those that communicate with others by more than one throat passage. Cul-de-sac, or dead-end, pores have only one throat passage connecting with another pore. Closed pores have no communication with other pores.

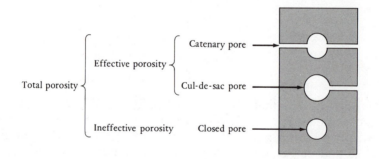

FIGURE 6.1   *The three basic types of pore.*

Catenary and cul-de-sac pores constitute effective porosity, in that hydrocarbons can emerge from them. In catenary pores hydrocarbons can be flushed out by a natural or artificial water drive. Cul-de-sac pores are unaffected by flushing, but may yield some oil or gas by expansion as reservoir pressure drops. Closed pores are unable to yield hydrocarbons (such oil or gas having invaded an open pore subsequently closed by compaction or cementation). The ratio of total to effective porosity is extremely important, being directly related to the permeability of a rock.

The size and geometry of the pores and the diameter and tortuosity of the connecting throat passages all affect the productivity of the reservoir. Pore geometry and continuity are difficult to analyze. In some studies

TABLE 6.1 A classification of the different types of porosity found in sediments

Time of formation	Type	Origin
Primary or depositional	Intergranular, or interparticle Intragranular, or intraparticle	Sedimentation
Secondary or postdepositional	Intercrystalline Fenestral	Cementation
	Vuggy Moldic	Solution
	Fracture	Tectonics, compaction, dehydration, diagenesis

pores are filled by a fluid, which then solidifies, and the rock is then dissolved by acid to reveal casts of the pores. Collins (1961) used molten lead in sandstones, and Wardlaw (1976) used epoxy resin on carbonates.

Several schemes have been drawn up to classify porosity types (e.g., Robinson, 1966; Levorsen, 1967; Choquette and Pray, 1970). Two main types of pore can be defined according to their time of formation (Murray, 1960). Primary pores are those formed when a sediment is deposited. Secondary pores are those developed in a rock some time after deposition (Table 6.1). Primary pores may be divided into two subtypes: interparticle (or intergranular) and intraparticle (or intragranular). Interparticle pores are initially present in all sediments. They are often quickly lost in clays and carbonate sands because of the combined effects of compaction and cementation. Much of the porosity found in sandstone reservoirs is preserved primary intergranular porosity (Fig. 6.2). Intraparticle pores are generally found within the skeletal grains of carbonate sands (Fig. 6.2) and are thus often cul-de-sac pores. Because of compaction and cementation they are generally absent in carbonate reservoirs.

Secondary pores are often caused by solution. Many minerals may be leached out of a rock, but, volumetrically, carbonate solution is the most significant. Thus solution-induced porosity is more common in carbonate reservoirs than in sandstone reservoirs. A distinction is generally made between moldic and vuggy porosity. Moldic porosity is fabric-selective (Plate 4 and Fig. 6.3), that is, only the grains or only the matrix has been leached out. Vugs, by contrast, are pores whose boundaries cross-cut grains, matrices, and/or earlier cement. Vugs thus tend to be larger than moldic pores (Fig. 6.4). With increasing size vuggy porosity changes into cavernous porosity. Cavernous pores are those large enough to cause the drill

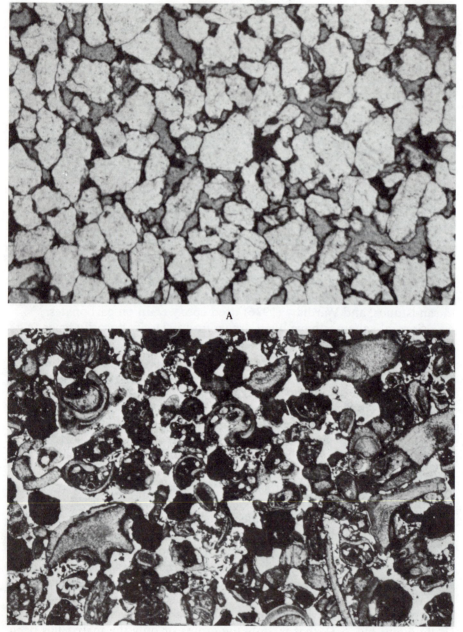

FIGURE 6.2   *Thin sections illustrating the different types of primary porosity.*
(A) Intergranular porosity in Middle Jurassic Brent Sandstone (North Sea, United
Kingdom). Field of view is approximately 1 cm. (B) Intergranular and intragranular
(within fossils) porosity in skeletal pelletal limestone. Field of view is about 1 cm.

FIGURE 6.3   *Portland Limestone (U. Jurassic) showing biomoldic porosity due to solution of shells. Portland Island, Dorset, United Kingdom.*

FIGURE 6.4   *Core of Zechstein (U. Permian) dolomite showing vuggy porosity (North Sea, United Kingdom).*

string to drop by half a meter or to contain one crouched mudlogger. Examples of cavernous porosity are known from the Arab D Jurassic limestone of the Abqaiq field of Saudi Arabia (McConnell, 1951) and from the Fusselman limestone of the Dollarhide field of Texas (Stormont, 1949), both having cavernous pores up to 5 m high.

Secondary porosity can also be caused not by solution but by the effects of cementation. Fenestral pores occur where there is a "primary or penecontemporaneous gap in rock framework larger than grain-supported interstices" (Tebbutt et al., 1965). Strictly speaking, therefore, such pores would appear to be primary rather than secondary. In fact, fenestral pores are characteristic of lagoonal pelmicrites in which dehydration has caused shrinkage and buckling of the laminae. This type of fabric has been described from Triassic lagoonal carbonates of the Alps, where it is termed *loferite* (Fischer, 1964), and also from Paleocene dolomite pellet muds from Libya (Conley, 1971).

Intercrystalline porosity, which refers to pores occurring between the crystal faces of crystalline rocks, is a far more important type of secondary porosity than is fenestral porosity. Most recrystallized limestones generally possess negligible porosity. Crystalline dolomites, on the other hand, are often highly porous, with a friable, sugary (saccharoidal) texture. This type of intercrystalline porosity occurs in secondary dolomites that have formed by the replacement of calcite. Dolomitization causes a 13 percent shrinkage of the original bulk volume, thereby increasing the porosity. Intercrystalline pores tend to be polyhedral, with sheet-like pore throats in contrast to the more common tubular ones (Wardlaw, 1976).

Fracture porosity is the last major type of pore to consider. It is extremely important not so much because it increases the storage capacity of a reservoir but because of the degree to which it may enhance permeability. Fractures are rare in unconsolidated, loosely cemented sediments, which respond to stress by plastic flow. They may occur in any brittle rock, not only sandstones and limestones but also shales and igneous and metamorphic rocks (Stearns and Friedman, 1972).

Because fractures are much larger than are most pores, they are seldom amenable to study from cores alone (Fig. 6.5). Furthermore, the process of coring itself may fracture the rock. Such artifacts must be distinguished from naturally occurring pores. Kulander et al. (1979) have outlined several significant criteria that may be used for this purpose. Fractures may also be recognized from wireline logs, seismic data, and the production history of a well. Cycle skip on sonic logs can be caused by fractures (and other phenomena), as discussed on page 69. A random *bag o' nails* motif on a dipmeter can be caused by fractures (and other phenomena), as discussed on page 74. Fractures can also be "seen" by the borehole TV log. Anomalously low velocities in seismic data may be due to fractures, although again this effect may be due to other types of porosity

FIGURE 6.5 *Fractured core of Gargaf Group sandstone (Cambro-Ordovician), Sirte basin, Libya.*

and/or the presence of gas. High initial production followed by rapid pressure drop and rapid decline in flow rate in production well tests is often indicative of fracture porosity.

Brittle rocks tend to fracture in three geological settings (Fig. 6.6). Fracture systems may dilate where strata are subjected to tension on the crests of anticlines and the nadirs of synclines. Detailed outcrop studies have demonstrated the correlation between fracture intensity and orientation and the structural style and trend of the anticline (e.g., Harris et al., 1960). Fractures often occur adjacent to faults, and in some cases a field may be directly related to a major fault and its associated fractures (Fig. 6.7). Fracture porosity is also found beneath unconformities, especially in carbonates, where the fractures may be enlarged by karstic solution. Frac-

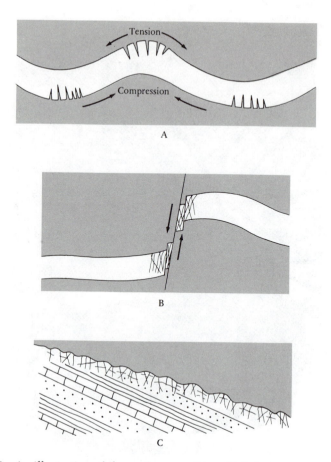

FIGURE 6.6   *An illustration of the various ways in which fracture porosity is commonly found. (A) Fractures may develop on the crests of anticlines and the nadirs of synclines. (B) Fractures may develop adjacent to faults. (C) Fractures may occur beneath unconformities.*

ture porosity also occurs in sandstones; for example, in the Buchan field of the North Sea, in which a fractured Devonian sandstone reservoir lies beneath the Cimmerian unconformity (Butler et al., 1976).

Porosity Measurement

Porosity may be measured in three ways: directly from cores, indirectly from geophysical well logs, or from seismic data. The last two of these methods are discussed on pages 70 and 96, respectively. The following account deals only with porosity measurement from cores. The main reason for cutting a core is to measure the petrophysical properties of the reservoir. The porosity of a core sample may be measured in the labora-

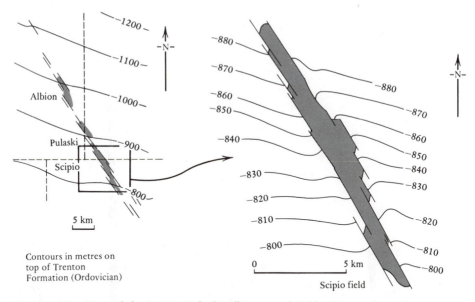

FIGURE 6.7   *Map of the Scipio-Pulaski-Albion trend field. This produces from fractured Trenton (Ordovician) dolomite. (B) is detail of boxed area in (A). (After Levorsen, 1967.)*

tory using several methods (Anderson, 1975; Monicard, 1981). For homogeneous rocks, like many sandstones, samples of only 30 mm³ or so may be cut or chipped from the core. For heterogeneous reservoirs, including many limestones, analysis of a whole core sample is generally necessary. Several of the more common porosity measuring procedures are briefly described in the following sections.

Washburn-Bunting Method

One of the earliest and simplest methods of porosity measurement is the gas expansion technique described by Washburn and Bunting in 1922. The basic apparatus is shown in Figure 6.8. Air within the pores of the sample is extracted when a vacuum is created by lowering and raising the mercury bulb.

The amount of air extracted can be measured in the burette. Then:

$$\text{Porosity (\%)} = \frac{\text{volume of gas extracted}}{\text{bulk volume of sample}} \times 100$$

Bulk volume must be determined independently by another method, which generally involves applying Archimedes principle of displacement to the sample when totally submerged in mercury.

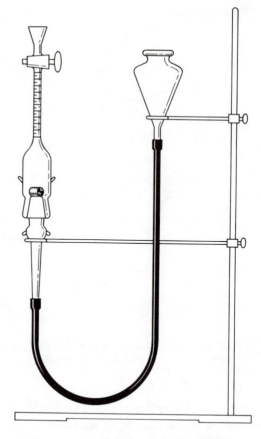

FIGURE 6.8 *Washburn-Bunting apparatus for measuring porosity by the gas expansion method.*

Boyles Law Method

Boyles law (pressure × volume = constant) can be applied to porosity measurement in two ways. One way is to measure the pore volume by sealing the sample in a pressure vessel, decreasing the pressure by a known amount, and measuring the increase in volume of the contained gas. Conversely, the grain volume can be measured and, if the bulk volume is known, porosity can be determined. The Ruska porosimeter is a popular piece of apparatus that is based on this technique (Fig. 6.9).

FIGURE 6.9 *Boyles law porosimeter.*

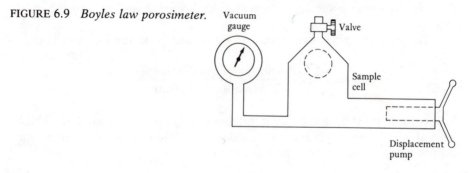

PERMEABILITY

The second essential requirement for a reservoir rock is permeability. Porosity alone is not enough; the pores must be connected. Permeability is the ability of fluids to pass through a porous material.

The original work on permeability was carried out by H. Darcy, who studied the flow rates of the springs at Dijon in France (Darcy, 1856). His work was further developed by Muskat and Botset (Botset, 1931; Muskat and Botset, 1931; Muskat, 1937). They formulated Darcy's law as follows:

$$Q = \frac{K(P_1 - P_2)A}{\mu L}$$

where
Q = rate of flow
K = permeability
$(P_1 - P_2)$ = pressure drop across the sample
A = cross-sectional area of the sample
L = length of the sample
μ = viscocity of the fluid

Figure 6.10 illustrates the basic arrangement for measuring permeability. The unit of permeability is the Darcy, which is defined as the permeability that allows a fluid of 1 centipoise (cP) viscosity to flow at a velocity of 1 cm/s for a pressure drop of 1 atm/cm. Since most reservoirs have permeabilities much less than a Darcy, the millidarcy (md) is commonly used. Average permeabilities in reservoirs are commonly in the range of 5 to 500 md. Permeability is generally referred to by the letter K.

Darcy's law is only valid when there is no chemical reaction between the fluid and the rock and when only one fluid phase completely fills the pores. The situation is far more complex for mixed oil and gas phases, although a Darcy-type equation is assumed to apply. Flow rate depends on the ratio of permeability to viscosity. Thus gas reservoirs may be able to flow at commercial rates with permeabilities of only a few millidarcies,

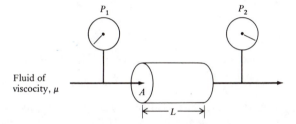

FIGURE 6.10 *Basic arrangement for the measurement of permeability.*

whereas oil reservoirs need minimal permeabilities in the order of tens of millidarcies. For this reason permeability is measured in the laboratory using an inert gas rather than a liquid. Under some conditions—small pore dimensions, very low gas densities, and relatively large mean free paths of gas molecules—the normal condition of zero velocity and no slip at solid surface may not be met. This is the Klinkenberg effect, and a correction is necessary to recalculate the permeability of the rock to air measured in the laboratory to the permeability it would exhibit for liquid or high-density gas (Klinkenberg, 1941). This correction may range from 1 or 2 percent for high-permeability rocks to as much as 70 percent for low-permeability rocks.

Permeability is seldom the same in all directions within a rock (see p. 238). Vertical permeability is generally far lower than permeability horizontal to the bedding. Permeability is thus commonly measured from plugs cut in both directions.

When a single fluid phase completely saturates the pore space, permeability is referred to as *absolute* or *specific* and is given the dimension (L^2). The *effective* permeability refers to saturations of less than 100 percent. The terms K_w, K_g, and K_o are used to designate the effective permeability with respect to water, gas, and oil, respectively. Effective permeability ranges between 0 and K at 100 percent saturation, but the sum of permeabilities to two or three phases is always less than 1. Relative permeability is the ratio of the effective permeability for a particular fluid at a given saturation to a base permeability. The relative permeability ranges from 0.0 to 1.0. Thus for oil:

$$K_{ro} = \frac{K_o}{K}$$

For gas:

$$K_{rg} = \frac{K_g}{K}$$

and for water:

$$K_{rw} = \frac{K_w}{K}$$

where K = absolute permeability (or permeability at irreducible wetting phase saturation)
K_r = relative permeability
K_g = effective permeability at 100 percent gas saturation
K_w = effective permeability at 100 percent water saturation
K_o = effective permeability at 100 percent oil saturation

FIGURE 6.11 *Typical curves for the relative permeabilities of oil and gas for different saturations.*

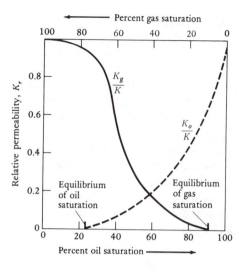

Figure 6.11 shows the typical relative permeability relationship for different saturations. It is necessary to consider the relative saturations, how the oil and gas are distributed, and the presence or absence of water. Most reservoirs are water-wet, having a film of water separating the pore boundaries from the oil. Few reservoirs are oil-wet, totally lacking a film of connate water at pore boundaries (Fig. 6.12). A mixed, or intermediate, wettability condition may also occur.

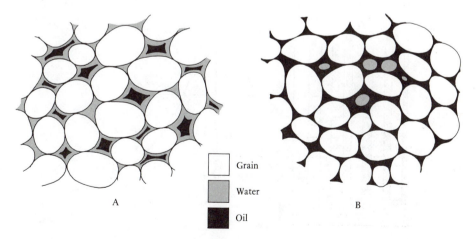

FIGURE 6.12 *The concept of wettability in reservoirs. (A) A water-wet reservoir (common). (B) An oil-wet reservoir (rare).*

CAPILLARY PRESSURE

Consideration of the wettability of pores leads to the concept of capillarity, the phenomenon whereby liquid is drawn up a capillary tube (Fig. 6.13). The capillary pressure is the difference between the ambient pressure and the pressure exerted by the column of liquid. Capillary pressure increases with decreasing tube diameter (Fig. 6.14). Translated into geological terms, the capillary pressure of a reservoir increases with decreasing pore size or, more specifically, pore throat diameter. Capillary pressure is also related to the surface tension generated by the two adjacent fluids—it increases with increasing surface tension. In water-wet pore systems the meniscus is convex with respect to water; in oil-wet systems it is concave (Fig. 6.15). For a mathematical analysis of capillary pressure, see Muskat (1949) and Dickey (1979).

FIGURE 6.13 *Capillary tube in a liquid-filled tank. The pressure on the water level (A) equals the pressure due to the hydrostatic head of water (h) minus the capillary pressure across the meniscus.*

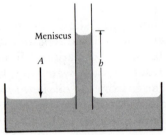

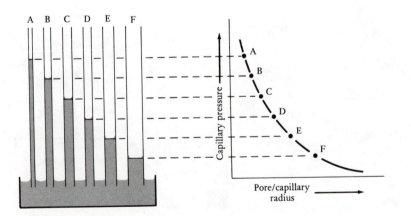

FIGURE 6.14 *Capillary tubes of various (exaggerated) diameters showing that the heights of the liquid columns are proportional to the diameters of the tubes.*

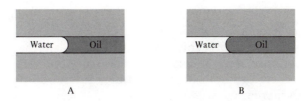

FIGURE 6.15 *Cross-sections of capillary tube/pores showing meniscus effect for oil-wet (A) and water-wet (B) reservoirs. Water-wet reservoirs are the general rule.*

Reservoirs are commonly subjected to capillary pressure tests in which samples with 100 percent of one fluid are injected with another (gas, oil, water, and mercury may be used). The pressure at which the injected fluid begins to invade the reservoir is the *displacement pressure*. As pressure increases, the proportions of the two fluids gradually reverse until the irreducible saturation point is reached, at which no further invasion by the second fluid is possible at any pressure. Data from these analyses are plotted up as capillary pressure curves (Fig. 6.16). Curve 1 is typical of a good quality reservoir—porous and permeable. Once the entry, or displacement, pressure has been exceeded, fluid invasion increases rapidly for a minor pressure increase until the irreducible water saturation is reached. At this point no further water can be expelled irrespective of pressure.

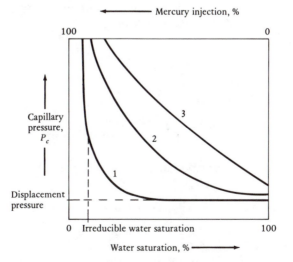

FIGURE 6.16 *Capillary pressure curves for various reservoirs. (1) A clean, well-sorted sand with uniform pore diameters; (2) an intermediate quality reservoir; (3) a poor quality reservoir with a wide range of pore diameters. For full explanation see text.*

Curve 2 is for a poorer quality reservoir with a higher displacement pressure and higher irreducible water content. Curve 3 is for a very poor quality reservoir, such as a poorly sorted sand with abundant matrix and hence a wide range of pore sizes. Displacement pressure and irreducible water saturation are therefore both high, and water saturation declines almost uniformly with increasing pressure.

RELATIONSHIP BETWEEN POROSITY, PERMEABILITY, AND TEXTURE

The texture of a sediment is closely correlated with its porosity and permeability. The texture of a reservoir rock is related to the original depositional fabric of the sediment, which is modified by subsequent diagenesis. This diagenesis may be negligible in many sandstones, but in carbonates it may be sufficient to obliterate all traces of original depositional features. Before considering the effects of diagenesis on porosity and permeability, the effects of the original depositional fabric on these two parameters must be discussed. The following account is based largely on studies by Krumbein and Monk (1942), Gaithor (1953), Rogers and Head (1961), Potter and Mast (1963), Chilingar (1964), Beard and Weyl (1973), and Pryor (1973). Pryor (1973) is particularly helpful.

The textural parameters of an unconsolidated sediment that may affect porosity and permeability are as follows:

$$\text{Grain shape} \begin{cases} \text{Roundness} \\ \text{Sphericity} \end{cases}$$

Grainsize

Sorting

$$\text{Fabric} \begin{cases} \text{Packing} \\ \text{Grain orientation} \end{cases}$$

These parameters are described and discussed in the following sections.

Relationship Between Porosity, Permeability, and Grain Shape

The two aspects of grain shape to consider are roundness and sphericity (Powers, 1953). As Figure 6.17 shows, these two properties are quite distinct. Roundness describes the degree of angularity of the particle. Sphericity describes the degree to which the particle approaches a spherical shape. Mathematical methods of analyzing these variables are available (e.g., Friedman and Sanders, 1978).

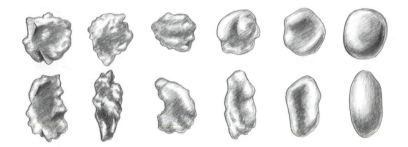

FIGURE 6.17 *Sand grains showing the difference between shape and sphericity.*

Data on the effect of roundness and sphericity on porosity and permeability is sparse. Fraser (1935) inferred that porosity might decrease with sphericity because spherical grains may be tighter packed than subspherical ones.

Relationship Between Porosity, Permeability, and Grainsize

Theoretically, porosity is independent of grainsize for uniformly packed and graded sands (Rogers and Head, 1961). In practice, however, coarser sands sometimes have higher porosities than do finer sands or vice versa (e.g., Lee, 1919; Sneider et al., 1977). This disparity may be due to separate but correlative factors such as sorting and/or cementation.

Permeability declines with decreasing grainsize because pore diameter decreases and hence capillary pressure increases (Krumbein and Monk, 1942). Thus a sand and a shale may both have porosities of 10 percent; whereas the former may be a permeable reservoir, the latter may be an impermeable cap rock.

Relationship Between Porosity, Permeability, and Grain Sorting

Porosity increases with improved sorting. As sorting decreases, the pores between the larger, framework-forming grains are infilled by the smaller particles. Permeability decreases with sorting for the same reason (Fraser, 1935; Rogers and Head, 1961; Beard and Weyl, 1973). As mentioned earlier, sorting sometimes varies with the grainsize of a particular reservoir sand, thus indicating a possible correlation between porosity and grainsize. Figure 6.18 summarizes the effects of sorting and grainsize on porosity and permeability in unconsolidated sand.

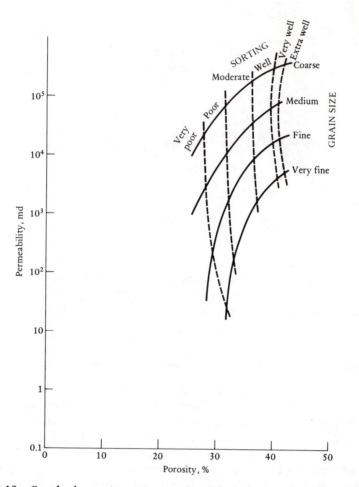

FIGURE 6.18   *Graph of porosity against permeability showing their relationship with grainsize and sorting for uncemented sands. (After Beard and Weyl, 1973, and Nagtegaal, 1978.)*

Relationship Between Porosity, Permeability, and Grain Packing

The two important characteristics of the fabric of a sediment are how the grains are packed and how they are oriented. The classic studies of sediment packing were described by Fraser (1935) and Graton and Fraser (1935). They showed that spheres of uniform size have six theoretical packing geometries. These geometries range from the loosest cubic style with a porosity of 48 percent down to the tightest rhombohedral style with a 26 percent porosity (Fig. 6.19).

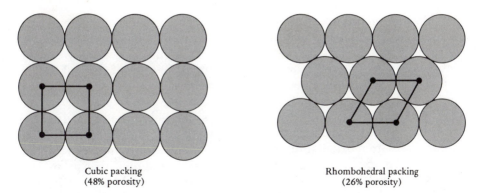

| Cubic packing | Rhombohedral packing |
| (48% porosity) | (26% porosity) |

FIGURE 6.19 *The loosest and tightest theoretical packings for spheres of uniform diameter.*

The significance of packing to porosity can be observed when trying to pour the residue of a packet of sugar into a sugar bowl. The sugar poured into the bowl has settled under gravity into a loose packing. Tapping the container causes the level of sugar to drop as the grains fall into a tighter packing, causing porosity to decrease and bulk density to increase.

Packing is obviously a major influence on the porosity of sediments, and several geologists have tried to carry out empirical, as opposed to theoretical, studies (e.g., Kahn, 1956; Morrow, 1971). Particular attention has been paid to relating packing to the depositional process (e.g., Martini, 1972). Like grain sphericity and roundness, packing is not amenable to extensive statistical analysis. Intuitively, one might expect sediments deposited under the influence of gravity, such as fluidized flows and turbidites, to exhibit looser grain packing than those laid down by traction processes. However, postdepositional compaction probably causes rapid packing adjustment and porosity loss during early burial.

Relationship Between Porosity, Permeability, and Grain Orientation

The preceding analysis of packing was based on the assumption that grains are spherical, which is generally untrue of all sediments except oolites. Most quartz grains are actually prolate spheroids, slightly elongated with respect to their *C* crystallographic axis (Allen, 1970). Sands also contain flaky grains of mica, clay, and other constituents. Skeletal carbonates have still more eccentric grain shapes. Thus the second element of fabric, namely orientation, is perhaps more significant to porosity and permeability than packing is. How grains are oriented has little effect on porosity, but a major effect on permeability.

Most sediments are stratified, the layering being caused by flaky grains, such as mica, shells, and plant fragments, as well as by clay laminae. Because of this stratification the vertical permeability is generally considerably lower than the horizontal permeability. The ratio of vertical to horizontal permeability in a reservoir is important because of its effect on coning as the oil and gas are produced (see p. 263).

Variation in permeability also occurs parallel to bedding. In most sands the grains generally show a preferential alignment within the horizontal plain. Grain orientation can be measured by various methods (Sippel, 1971). Studies of horizontally bedded sands have shown that grains are elongated parallel to current direction (e.g., Shelton and Mack, 1970; Von Rad, 1971; Martini, 1971, 1972). For cross-bedded sands the situation is more complex because grains may be aligned parallel to the strike of foresets due to gravitational rolling. Figure 6.20 shows that permeability will be greatest parallel to grain orientation, since this orientation is the fabric alignment with least resistance to fluid movement (Schieddegger, 1957).

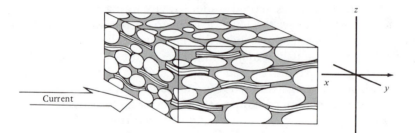

FIGURE 6.20   *Block diagram of sand showing layered fabric with grains oriented parallel to current. Generally, $K_x > K_y > K_z$.*

Studies of the relationship between fabric and permeability variation have produced different results on both small and large scales. Potter and Pettijohn (1977) have reviewed the conflicting results of a number of case histories, some of which show a correlation with grain orientation and some of which do not. It is necessary to consider not only small-scale permeability variations caused by grain alignment but also the larger variations caused by sedimentary structures.

Grainsize differences cause permeability variations far greater than those caused by grain orientation. Thus in the cross-bedded eolian sands of the Leman field (North Sea), horizontal permeabilities measured parallel to strike varied from as much as 0.5 to 38.5 md between adjacent foresets. This range in permeabilities is attributable to variations in grainsize and sorting. Similarly, since eolian cross-beds generally show a decreas-

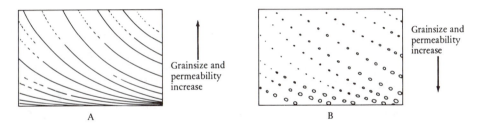

FIGURE 6.21 *Permeability variations for (A) downward-fining and (B) downward-coarsening avalanche cross-beds.*

ing grainsize from foreset to toeset, permeability diminishes downward through each cross-bedded unit (Van Veen, 1975). Conversely, in many aqueously deposited cross-beds, avalanching causes grainsize to increase downward. Thus permeability also increases down each foreset for the reason previously given (Fig. 6.21).

On the still larger scale of whole sand bodies, grainsize-related permeabilities often have considerable variations. In a study of a Holocene sand-filled channel in Holland, permeability increased from 25 md at the margin to 270 md in the center some 175 m away (Weber et al., 1972). Pumping tests showed that the drawdown was nearly concentric to the borehole, which suggests that the permeability had little preferred direction within the horizontal plane (Fig. 6.22).

EFFECTS OF DIAGENESIS ON RESERVOIRS

The preceding section examined the control of texture on the petrophysics of unconsolidated sediment. Once burial begins, however, many changes take place, most of which diminish the porosity and permeability of a potential reservoir. These changes, collectively referred to as diagenesis, are numerous and complex. The following account specifically focuses on the effect of diagenesis on reservoir quality. More detailed accounts are found in Friedman and Sanders (1978); Blatt, Middleton, and Murray (1980); and Selley (1982). For convenience the effects of diagenesis are considered separately for sandstone and carbonate reservoirs.

Effects of Diagenesis on Sandstone Reservoirs

The effects of diagenesis on sandstone reservoirs is first discussed on a general scale and then in detail. The first of these effects is due to burial; the second to cementation and solution.

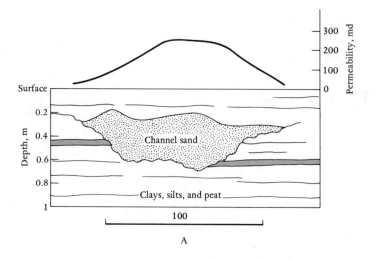

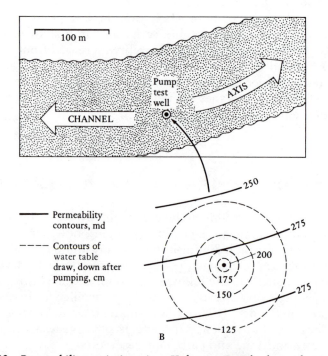

FIGURE 6.22 *Permeability variations in a Holocene Dutch channel sand. (A) A cross-section showing how permeability increases toward the channel axis. (B) Maps showing how contours of the water level drop are concentric around the pump test well. (After Weber et al., 1972.)*

Porosity Gradients in Sandstone Reservoirs

Studies of Recent sands, such as those of Pryor (1973), show that they have porosities of some 40 to 50 percent and permeabilities of tens of Darcies. Most sandstone reservoirs, however, have porosities in the range of 10 to 20 percent and permeabilities measurable in millidarcies. Although fluctuations do occur, the porosity and permeability of reservoirs decrease with depth. This relationship is of no consequence to the production geologist or engineer concerned with developing a field, but it is important for the explorationist who has to decide the greatest depth at which commercially viable reservoirs may occur.

The porosity of a sandstone at a given depth can be determined if the porosity gradient and primary porosity are known:

$$\phi^D = \phi^P - GD$$

where ϕ^D = porosity at a given depth
ϕ^P = primary porosity at the surface
G = porosity gradient (% ϕ/km)
D = burial depth

Reviews of porosity gradients and the factors controlling them have been given by Selley (1978) and Magara (1980).

The main variables that affect porosity gradients are the mineralogy and texture of the sediments and the geothermal and pressure regimes to which they are subjected. The more mineralogically mature a sand is, the better its ability to retain its porosity (Dodge and Loucks, 1979). Taking the two extreme cases, chemically unstable volcaniclastic sands tend to lose porosity fastest, and the more stable pure quartz sands tend to have the lowest gradients (Fig. 6.23).

Texture also affects the gradient: poorly sorted sands with abundant clay matrix compact more and lose porosity faster than do clean, well-sorted sand (Rittenhouse, 1971). A similar effect is noted in micaceous sand, as, for example, in the Jurassic of the North Sea (Hay, 1977).

Geothermal gradients also affect the porosity gradient of sand, as shown in Figure 6.24. Because the rate of a chemical reaction increases with temperature, the higher the geothermal gradient, the faster the rate of porosity loss. Pressure gradient also affects porosity. Abnormal pressure preserves porosity, presumably by decreasing the effect of compaction (Atwater and Miller, 1965). Oil and gas have been commonly observed to preserve porosity in sands (e.g., Fuchtbauer, 1967). Once petroleum enters a trap, the circulation of connate water is terminated and further cementation inhibited.

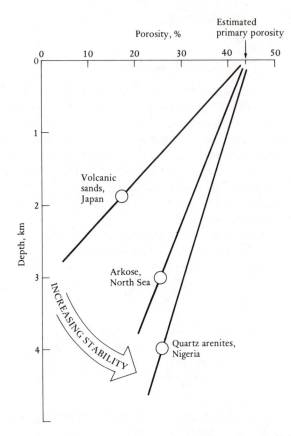

FIGURE 6.23 *Graph showing how the porosity gradient decreases with increasing mineralogical maturity. (After Nagtegaal, 1978.)*

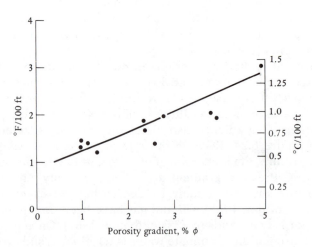

FIGURE 6.24 *Graph of geothermal gradient versus porosity for various sandstones showing how they tend to increase together. (From Selley, 1978.)*

242

Pleistocene	Dada Conglomerate/Manchar Fm.		D	
Plio-Pleistocene	Siwaliks	Chaudwan Fm. Litra Fm. Vihowa Fm.	CH L V	Bedded conglomerates etc.
?Miocene	Chitarwata Fm. (Gaj/Nari/Sibi)		$CHIT$	Sst./Shale
Eocene	Kirther Fm. K	Drazinda & Domanda Pirkoh Habib Rahi	E PK HR	Gypsiferous shale Pale lst. Platy lst.
	Ghazij Fm.		GH	Shales, lst., etc.
Paleocene	Dungan/Ranikot		DU	Lst.

~U~	Unconformity	—F	Faults
Major anticline		—∠	Lineaments
Major syncline		(--)	?Trace of buried fold

PLATE 4 *Slabbed core of Libyan Paleocene limestone showing pelmoldic and vuggy porosity. Field of view is 3 cm.*

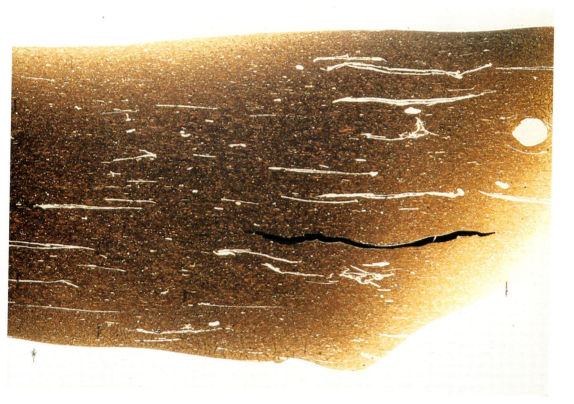

PLATE 1   *Photomicrograph of Kimmeridge Coal oil shale (Upper Jurassic), Dorset, United Kingdom, showing amorphous kerogen (amber), black humic detritus, and thin-shelled bivalves.*

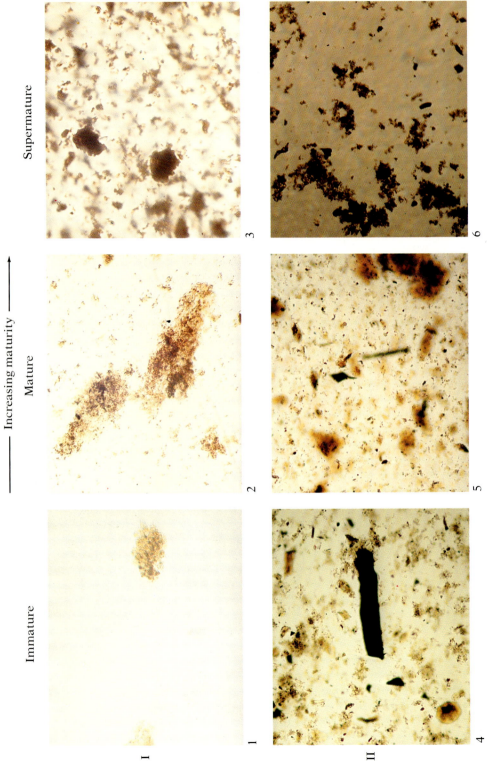

Immature   Mature   Supermature

Increasing maturity →

Type of kerogen

I

II

1

2

3

4

5

6

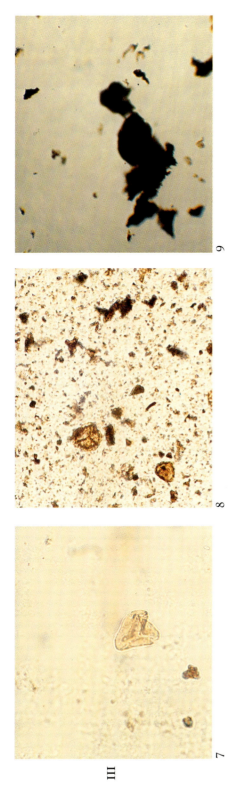

III

7

8

9

PLATE 2 *Photographs of different types of kerogen at different levels of thermal maturation. (Courtesy of R. R. F. Kinghorn and F. Rahman.)*

Photograph Number	1	2	3	4	5	6	7	8	9
Locality	Gurpi Fm. Zagros, Iran	Kazdumi Fm. Zagros, Iran	Garan Fm. Zagros, Iran	Kimmeridge Clay Dorest, UK	Kimmeridge Clay Dorset, UK	Sembar Fm. Lower Indus, Pakistan	Shamser Fm. Alborz, Iran	Coal Measures South Wales	Coal Measures South Wales
Type of Kerogen	I	I	I	II	II	II	III	III	III
Maturation Level	Immature	Mature	Supermature	Immature	Mature (just)	Supermature	Immature	Mature	Supermature
$R_o(\%)$	0.4	0.9	2.0	0.4	0.5	2.0	0.4	1.2	3.0
Magnification	X25	X25	X62	X25	X25	X62	X25	X25	X50

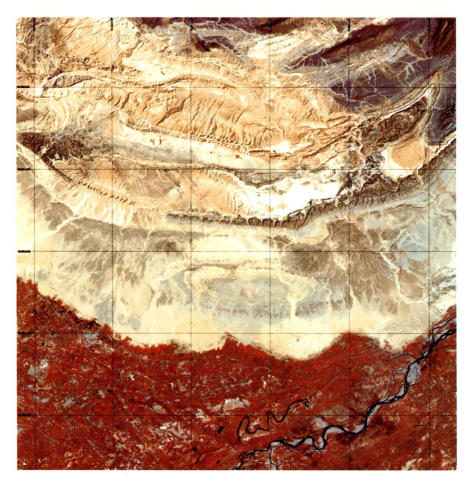

PLATE 3   *Enhanced LANDSAT image and interpretation of part of Pakistan. When used in conjunction with recordings of gravity and magnetics, multispectral sensing may delineate anomalies that deserve further attention on the ground. (Courtesy of Hunting Geology and Geophysics Limited.)*

PLATE 5   *Secondary porosity indicated by large, irregular pores and loose grain packing in Brent Sandstone (Jurassic), North Sea, United Kingdom. Field of view is 0.5 cm.*

PLATE 6   *Thin section of coral walls showing porosity obliterated by a sparite cement in Wenlock Limestone (Silurian), Wales, United Kingdom. Field of view is 0.5 cm.*

PLATE 7 *Photomicrograph of secondary crystalline dolomite with intercrystalline porosity. Zechstein Group (Upper Permian), North Sea, United Kingdom.*

Porosity Loss by Cementation

On a regional scale the porosity of sandstones is largely controlled by the factors previously discussed (mineralogy, texture, geothermal and pressure gradient). Within a reservoir itself porosity and permeability vary erratically with primary textural changes and secondary diagenetic ones. The diagenesis of sandstones is a major topic beyond the scope of this text. It is treated in depth by Blatt et al., (1980), Chilingarian and Wolf (1975), and Friedman and Sanders (1978). The following account deals concisely with those aspects particularly relevant to petroleum geology. Diagenetic changes in a sandstone reservoir include cementation and solution, which are now discussed in turn.

A small degree of cementation is beneficial to a sandstone reservoir because it prevents sand from being produced with the oil. The presence of sand in the oil not only damages the reservoir itself but also the production system. Extensive cementation is deleterious, however, because it diminishes porosity and permeability. Many minerals may grow in the pores of a sandstone, but only three are of major significance: quartz, calcite, and the authigenic clays.

Quartz is a common cement. It generally grows as optically continuous overgrowths on detrital quartz grains (Fig. 6.25A). The solubility of silica increases with pH, so silica cements occur where acid fluids have moved through the pores.

Calcium carbonate is another common cement. It generally occurs as calcite crystals, which, as they grow from pore to pore, may form a poikilitic fabric of crystals enclosing many sand grains (Fig. 6.25B). The grains frequently appear to "float" in the crystals. Detailed observation often shows that grain boundaries are corroded, suggesting that some replacement has occurred. Calcite solubility is the reverse of silica solubility; that is, it decreases with pH. Thus calcite cementation is the result of alkaline fluids moving through the pores.

The presence of clay in a reservoir obviously destroys the porosity and permeability. Clay may be present either as a detrital matrix or as an authigenic cement. As clays recrystallize and alter during burial, this distinction is not always easy to make. The kaolinitic, illitic, and montmorillonitic clays have different effects on reservoirs and different sources of formation.

Kaolin generally occurs as well-formed, blocky crystals within pore spaces (Fig. 6.25C). This make-up diminishes the porosity of the reservoir, but may have only a minor effect on permeability. Kaolin forms and is stable in the presence of acid solutions. Therefore it occurs as a detrital clay in continental deposits and as an authigenic cement in sands that have been flushed by acidic waters, such as those of meteoric origin.

Illitic clay is quite different from kaolin. Authigenic illite grows as fibrous crystals, which typically occur as fur-like jackets on the detrital

A

FIGURE 6.25   *Various types of sandstone cements. (A) Silica cement growing in optical continuity on detrital quartz grains. (B) Calcite cement with poikilitic fabric of large crystals enclosing corroded quartz grains. (C) Authigenic kaolin crystals within pores. (D) Authigenic illite showing fibrous habit. (A) and (B) are thin sections photographed in ordinary and polarizing light, respectively. (C) and (D) are scanning electron micrographs.*

grains (Fig. 6.25D). These structures often bridge over the throat passages between pores in a tangled mass. Thus illitic cement may have a very harmful effect on the permeability of a reservoir. This effect is demonstrated in Figure 6.26, which is taken from studies of the Lower Permian Rotliegende sandstone of the southern North Sea by Stalder (1973) and Seemann (1979). Illitic clays typically form in alkaline environments. They are thus the dominant detrital clay of most marine sediments and occur as an authigenic clay in sands through which alkaline connate water has moved.

The montmorillonitic, or smectitic, clays are formed from the alteration of volcanic glass and are found in continental or deep marine deposits. They have the ability to swell in the presence of water. Reservoirs with montmorillonite are thus very susceptible to formation damage if drilled

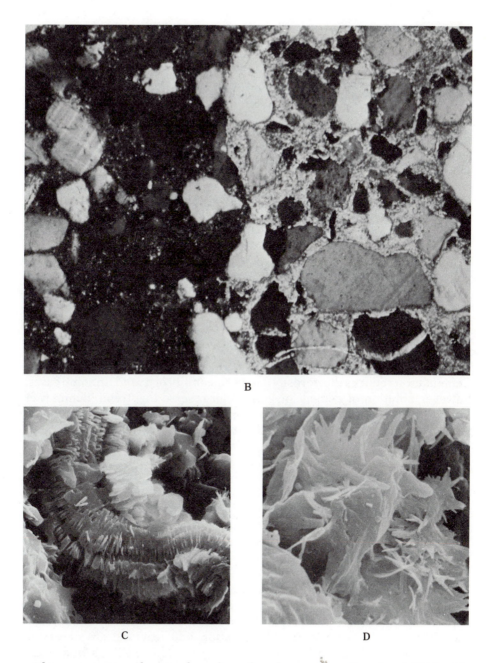

B

C   D

with a conventional water-based mud and must therefore be drilled with
an oil-based mud. When production begins, water displaces the oil, causing the montmorillonitic clays to expand and destroy the permeability of
the lower part of the reservoir.

FIGURE 6.26 *Porosity-permeability data for illite (B and D) and kaolin (A and C) cemented Rotliegende reservoir sands of the North Sea. (After Stalder, 1973, and Seemann, 1979.)*

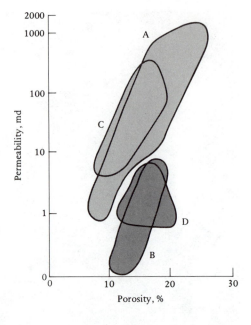

Porosity Enhancement by Solution

Cementation reduces the porosity and permeability of a sand. In some cases, however, solution of cement or grains can reverse this trend. Schmidt et al., (1977) have outlined the petrographic criteria for the recognition of secondary solution porosity in sands. It generally involves the leaching of carbonate cements and grains, including calcite, dolomite, siderite, shell debris, and unstable detrital minerals, especially feldspar. Leached porosity in sands is generally associated with kaolin, which both replaces feldspar and occurs as an authigenic cement.

The fact that carbonate is leached out of the sand and the predominance of kaolin indicate that the leaching was caused by acidic solutions. There are two sources of acidic leaching: epidiagenesis or, more specifically, weathering due to surface waters (Fairbridge, 1967) and decarboxylation of coal or kerogen (Schmidt et al., 1977).

Meteoric water rich in carbonic and humic acids weathers sandstones at the earth's surface. Thus many geologists from Hea (1971) to Al-Gailani (1981) have studied the petrography of sandstones beneath unconformities (Plate 5). In many cases kaolinization and leaching are observed, although secondary porosity is only preserved within hydrocarbon reservoirs. Usually, the weathering-induced porosity is destroyed during reburial by normal diagenesis. The preservative effect of hydrocarbons on porosity has already been noted. Examples of subunconformity sand reservoirs with secondary porosity include the Sarir sand of Libya (which formed the basis for Hea's study), the Cambro-Ordovician sand of Hassi-Messaoud, Algeria (Balducci and Pommier, 1970), the sub-Cimmerian unconformity Jurassic sands of the North Sea (Hancock, 1978), and Prudoe Bay, Alaska.

An alternative source for the acid fluid is the decarboxylation of coal or kerogen, causing the expulsion of solutions of carbonic acid (Schmidt et al., 1977). According to this theory, acidic solutions are expelled from a maturing source rock ahead of petroleum migration. These acid fluids generate secondary solution porosity in the reservoir beds. This mechanism explains solution porosity, which has been observed in sandstones that have not undergone uplift and subaerial exposure.

Timing of Migration and Diagenesis

Porosity and permeability are often higher in hydrocarbon reservoirs than in the underlying water zones. As discussed in the preceding section, this discrepancy may be due to epidiagenesis where the reservoir lies beneath an unconformity. In such instances the changes in porosity and permeability may be above or beneath the oil-water contact.

As already noted (p. 241), the presence of hydrocarbons inhibits cementation by preventing connate water from continuing to move through the trap. Flow may continue in the underlying water zone, however, so porosity and permeability are reduced by cementation after oil or gas has migrated into the trap.

Postmigration silica cement in water zones beneath traps has been described from the Silurian Clinton Sand of Ohio (Heald, 1940) and the Jurassic reservoirs of the Gifhorn trough, Germany (Philip et al., 1964). Calcite and anhydrite postdate migration in the Permian Lyons sand of the Denver basin (Levandowski et al., 1973). Many other instances of postmigration cementation are known. In extreme cases postmigration cementation may be so pervasive as to act as a seat seal, giving rise to *frozen-in* paleotraps, as in the Raudhatain field of Kuwait (Al-Rawi, 1981).

Because of postmigration cementation it is important to cut cores below the oil-water contact of a field. If a water drive production mechanism is anticipated, then the permeability of the aquifer must be known. Data must not be extrapolated from the reservoir downward. Figure 6.27 summarizes the potential diagenetic pathways of sandstones.

Effects of Diagenesis on Carbonate Reservoirs

The carbonate rocks include the limestones, composed largely of calcite ($CaCO_3$), and the dolomites, composed largely of the mineral of the same name [$CaMg(CO_3)_2$]. Carbonates are equally important reservoirs as are sandstones, but their development and production present geologists and

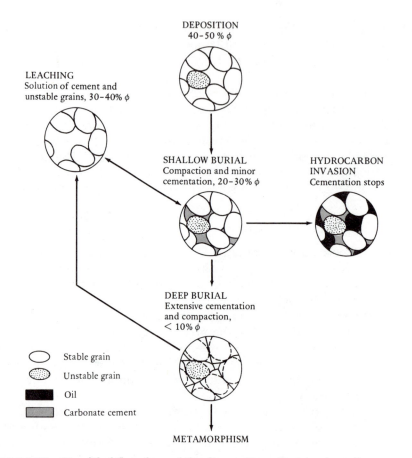

DEPOSITION
40-50 % ϕ

LEACHING
Solution of cement and
unstable grains, 30-40% ϕ

SHALLOW BURIAL
Compaction and minor
cementation, 20-30% ϕ

HYDROCARBON
INVASION
Cementation stops

DEEP BURIAL
Extensive cementation
and compaction,
< 10% ϕ

Stable grain

Unstable grain

Oil

Carbonate cement

METAMORPHISM

FIGURE 6.27   *Simplified flowchart of the diagenetic pathways of sandstones.*

engineers with a different set of problems. Silica is chemically more stable
than calcite. Thus the effects of diagenesis are more marked in limestones
than in sandstones. With sandstone reservoirs the main problem is to establish the original sedimentary variations, working out the depositional
environment and paleogeography and using these to predict reservoir
variation as the field is drilled. The effects of diagenesis are generally sub-
ordinate to primary porosity variations, which is seldom true of carbonate
reservoirs. When first deposited, carbonate sediments are highly porous
and permeable and are inherently unstable in the subsurface environment.
Carbonate minerals are thus dissolved and reprecipitated to form limestones whose porosity and permeability distribution are largely secondary
in origin and often unrelated to the primary porosity. Thus with carbonate
reservoirs facies analysis may only aid development and production in

those rare cases where the diagenetic overprint is minimal. Having discussed the general effects of diagenesis on carbonate reservoirs, it is now appropriate to consider them in more detail.

Major texts covering this topic have been published by Bathurst (1975), Wilson (1975), Chilingar et al., (1972), and Reeckmann and Friedman (1982). Shorter accounts that specifically relate carbonate diagenesis to porosity evolution have been given by Purser (1978) and Longman (1980). The following description is based on these references, concentrating on those aspects of diagenesis that significantly affect reservoir quality. Carbonate reefs, sands, and muds will be considered separately.

Diagenesis and Petrophysics of Reefs

Reefs (used in the broadest sense of the term) are unique among the sediments because they actually form as rock rather than from the lithification of unconsolidated sediment. Modern reefs are reported with porosities of 60 to 80 percent. Because reefs are formed already lithified, they do not undergo compaction as do most sediments. Porosity may thus be preserved. However, porosity is only likely to be preserved when the reef is maintained in a static fluid environment, such as when rapidly buried by impermeable muds. Where a steady flow of alkaline connate water passes through the reef, porosity may gradually be lost by the formation of a mosaic sparite cement (Plate 6). Where acid meteoric water flows through the reef, the reef may be leached and its porosity enhanced.

Before a reef is finally buried, sea level may have fluctuated several times. Thus several diagenetic phases of both cementation and solution may have been superimposed on the primary fabric and porosity of the reef. It is not surprising, therefore, that there may be no correlation between the reservoir units that an engineer delineates in a reef trap and the facies defined by geologists. This phenomenon is illustrated by the Intisar reefs of Libya, described in Chapter 7 (p. 309).

Diagenesis and Petrophysics of Lime Sands

Like reefs, lime sands may have high primary porosities (Enos and Sawatsky, 1979). Because they are unconsolidated, however, they begin to lose porosity quickly upon burial. Mud compacts and shells fracture as the overburden pressure increases. Like reefs, lime sands may be flushed by various fluids. Acid meteoric waters may leach lime sands and increase porosity, whereas alkaline fluids may cause them to become cemented. Early cementation takes various distinctive forms, which may be recognized microscopically. These forms include fibrous, micritic, and microcrystalline cements formed in the spray zone (beach rock) and on the sea floor (hard ground). These early cements diminish porosity somewhat and may diminish permeability considerably where they bridge pore throats. However, early cementation may enhance reservoir potential because it lithifies the sediment, and, as with reefs, the effect of compaction is then diminished. Later diagenesis depends on the fluid environment. Acidic meteoric water may enhance porosity by leaching; alkaline connate water may cause a mosaic of sparite crystals to infill the pores. Where there is no fluid flow, porosity may be preserved and hydrocarbons may invade the reservoir.

Carbonate sands that have not undergone early shallow cementation have less chance of becoming reservoirs. As the sands are buried, they lose porosity rapidly by compaction; although again they are susceptible to secondary leaching when invaded by acid meteoric water during later epidiagenesis. Figure 6.28 summarizes the diagenetic pathways for carbonate sands.

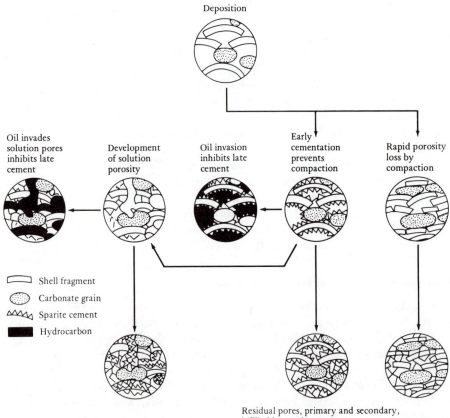

FIGURE 6.28   *Simplified flowchart of the diagenetic pathways of lime sand.*

Diagenesis and Petrophysics of Lime Muds

Recent carbonates are largely composed of the mineral aragonite, the orthorhombic variety of calcium carbonate. Aragonite is unstable in the subsurface, where it reverts to the hexagonal isomorph calcite. This polymorphic reaction causes an 8 percent increase in bulk volume, which results in a loss of porosity. Coupled with compaction, this porosity loss means that most ancient lime mudstones are hard, tight, splintery rocks. They have no reservoir potential unless fractured or dolomitized.

A significant exception to this general rule occurs when the lime mud is calcitic rather than aragonitic. The Coccolithic oozes, which occur in Cretaceous and younger rocks, are of original calcitic composition. Lacking the aragonite-calcite reaction, these lime muds remain as friable chalks with porosities of 30 to 40 percent. Ordinarily, they have very low permeabilities because of their small pore diameters. Chalks are not usually considered to be reservoirs and may even act as cap rocks. In certain circumstances, however, they can act as reservoirs. For example, fractures may enhance their permeability as in the Austin Chalk (U. Cretaceous) of Texas. In the similar Cretaceous Chalk of the North Sea, reservoirs occur in the Ekofisk and associated fields of offshore Norway (see p. 299). Here overpressure has inhibited porosity loss by compaction, whereas fracturing over salt domes has increased permeability (Scholle, 1977). Because of their uniformity, porosity gradients of lime muds can be constructed analogously to those drawn for sandstones (Scholle, 1981). These porosity gradients can seldom be prepared for the more heterogeneous calcarenites, although there are exceptions (see Schmoker and Halley, 1982).

Dolomite Reservoirs

The second main group of carbonate reservoirs include the dolomites. Dolomite can form from calcite (dolomitization) and vice versa (dedolomitization, or calcitization) as shown in the following reaction:

$$2CaCO_3 + Mg^{2+} \rightleftharpoons CaCO_3MgCO_3 + Ca^{2+}$$

The direction in which this reaction moves is a function not only of the Mg-Ca ratio but also of salinity (Fig. 6.29).

A distinction is made between primary and secondary dolomites. Primary dolomites are generally bedded and often form laterally continuous units. Characteristically, they occur in *sabkha* (salt marsh) carbonate sequences, passing down into fecal pellet lagoonal muds and up into supratidal evaporites. Such dolomites are generally cryptocrystalline, often with a chalky texture. They may thus be porous, but, because of their narrow pore diameter, lack permeability. Examples of primary dolomites in sab-

FIGURE 6.29 *Graph of
Mg-Ca ratio plotted against
salinity showing stability
fields of aragonite, calcite,
and dolomite. Note how
dolomitization may be
caused by a drop in salinity
even for low Mg-Ca ratios.
(After Folk and Land, 1974.)*

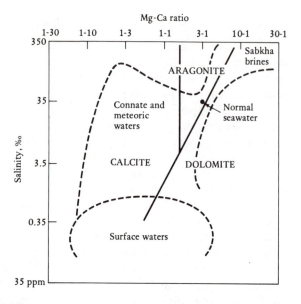

kha cycles have been described from the Jurassic Arab-Darb Formation of
Abu Dhabi (Wood and Wolfe, 1969). Such dolomites are believed to be
primary, or at least penecontemporaneous, although the exact chemistry
of their formation is still debated.

Secondary dolomites cross-cut bedding often occurring in irregular
lenses or zones frequently underlying unconformities or forming enve-
lopes around faults and fractures (Fig. 6.30). These secondary dolomites
are commonly crystalline and sometimes possess a friable saccharoidal
texture (Plate 7). Porosity is of secondary intercrystalline type and may
exceed 30 percent. Unlike primary dolomites, secondary dolomites are
permeable. When calcite is replaced by dolomite, bulk volume reduces by
some 13 percent, and hence porosity increases correspondingly. Thus secondary dolomites are often important reservoirs, as in the Devonian reefs
of Alberta (Toomey et al., 1970). Regionally extensive subunconformity
dolomites occur, for example, on the Wisconsin and San Marcos arches
(Badiozamani, 1973; Rose, 1972).

Carbonate Diagenesis and Petrophysics: Summary

The preceding section gave a simplified account of the complexities of carbonate diagenesis and porosity evolution. Many scholarly studies of these
topics have been carried out in universities and oil companies, but even in
the latter it is always germane to ask what contribution a particular study
has made to the *prediction* of reservoir quality. Most carbonate diagenetic
studies tend to be essentially descriptive.

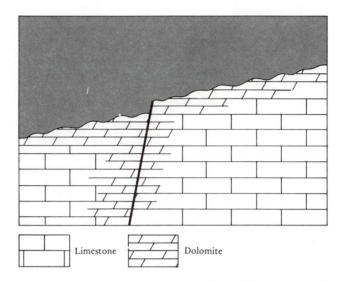

FIGURE 6.30 *Cross-section showing how secondary dolomites are often related to faults and unconformities.*

Jardine et al., (1977) and Wilscn (1981) have reviewed the relationships between carbonate reservoir quality, facies, and diagenesis. Wilson (1981) recognizes the following six major carbonate reservoir types, ranging from those whose porosity is largely facies controlled to those with extensive diagenetic overprints.

1. Shoals with primary porosity still preserved; for example, the Arab D of Saudi Arabia and the Smackover of Louisiana.

2. Buildups with primary porosity preserved; for example, some Alberta Devonian reefs and the Golden Lane atoll of Mexico.

3. Forereef talus with primary porosity preserved; for example, Poza Rica and Reforma fields of Mexico.

4. Pinchout traps where grainstones with preserved porosity or intercrystalline dolomitic porosity are sealed up-dip by sabkha evaporites; for example, the San Andres of West Texas.

5. Subunconformity reservoirs. Porosity is largely secondary due to solution, fracturing, and dolomitization; for example, the Natih and Fahud fields of Oman.

6. Calcitic chalks with permeability enhanced by fractures; for example, the Ekofisk and associated fields of the North Sea.

Finally, Figure 6.31 shows the relationship between porosity, permeability, and the various types of pore found in carbonate and other reservoirs.

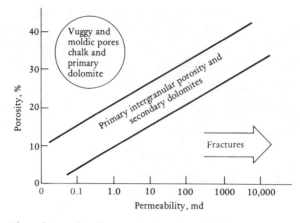

FIGURE 6.31   *The relationship between porosity and permeability for various types of rock and pore systems.*

Atypical and Fractured Reservoirs

Some 90 percent of the world's oil and gas occur in sandstone or carbonate reservoirs. The remaining 10 percent occur in what may therefore be termed atypical reservoirs, which range from various types of basement to fractured shale. As discussed earlier, theoretically any rock can be a petroleum reservoir if it is both porous and permeable. Atypical reservoirs may form by two processes: weathering and fracturing.

The role of weathering in the formation of solution porosity in sandstones and carbonate has already been noted (p. 246). Weathering of certain other rocks has a similar effect. Minerals weather at different rates; so polyminerallic crystalline rocks can form a porous veneer as the unstable minerals weather out, leaving a granular porous residue of stable mineral grains. This situation usually occurs in granites and gneisses, where the feldspars leach out to leave an unconsolidated quartz sand. This *granite wash* is a well site geologist's nightmare. The transition from arkose via in situ granite wash to unweathered granite may be long and gentle.

A number of fields produce from weathered basement reservoirs. In the Panhandle-Hugoton field of Texas and Oklahoma the productive granite wash zone is some 70 m thick, although this zone ranges from red shale, via arkose, into fractured granite (Pippin, 1970). The Augila field of the Sirte basin, Libya, also produces from weathered and fractured granite. Some wells in this field flowed at more than 40,000 BOPD (Williams, 1972). The Long Beach and other fields of California produce from fractured Franciscan (Jurassic) schists (Truex, 1972).

Fracturing can turn any brittle rock into a reservoir (p. 224). Porous but impermeable rocks can be rendered permeable by fracturing, as, for example, in chalk and shale. Even totally nonporous rocks may become reservoirs by fracturing. Fractures can be identified from wireline logs, from visual inspection of cores, and from production tests. Parameters of great importance include fracture intensity, which controls porosity and permeability, and fracture orientation, which affects the isotropy of the reservoir (Pirson, 1978, pp. 180–206).

Fracture intensity may be expressed by the fracture intensity index (FII):

$$FII = \frac{(\phi_t - \phi_m)}{(1 - \phi_m)}$$

where ϕ_t = total porosity
       ϕ_m = porosity not due to fractures

Fractured reservoirs present particular problems of production, the main danger being that oil in the fractures may be rapidly produced and replaced by water, thus preventing recovery from the rest of the pores (Aguilera and Van Poollen, 1978; Reiss, 1980). Examples of reservoirs that produce almost entirely from fractures include the Miocene Monterey cherts of California, which produce oil in the Santa Maria and other fields. Recovery rates are low, in the order of 15,400 barrels per acre (Eggleston, 1948; Regan and Hughes, 1949). Shales, which are generally only cap rocks, may themselves act as reservoirs when fractured.

Some shale source rocks produce free oil from fractures with no associated water. An example of this phenomenon is the Cretaceous Pierre shale of the Florence field, Colorado (McCoy et al., 1951). Production is erratic, although one well in this field is reported to have produced one and a half million barrels of oil. Gas production from fractured shale is more common, with many fields producing from Paleozoic shales in Kentucky, Kansas, and elsewhere in the eastern United States.

Further details on anomalous and fractured reservoirs can be found in Dott and Reynolds (1969), Reiss (1980), and Chung-Hsiang P'An (1982).

RESERVOIR CONTINUITY

Once an oil or gas field has been discovered, its reserves and the optimum method for recovering them must be established. A detailed knowledge of reservoir continuity is a prerequisite for solving both these problems. Few traps contain reservoirs that are uniform in thickness, porosity, and permeability; most are heterogeneous to varying degrees. Thus a reservoir is commonly divided into the gross pay and the net pay intervals. The gross pay is the total vertical interval from the top of the reservoir down to the petroleum-water contact. The net pay is the cumulative vertical thickness from which petroleum may actually be produced. In many fields the net pay may be considerably less than the gross pay. The difference between gross and net pay is due to two factors: the primary porosity with which the reservoir was deposited and the diagenetic processes that have destroyed this original porosity by cementation or enhanced it by solution.

Many attempts have been made to quantify reservoir continuity. These efforts have largely been directed toward sand body geometry, where the effects of diagenesis are generally minimal.

Lateral Continuity

Schemes to classify and quantify the lateral continuity of sands have been produced by Potter (1962), Polasek and Hutchinson (1967), Pryor and Fulton (1976), and Harris and Hewett (1977). Figure 6.32 is based on the scheme proposed by Potter (1962). Two major groups of sand may be recognized according to their lateral continuity: sheets and elongate bodies. Sheets are more or less continuous with length-width ratios of 1-1.

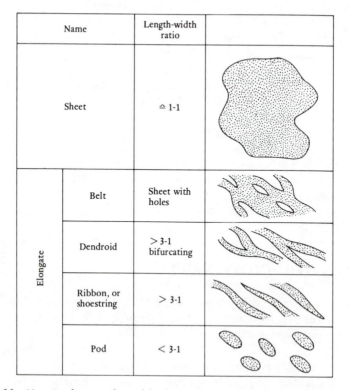

Name		Length-width ratio	
Sheet		≏ 1-1	
Elongate	Belt	Sheet with holes	
	Dendroid	> 3-1 bifurcating	
	Ribbon, or shoestring	> 3-1	
	Pod	< 3-1	

FIGURE 6.32   *Nomenclature of sand body geometry. (After Potter, 1962.)*

Sheet sands occur in many environments, ranging from turbidite fans and crevasse-splays to coalesced channel sands in braided alluvial plains. Sheets may be discontinuous because of nondeposition or later erosion, with the sands being locally replaced by shales. Discontinuous sheets grade into belt sands, which, although still extensive, contain many local elongate vacuoles.

Of the elongate sands with length-width ratios greater than 3-1, the best known are ribbons, or shoestrings (Rich, 1923), which are generally deposited in barrier bar environments. Dendroids are bifurcating shoestrings and include both fluvial tributary channel sands and deltaic and submarine fan distributary channels. Pods are isolated sands with length-width ratios of less than 3-1. They include some tidal current sands and some eolian dunes.

Vertical Continuity

Considered vertically, sands may be differentiated into isolated and stacked reservoirs (Harris and Hewitt, 1977). Stacked sands are those that are in continuity with one another either laterally or vertically. The latter are often referred to as multistorey sands (Fig. 6.33).

Pryor and Fulton (1976) have produced a more rigorous analysis of sand body continuity with a lateral continuity index (LCI) and a vertical continuity index (VCI). The LCI is calculated by constructing a series of cross-sections, measuring maximum and minimum sand body continuities, and dividing that length by the length of the section. The VCI is calculated by measuring maximum sand thickness for each well and dividing it into the thickness of the thickest sand. Pryor and Fulton (1976) applied these indices to detailed core data on the Holocene sands of the Rio Grande. Table 6.2 summarizes their data.

Many fields have reservoirs with alternating productive sand and nonproductive shale. In these cases detailed correlation is essential both to

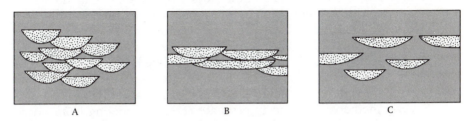

FIGURE 6.33 *Descriptive terms for vertical sand body continuity: (A) vertically stacked (multistorey); (B) laterally stacked; (C) isolated. (After Harris and Hewitt, 1977.)*

TABLE 6.2   Maximum lateral continuity indices (LCI) and vertical continuity indices (VCI) for Holocene sand bodies of the Rio Grande delta.

| Environment | Average LCI | | Average VCI | Average sand thickness, m |
	Perpendicular	Parallel		
Fluvial	0.49	0.83	0.75	5–7
Fluviomarine	1.0	1.0	0.84	7–8
Pro-delta	0.3	0.17	0.56	1.5

From Pryor and Fulton, 1976. Reprinted with permission.

select the location of wells during development drilling and to predict how fluids will move as the field is produced. These fluids include not only naturally occurring oil and gas but any water or gas that may be injected to maintain pressure and enhance production.

Figure 6.34 illustrates just how hetereogeneous a reservoir can be. The Handil field of Indonesia is essentially an anticlinal trap with some faulting. The gross pay is some 1600 m thick; however, it is split up into many separate accumulations with their own gas caps and oil-water contacts. Separate accumulations are partly due to faults, but also occur because the formation has a sand-shale ratio of 1-1, with individual units being 5 to 20 m thick. Subsurface facies analysis using cores and logs (as discussed in Chapter 3, p. 79) shows that the productive sands include discontinuous sheets and pods of marine sands interbedded with, and locally cut into by, channel shoestrings (Verdier et al., 1980). This arrangement is not an isolated case; many other shallow marine reservoirs exhibit a similar complexity, notably those of the Niger delta (Weber, 1971).

RESERVE CALCULATIONS

Estimates of possible reserves in a new oil or gas field can be made before a trap is even drilled. The figures used are only approximations, but they may give some indication of the economic viability of the prospect. As a proven field is developed and produced, its reserves are known with greater and greater accuracy until they are finally depleted. The calculation of reserves is more properly the task of the petroleum engineer, but since this task is based on geological data, it deserves consideration here.

Several methods are used to estimate reserves, ranging from crude approximations made before a trap is tested to more sophisticated calculations as hard data become available.

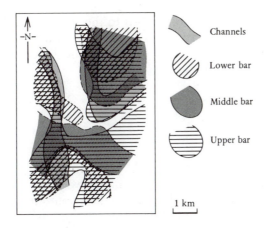

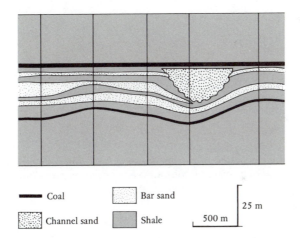

FIGURE 6.34 *Map and cross-section of part of the Handil field, Indonesia, indicating the degree of vertical and lateral continuity of reservoirs found in shallow fluviomarine sands. (After Verdier et al., 1980.)*

Preliminary Volumetric Reserve Calculations

A rough estimate of reserves prior to drilling a trap can be calculated as follows:

$$\text{Recoverable oil reserves (bbl)} = Vb \cdot F$$

where Vb = bulk volume
F = recoverable oil (bbl/acre-ft)

This formula assumes that the trap is full to the spill point. The bulk volume is calculated from the area, and closure is estimated from seismic data. A planimeter is used to measure the area of the various contours (Fig. 6.35). The volume is then calculated as follows:

$$V = h\left(\frac{a_o}{2} + a_1 + a_2 + a_3 + \cdots + a_{n-1} + \frac{a_n}{2}\right)$$

where V = volume
   h = contour interval
   a_o = area enclosed by the oil-water contact
   a_1 = area enclosed by the first contour
   a_n = area enclosed by the nth contour

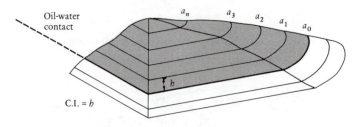

FIGURE 6.35 *Isometric sketch of a trap showing how a planimeter survey measures the areas above selected contours. For additional explanation see text.*

Obviously, the closer the measured contour intervals, the more accurate the end result.

The recoverable oil per acre-foot is the most difficult figure to assess unless local information is available from adjacent fields. It may be calculated according to an estimate of the average porosity of the reservoir (say 10 to 30 percent) and an estimate of the recovery factor (generally, 30 percent or more for sands and 10 to 20 percent for carbonate reservoirs). The recovery factor will vary according to well spacing, reservoir permeability, fluid viscosity, and the effectiveness of the drive mechanism. Generally, several hundred bbl/acre-ft of oil may be recovered.

This method of calculating reserves is very approximate, but may be the only one available before bidding for acreage. It may also be useful for ranking the order in which to drill a number of prospects.

Postdiscovery Reserve Calculations

Once a field has been discovered, accurate reservoir data become available and a more sophisticated formula may be applied:

$$\text{Recoverable oil (bbl)} = \frac{7758 V \phi (1 - S_w) R}{FVF}$$

where
V = the volume (area $\times$ thickness)
7758 = conversion factor from acre feet to barrels
ϕ = porosity (average)
S_w = water saturation (average)
R = recovery factor (estimated)
FVF = formation volume factor

These variables will now be discussed in more detail. As wells are drilled on the field, the seismic interpretation becomes refined so that an accurate structure contour map can be drawn. Log and test data establish the oil-water contact and hence the thickness of the hydrocarbon column (Dahlberg, 1979).

The porosity is calculated from wireline logs calibrated from core data, and the water saturation is calculated from the resistivity logs. The recovery factor is hard to estimate even if the performances of similar reservoirs in adjacent fields are available. Approximate values have been given previously.

The formation volume factor (FVF) converts a stock tank barrel of oil to its volume at reservoir temperatures and pressures. It depends on oil composition, but this dependence can generally be approximated by calculating the FVF's dependence on the solution gas-oil ratio (GOR) and oil density (API gravity). The formation volume factor ranges from 1.08 for low GORs and heavy crudes to values of more than 2.0 for volatile oils and high GORs.

The gas-oil ratio is calculated from the following equation:

$$\text{Gas-oil ratio (in the reservoir)} = \frac{Q_g}{Q_o} = \frac{\mu_o K_g}{\mu_g K_o}$$

where Q = flow rate   } At reservoir temperatures
μ = viscosity   } and pressures
K = effective permeability
g = gas
o = oil

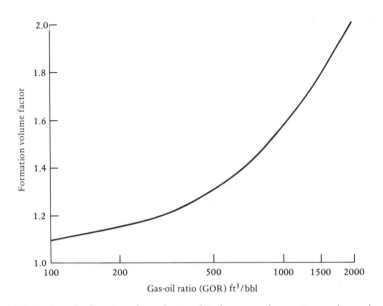

FIGURE 6.36 *Graph showing the relationship between formation volume factor and gas-oil ratio.*

Figure 6.36 shows the relationship between formation volume factor and gas-oil ratio. A more accurate measurement of FVF can be made in the laboratory. Ideally, a sample of the reservoir fluid is collected in a pressure bomb and reheated to the reservoir temperature. Temperature and pressure are reduced to surface temperature and pressure so that the respective volumes of gas and oil can be measured accurately. This procedure is referred to as pressure-volume-temperature (PVT) analysis.

As a field is produced, several changes take place in the reservoir. Pressure drops and the flow rate diminishes. The gas-oil ratio may also vary, although this depends on the type of drive mechanism, as discussed in the next section. These changes are sketched in Figure 6.37. The changes may be formulated in the material balance equation, which, as its name suggests, equates the volumes of oil and gas in the reservoir at virgin pressure with those produced and remaining in the reservoir at various stages in its productive life. At its simplest level, the material balance equation is a variant of the law of conservation of mass:

$$
\begin{matrix}
\text{Weight of} \\
\text{hydrocarbons} \\
\text{originally} \\
\text{in reservoir}
\end{matrix}
=
\begin{matrix}
\text{weight of} \\
\text{produced} \\
\text{hydrocarbons}
\end{matrix}
+
\begin{matrix}
\text{weight of} \\
\text{hydrocarbons} \\
\text{retained} \\
\text{in reservoir}
\end{matrix}
$$

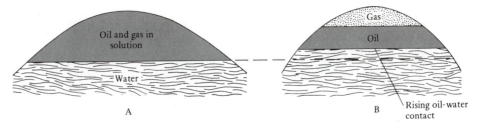

FIGURE 6.37 *The changes within a reservoir that may be caused by production. (A) The situation at virgin pressure before production. (B) The situation with depleted pressure after commencement of production.*

If all the weights are expressed as stock tank (ST) barrels of oil and stock tank cubic feet (SCF) of gas, then weight can be substituted by volume.

The mass balance equation has several complex variations. Although its solution is a job for the petroleum engineer, much of the data on which it is based are provided by geologists.

PRODUCTION METHODS

As mentioned in the review of the mass balance equation in the previous section, hydrocarbons may be produced by several mechanisms. Three natural drive mechanisms can cause oil or gas to flow up the well bore: water drive, gas drive, and gas solution (or dissolved gas) drive.

Water Drive

In water drive production, oil or gas trapped within a reservoir may be viewed as being sealed within a water-filled U-tube (Fig. 6.38). When the tap is opened, oil and gas will flow from the reservoir because of the hydrostatic head of water (Fig. 6.39). Not all aquifers have a continuous recharge of the earth's surface. The degree of water encroachment and pressure maintenance will depend on the size and productivity of the aquifer. Note that as the field is produced, water invades the lower part of the trap to displace the oil (Fig. 6.40); only in the most uniform reservoirs does the oil-water contact rise evenly. Because adjacent beds seldom have the same permeability, water encroachment advances at different rates, giving rise to fingering. A further complication may be caused by coning of the water adjacent to boreholes. The extent of coning depends on the rate of production and the ratio between vertical and horizontal permeability.

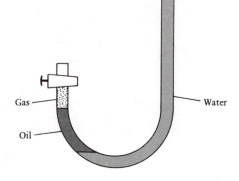

When a field with a water drive mechanism is produced, the reservoir pressure drops in inverse proportion to the effectiveness of the recharge from the aquifer. Generally, little change occurs in the gas-oil ratio. With an effective water drive, the flow rate remains constant during the life of the fluid, but oil production declines inversely with an increase in water production. These trends are illustrated in Figure 6.41. The water drive mechanism is generally the most effective, with a recovery factor of up to 60 percent.

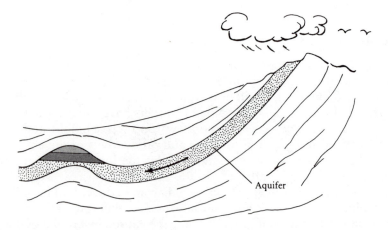

FIGURE 6.39   *A typical geological setting of water drive mechanism. As the field is*
produced, petroleum is displaced by water from the aquifer.

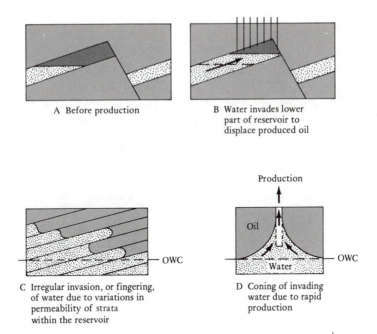

A Before production

B Water invades lower
part of reservoir to
displace produced oil

C Irregular invasion, or fingering,
of water due to variations in
permeability of strata
within the reservoir

D Coning of invading
water due to rapid
production

FIGURE 6.40 *Water drive mechanism before production (A) and during production (B). (C) and (D) illustrate potential irregularities in the rising oil-water contact.*

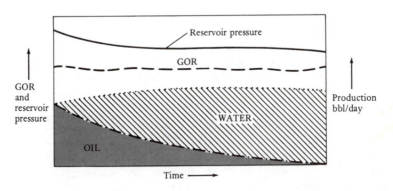

FIGURE 6.41 *Typical production history for a water drive field. For explanation see text.*

Gas Cap Drive

A second producing mechanism is the gas cap drive, in which the field contains both oil and gas zones. As production begins, the drop in pressure causes gas dissolved in the oil to come out of the solution. This new gas moves up to the gas cap and, in so doing, expands to occupy the pores vacated by the oil. A transitional zone of degassing thus forms at the gas-oil contact. Drawdown zones may develop adjacent to boreholes in a manner analogous to, but the reverse of, coning at the oil-water contact (Fig. 6.42).

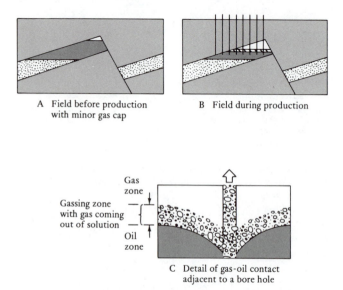

A   Field before production   B   Field during production
    with minor gas cap

C   Detail of gas-oil contact
    adjacent to a bore hole

FIGURE 6.42   *The gas expansion drive mechanism. A transitional zone develops at the gas-oil contact as pressure drops and the gas separates out from the oil. Note the drawdown effect, which may develop adjacent to boreholes (C).*

The production history of gas cap drive fields is very different from that of water drive fields. Pressure and oil production drop steadily, while the ratio of gas to oil naturally increases (Fig. 6.43). The gas cap drive mechanism is generally less effective than the water drive mechanism, with a recovery factor of 20 to 50 percent.

Dissolved Gas Drive

The third type of production mechanism is the dissolved gas drive, sometimes called the solution gas drive. This type of drive occurs in oil fields that initially have no gas cap. Production is analogous to a soda fountain.

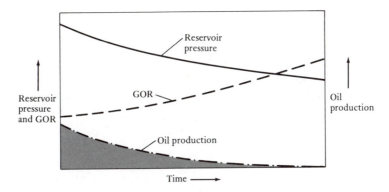

FIGURE 6.43 *The typical production history of a field with a gas drive mechanism. For explanation see text.*

As production begins, pressure drops and gas bubbles form in the oil and expand, forcing the oil out of the pore system and toward the boreholes. As the gas expands, it helps to maintain reservoir pressure. Initially, the gas bubbles are separated. As time passes, they come together and may form a continuous free-gas phase, which may accumulate as a gas cap in the crest of the reservoir. This point is termed the *critical gas saturation*, and care should be taken to prevent it from being reached. It may be avoided either by maintaining a slow rate of production or by reinjecting the produced gas to maintain the original reservoir pressure. Figure 6.44 illustrates what happens in a reservoir with a gas expansion drive, and Figure 6.45 shows the typical production history. This drive mechanism is generally considered to be the least efficient of the three, with a recovery factor in the range of 7 to 15 percent.

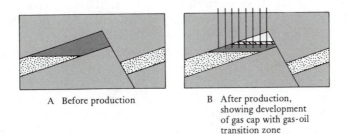

A Before production

B After production, showing development of gas cap with gas-oil transition zone

FIGURE 6.44 *A field with a gas expansion drive mechanism. If the pressure drops sufficiently to cross the critical gas saturation point, then a free-gas cap may form (B).*

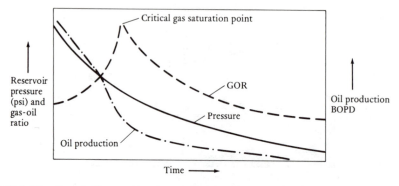

FIGURE 6.45 *Graph illustrating the production history of a field with a gas expansion drive. Note how the GOR increases until the critical gas saturation point is reached; then it drops off sharply as the gas separates out.*

Artificial Lift and Enhanced Recovery

Not all fields produce by natural drive mechanisms, and even these natural drive mechanisms do not recover all the oil. Artificial methods are used to produce oil from fields lacking natural drive, and enhanced recovery methods increase the recoverable reserves.

A well will flow oil to the surface if the static pressure at the bottom of the well exceeds the pressure of the column of mud and the frictional effect of the borehole. In many shallow fields with low reservoir pressure and in depleted fields, the wells will not flow spontaneously. In such instances the oil is pumped to the surface using the nodding donkey or bottom hole pumps described earlier (p. 49).

Once a field becomes depleted, it may be abandoned or subjected to a secondary recovery program to step up production. In due course even a tertiary recovery program may be initiated. Today the practice is to initiate an enhanced recovery program when a field first goes on stream. There are many enhanced recovery techniques, and this rapidly expanding field is beyond the scope of this text (see Van Poollen and Associates, 1980). Some of the methods are briefly discussed as follows.

One of the main objectives of enhanced recovery is to maintain or reestablish the original reservoir pressure. This objective may be accomplished in one of several ways. Gas may be injected, either petroleum gas from the same or adjacent field or naturally occurring or industrially manufactured inert gases, such as carbon dioxide or nitrogen. Alternatively, liquids may be injected; these liquids may be seawater or connate waters from adjacent strata. Careful chemical analysis and treatment of injected waters is essential to prevent and monitor unwanted chemical changes taking place within the reservoir. Some waters damage permeability by precipitating salts in pore spaces and causing clays to swell (Giraud and Neu, 1971). Seawater must be treated to destroy bacteria, since they degrade oil and increase its viscosity. More elaborate enhanced recovery methods involve the injection of detergents (micellar floods) to emulsify heavy oil and move it to the surface. For these and further details of enhanced recovery, see Latil (1980) and Van Poollen and Associates (1980).

SELECTED BIBLIOGRAPHY

For details of the petrography of reservoir rocks, see:

BATHURST, R. G. C. 1975. *Carbonate Sediments and Their Diagenesis*, Second Edition. Amsterdam: Elsevier, 658 pp.

FRIEDMAN, G. H. and SANDERS, J. E. 1978. *Principles of Sedimentology*. New York: Wiley, 792 pp. This major text deals not only with petrography but also with wider aspects of reservoir rocks.

PETTIJOHN, F. J.; POTTER, P. E.; and SIEVER, R. J. 1972. *Sand and Sandstone*. Berlin: Springer-Verlag, 618 pp.

SELLEY, R. C. 1982. *Introduction to Sedimentology*, Second Edition. London: Academic Press, 417 pp.

Friedman and Sanders (1978) deal with carbonates and sandstones in great detail. Pettijohn et al. (1972) and Bathurst (1975) cover sandstones and carbonates in detail, respectively. Selley (1982) covers carbonates and sandstones at a less advanced level, but deals with the application of petrography to porosity and permeability predictions.

For further details on the engineering aspects of reservoirs, see:

CLARK, N. J. 1969. *Elements of Petroleum Reservoirs*. Dallas: Soc. Petrol. Eng., 250 pp.

TIMMERMAN, E. H. 1982. *Practical Reservoir Engineering*, Vols. I and II. Tulsa: Penn Well, 365 and 367 pp.

REFERENCES

AGUILERA, R. and VAN POOLLEN, H. K. 1978. Geologic aspects of naturally fractured reservoirs explained. *Oil & Gas J.*, 18 Dec., 47–51.

AL-GAILANI, M. B. 1981. Authigenic mineralization at unconformities: implication for reservoir characteristics. *Sedim. Geol.*, 29, 89–115.

ALLEN, J. R. L. 1970. The systematic packing of prolate spheroids with reference to concentration and dilatency. *Geol. Mijnb.*, 49, 211–220.

AL-RAWI, M. 1981. Geological interpretation of oil entrapment in the Dubair Formation, Raudhatain Field. *Soc. Petrol. Eng.*, Preprint No. 9591, 149–158.

ANDERSON, G. 1975. *Coring and Core Analysis Handbook*. Tulsa: Petroleum Publishing Co., 200 pp.

ATWATER, G. I. and MILLER, E. E. 1965. The effect of decrease in porosity with depth on future development of oil and gas reserves in South Louisiana. Abstract. *Am. Assoc. Petrol. Geol. Bull.*, *49*, 334.

BADIOZAMANI, K. 1973. The Dorag dolomitization model-application to the Middle Ordovician of Wisconsin. *J. Sedim. Petrol.*, *43*, 965–984.

BALDUCCI, A. and POMMIER, G. 1970. Cambrian oil field of Hassi Messaoud, Algeria. In: *Geology of Giant Petroleum Fields*. M. T. Halbouty (ed.). Tulsa: Am. Assoc. Petrol. Geol., Mem. No. 14, 477–488.

BATHURST, R. G. C. 1975. *Carbonate Sediments and Their Diagenesis*, Second Edition. Amsterdam: Elsevier, 658 pp.

BEARD, D. C. and WEYL, P. K. 1973. The influence of texture on porosity and permeability of unconsolidated sand. *Am. Assoc. Petrol. Geol. Bull.*, *57*, 349–369.

BLATT, H.; MIDDLETON, G.; and MURRAY, R. 1980. *Origin of Sedimentary Rocks*, Second Edition. New Jersey: Prentice-Hall, 782 pp.

BOTSET, H. G. 1931. The measurement of permeability of porous alundum discs of water and oils. *Rev. Sci. Instr.*, *2*, 84–95.

BRADY, T. J.; CAMPBELL, N. D. J.; and MAHER, C. E. 1980. Intisar 'D' oil field, Libya. In: *Giant Oil and Gas Fields of the Decade: 1968–1978*. M. T. Halbouty (ed.). Tulsa: Am. Assoc. Petrol. Geol., 543–564.

BUTLER, M.; PHELAN, M. J.; and WRIGHT, A. W. 1976. The Buchan field, evaluation of a fractured sandstone reservoir. *Trans. 4th Europ. Symp.* London: Soc. Prof. Well Log Anal., 18 pp.

CALHOUN, J. C. 1960. *Fundamentals of Reservoir Engineering*. Norman: Univ. of Oklahoma, 426 pp.

CHILINGAR, G. V. 1964. Relationship between porosity, permeability and grainsize distribution of sands and sandstones. In: *Deltaic and Shallow Marine Deposits*. L. M. J. U. Van Straaten (ed.). Amsterdam: Elsevier, 71–75.

CHILINGAR, G. V.; MANNON, R. W.; and RIEKE, H. 1972. *Oil and Gas Production from Carbonate Rocks*. Amsterdam: Elsevier, 408 pp.

CHILINGARIAN, G. V. and WOLF, K. H. 1975. *Compaction of Coarse-Grained Sediments*, vol. 1. Amsterdam: Elsevier, 552 pp.

CHOQUETTE, P. W. and PRAY, L. C. 1970. Geologic nomenclature and classification of porosity in sedimentary carbonates. *Am. Assoc. Petrol. Geol. Bull.*, *54*, 207–250.

CHUNG-HSIANG P'AN. 1982. Petroleum in basement rocks. *Am. Assoc. Petrol. Geol. Bull.*, *66*, 1597–1643.

COLLINS, R. E. 1961. *Flow of fluids through porous media*. New York: Reinhold, 406 pp.

CONLEY, C. D. 1971. Stratigraphy and lithofacies of Lower Paleocene rocks, Sirte Basin, Libya. In: *Symposium on the Geology of Libya*. C. Gray (ed.). Tripoli: Univ. of Libya, 127–140.

DAHLBERG, E. C. 1979. Hydrocarbon reserve estimation from contour maps. *Can. Petrol. Geol. Bull.*, *27*, 94–99.

DAKE, L. P. 1978. *Fundamentals of Reservoir Engineering*. Amsterdam: Elsevier, 444 pp.

DARCY, H. 1856. *Les fontaines publiques de la Ville de Dijon*. Paris: V. Dalmont, 674 pp.

DICKEY, P. A. 1979. *Petroleum Development Geology.* Tulsa: Petroleum Publishing Co., 398 pp.

DODGE, M. M. and LOUCKS, R. G. 1979. Mineralogic composition and diagenesis of Tertiary sandstones along Texas Gulf Coast. *Am. Assoc. Petrol. Geol. Bull., 63,* 440.

DOTT, R. H. and REYOLDS, M. J. 1969. *Source book for petroleum geology.* Tulsa: Am. Assoc. Petrol. Geol., Mem. No. 5, 471 pp.

EGGLESTON, W. S. 1948. Summary of oil production from fractured rock reservoirs in California. *Am. Assoc. Petrol. Geol. Bull., 32,* 1352–1355.

ENOS, P. and SAWATSKY, L. H. 1979. Pore space in Holocene carbonate sediments. *Am. Assoc. Petrol. Geol. Bull., 63,* 445.

FAIRBRIDGE, R. 1967. Phases of diagenesis and authigenesis. In: G. Larsen and G. V. Chilingar (eds.). *Diagenesis in Sediments.* Amsterdam: Elsevier, 19–89.

FISCHER, A. G. 1964. The Lofer cyclothems of the Alpine Triassic. In: *Symposium on Cyclic Sedimentation.* D. Merriam (ed.). *Kans. Univ. Geol. Surv. Bull., 169,* 107–150.

FOLK, R. L. and LAND, L. S. 1974. Mg/Ca ratio and salinity: two controls over crystallization of dolomite. *Am. Assoc. Petrol. Geol. Bull., 59,* 60–68.

FRASER, H. J. 1935. Experimental study of the porosity and permeability of clastic sediments. *J. Geol., 43,* 910–1010.

FRIEDMAN, G. M. and SANDERS, J. E. 1978. *Principles of Sedimentology.* New York: Wiley, 792 pp.

FUCHTBAUER, H. 1967. Influence of different types of diagenesis on sandstone porosity. *Proc. 7th World Petrol. Cong.,* 353–369.

GAITHOR, A. 1953. A study of porosity and grain relationships in sand. *J. Sedim. Petrol., 23,* 186–195.

GIRAUD, A. and NEU, J. M. 1971. Injection d'eau: interaction entre l'eau injectee et le reservoir. *Revue Institute Francaise du Petrole, 26,* 1009–1028.

GRATON, L. C. and FRASER, H. J. 1935. Systematic packing of spheres, with particular reference to porosity and permeability. *J. Geol., 43,* 785–909.

HANCOCK, N. 1978. Diagenetic modelling in the middle Jurassic Brent Sand of the northern North Sea. London: *Europ. Offsh. Petrol. Conf.,* 275–280.

HARRIS, D. G. and HEWITT. 1977. Synergism in reservoir management. The geologic perspective. *J. Petrol. Tech.,* July, 761–770.

HARRIS, J. F.; TAYLOR, G. L.; and WALPER, J. L. 1960. Relation of deformational fractures in sedimentary rocks to regional and local structure. *Am. Assoc. Petrol. Geol. Bull., 44,* 1853–1873.

HAY, J. T. C. 1977. The Thistle oilfield. In: *Mesozoic Northern North Sea Symposium.* Oslo: Norweg. Petrol. Soc., Paper II, 20 pp.

HEA, J. P. 1971. Petrography of the Paleozoic—Mesozoic Sandstones of the Southern Sirte basin, Libya. In: *The Geology of Libya.* C. Gray (ed.). Tripoli: Univ. of Libya, 107–125.

HEALD, K. C. 1940. Essentials for oil pools. In: *Elements of the Petroleum Industry.* E. L. De Golyer (ed.). Am. Inst. Mining and Metall. Eng., 26–62.

JAMISON, H. C.; BROCKETT, L. D.; and McINTOSH, R. A. 1980. Prudoe Bay: A 10-year perspective. In: *Giant Oil and Gas Fields of the Decade: 1968–1978.* M. T. Halbouty (ed.). Tulsa: Am. Assoc. Petrol. Geol., 289–314.

JARDINE, D.; ANDREWS, D. P.; WISHART, J. W.; and YOUNG, J. W. 1977. Distribution and continuity of carbonate reservoirs. *J. Petrol. Tech.,* July, 873–885.

KAHN, J. S. 1956. The analysis and distribution of the properties of packing in sand size sediments. *J. Geol.*, *64*, 385–395.

KLINKENBERG, L. J. 1941. The permeability of porous media to liquids and gases. *Drill and Prod. Prac.*, 200–213.

KRUMBEIN, W. C. and MONK, G. D. 1942. Permeability as a function of the size parameters of unconsolidated sands. *Am. Inst. Min. and Metal. Eng.*, *Tech. Pub.* 1492, 1–11.

KULANDER, B. R.; BARTON, C. C.; and DEAN, S. L. 1979. Fractographic distinction of coring—induced fractures from natural cored fractures. *Am. Assoc. Petrol. Geol. Bull.*, *63*, 482.

LANGRES, G. L.; ROBERTSON, J. O.; and CHILINGAR, G. V. 1972. *Secondary Recovery and Carbonate Reservoirs*. Amsterdam: Elsevier, 250 pp.

LATIL, M. 1980. *Enhanced Oil Recovery*. London: Graham and Trotman, 252 pp.

LEE, C. H. 1919. Geology and groundwaters of the western part of the San Diego County, California. Washington: *Water Supply Invig. Papers*, 446–121 p.

LEVANDOWSKI, D. W.; KALEY, M. E.; SILVERMAN, S. R.; and SMALLEY, R. G. 1973. Cementation in Lyons Sandstone and its role in oil accumulation, Denver Basin, Colorado. *Am. Assoc. Petrol. Geol. Bull.*, *57*, 2217–2244.

LEVORSEN, A. I. 1967. *The Geology of Petroleum*. Oxford: Freeman, 724 pp.

LONGMAN, M. W. 1980. Carbonate diagenetic textures from nearsurface diagenetic environments. *Am. Assoc. Petrol. Geol. Bull.*, *64*, 461–487.

LOUCKS, R. G.; DODGE, M. M.; and GALLOWAY, W. W. 1979. Reservoir quality in Tertiary sandstones along Texas Gulf Coast. *Am. Assoc. Petrol. Geol. Bull.*, *63*, 488.

MAGARA, K. 1980. Comparison of porosity-depth relationships of shale and sandstone. *J. Petrol. Geol.*, *3*, 175–185.

MARTINI, I. P. 1971. Grain size orientation and paleocurrent systems in the Thorold and Grimsby Sandstones (Silurian), Ontario and New York. *J. Sedim. Petrol.*, *41*, 425–434.

MARTINI, I. P. 1972. Studies of microfabrics: an analysis of packing in the Grimsby Sandstone (Silurian) Ontario and New York State. Montreal: 24. *Internat. Geol. Cong.*, Section 6. *Stratigraphy and Sedimentology*, 415–423.

McCONNELL, P. C. 1951. Drilling and production techniques that yield nearly 850,000 barrels per day in Saudi Arabia's fabulous Abqaiq field. *Oil & Gas J.*, 20 Dec., 197.

McCOY, A. W.; SIELAFF, R. L.; DOWNS, G. R.; BASS, N. W.; and MAXSON, J. H. 1951. Types of oil and gas traps in the Rocky Mountains. *Am. Assoc. Petrol. Geol. Bull.*, *35*, 1000–1037.

MONICARD, R. P. 1981. *Properties of reservoir rocks: core analysis*. Paris: Ed. Technip, 184 pp.

MORROW, N. R. 1971. Small scale packing heterogeneities in porous sedimentary rocks. *Am. Assoc. Petrol. Geol. Bull.*, *55*, 514–522.

MURRAY, R. C. 1960. Origin of porosity in carbonate rocks. *J. Sedim. Petrol.*, *30*, 59–84.

MUSKAT, M. 1937. *Flow of Homogenous Fluids Through Porous Media*. New York: McGraw-Hill, 763 pp.

MUSKAT, M. 1949. *Physical Principles of Oil Production*. New York: McGraw-Hill (reissued by IHRDC, Boston, 1981), 922 pp.

MUSKAT, M. and BOTSET, H. G. 1931.   Flow of gas through porous materials. *Physics 1*, 27–47.

NAGTEGAAL, P. J. C. 1978.   Sandstone-framework instability as a function of burial diagenesis. *J. Geol. Soc. Lond.*, *135*, 101–105.

PHILIP, W.; DROUG, H. J.; HADDENBACH, H. G.; and JANKOWSKY, W. 1964.   The history of migration in the Gifhorn trough (N. W. Germany). Frankfurt: *Proc. 6th World Petrol. Cong.*, 547–742.

PIPPIN, L. 1970.   Panhandle-Hugoton field, Texas-Oklahoma-Kansas—the first fifty years. In: *Geology of Giant Petroleum Fields*. M. T. Halbouty (ed.). Tulsa: Am. Assoc. Petrol. Geol., Mem. No. 14, 204–222.

PIRSON, S. J. 1978.   *Geologic Well Log Analysis*, Second Edition. Houston: Gulf Publishing Co., 385 pp.

POLASEK, T. L. and HUTCHINSON, C. A. 1967.   Characterization of non-uniformities within a sandstone reservoir from a fluid mechanics point of view. *Proc. 7th World Petrol. Cong.*, *2*. Amsterdam: Elsevier, 397–407.

POTTER, P. E. 1962.   Late Mississippian sandstones of Illinois Basin. *Illinois Geol. Surv. Circ.*, 340.

POTTER, P. E. and MAST, R. F. 1963.   Sedimentary structures, sandshape fabrics and permeability, I. *J. Geol.*, *71*, 441–471.

POTTER, P. E. and PETTIJOHN, F. J. 1977.   *Paleocurrents and Basin Analysis*. New York: Springer-Verlag, 425 pp.

POWERS, M. C. 1953.   A new roundness scale for sedimentary particles. *J. Sedim. Petrol.*, *23*, 117–119.

PRYOR, W. A. 1973.   Permeability—porosity patterns and variations in some Holocene Sand Bodies. *Am. Assoc. Petrol. Geol. Bull.*, *57*, 162–189.

PRYOR, W. A. and FULTON, K. 1976.   Geometry of reservoir type sand bodies in the Holocene Rio Grande Delta and comparison with ancient reservoir analogs. *Soc. Petrol. Eng.*, Preprint 7045, 81–92.

PURSER, B. H. 1978.   Early diagenesis and the preservation of porosity in Jurassic limestones. *J. Petrol. Geol.*, *1*, 83–94.

REECKMANN, A. and FRIEDMAN, G. M. 1982.   *Exploration for Carbonate Petroleum Reservoirs*. New York: Wiley, 213 pp.

REGAN, L. J. and HUGHES, A. W. 1949.   Fractured reservoirs of Santa Maria district, California. *Am. Assoc. Petrol. Geol. Bull.*, *35*, 32–51.

REISS, L. H. 1980.   *The Reservoir Engineering Aspects of Fractured Formations*. Paris: Ed. Tecnip, 120 pp.

RICH, J. L. 1923.   Shoestring sands of eastern Kansas. *Am. Assoc. Petrol. Geol. Bull.*, *7*, 103–113.

RITTENHOUSE, G. 1971.   Mechanical compaction of sands containing different percentages of ductile grains: a theoretical approach. *Am. Assoc. Petrol. Geol. Bull.*, *52*, 92–96.

ROBINSON, R. B. 1966.   Classification of reservoir rocks by surface texture. *Am. Assoc. Petrol. Geol. Bull.*, *50*, 547–559.

ROGERS, J. J. and HEAD, W. B. 1961.   Relationship between porosity median size, and sorting coefficients of synthetic sands. *J. Sedim. Petrol.*, *31*, 467–470.

ROSE, P. R. 1972.   Edwards Group, surface and subsurface, central Texas. Austin: *Bur. Econ. Geol. Univ. Tex.*, 198 pp.

SCHEIDDEGGER, A. E. 1960.   *The Physics of Flow Through Porous Media*. New York: MacMillan, 313 pp.

SCHMIDT, V.; McDONALD, D. A.; and PLATT, R. L. 1977. Pore geometry and reservoir aspects of secondary porosity in sandstones. *Can. Soc. Petrol. Geol. Bull., 25,* 271–290.

SCHMOKER, J. W. and HALLEY, R. B. 1982. Carbonate porosity versus depth. A predictable relation for south Florida. *Am. Assoc. Petrol. Geol. Bull., 66,* 2561–2570.

SCHOLLE, P. A. 1977. Chalk diagenesis and its relation to petroleum exploration: oil from chalks, a modern miracle. *Am. Assoc. Petrol. Geol. Bull., 61,* 982–1009.

SCHOLLE, P. A. 1981. Porosity prediction in shallow vs. deepwater limestones. *J. Petrol. Tech.,* Nov., 2236–2242.

SEEMANN, U. 1979. Diagenetically formed interstitial clay minerals as a factor in Rotliegende Sandstone reservoir quality in the North Sea. *J. Petrol. Geol., 1 (3),* 55–62.

SELLEY, R. C. 1978. Porosity gradients in North Sea oil-bearing sandstones. *J. Geol. Soc. Lond., 135,* 119–132.

SELLEY, R. C. 1982. *Introduction to Sedimentology,* Second Edition. London: Academic Press, 417 pp.

SHELTON, J. W. and MACK, D. E. 1970. Grain orientation in determination of paleocurrents and sandstone trends. *Am. Assoc. Petrol. Geol. Bull., 54,* 1108–1119.

SIPPEL, R. F. 1971. Quartz grain orientations, I. (the photometric method). *J. Sedim. Petrol., 41,* 38–59.

SNEIDER, R. M.; RICHARDSON, F. H.; PAYNTER, D. D.; EDDY, R. E.; and WYANT, I. A. 1977. Predicting reservoir rock geometry and continuity in Pennsylvanian reservoirs, Elk City field, Oklahoma. *J. Petrol. Tech., 29,* 851–866.

STALDER, P. J. 1973. Influence of crystallographic habit and aggregate. Structure of authigenic clay minerals on sandstone permeability. *Geol. Mijnb., 52,* 217–220.

STEARNS, D. W. and FRIEDMAN, M. 1972. Reservoirs in fractured rocks. In: *Stratigraphic Oil and Gas Fields—Classification, Exploration Methods, and Case Histories.* Am. Assoc. Petrol. Geol., Mem. No. 16, 82–106.

STORMONT, D. H. 1949. Huge caverns encountered in Dollarhide Field. *Oil & Gas J.* 7 April, 66–68, 94.

TEBBUTT, G. E.; CONLEY, C. D.; and BOYD, D. W. 1965. Lithogenesis of a distinctive carbonate fabric. *Wyoming Univ. Contrib. Geol., 4, (1),* 13 pp.

TOOMEY, D. F.; MOUNTJOY, E. W.; and MACKENZIE, W. S. 1970. Upper Devonian (Frasnian) algae and foraminifera from the Ancient Wall carbonate complex, Jasper National Park, Alberta, Canada. *Can. J. Earth Sci., 7,* 946–981.

TRUEX, J. N. 1972. Fractured shale and basement reservoir Long Beach Unit, California. *Am. Assoc. Petrol. Geol. Bull., 56,* 1931–1938.

VAN POOLLEN and ASSOCIATES. 1980. *Fundamentals of Enhanced Recovery.* Tulsa: Pennwell, 155 pp.

VAN VEEN, F. R. 1975. Geology of the Leman gas-field. In: *Petroleum and the Continental Shelf of North-West Europe,* vol. I. Geology. A. E. Woodland (ed.). Barking: Applied Science Publishers, 223–232.

VERDIER, A. C.; OKI, T.; and ATIK, S. 1980. Geology of the Handil field (East Kalimantan, Indonesia). In: *Giant Oil and Gas Fields of Decade 1968–1978.* M. T. Halbouty (ed.). Am. Assoc. Petrol. Geol., 399–422.

VON RAD, U. 1971.   Comparison between 'magnetic' and sedimentary fabric in graded and cross-laminated sand layers, Southern California. *Geol. Runsch.,* *60,* 331–354.

WARDLAW, N. C. 1976.   Pore geometry of carbonate rocks as revealed by pore casts and capillary pressure. *Am. Assoc. Petrol. Geol. Bull., 60,* 245–257.

WASHBURN, E. W. and BUNTING, E. N. 1922.   Determination of porosity by the method of gas expansion. *J. Am. Cer. Soc., 5,* 48–112.

WEBER, K. J. 1971.   Sedimentological aspects of oil fields in the Niger delta. *Geol. Mijnb., 50,* 559–576.

WEBER, K. J.; EIJPE, R.; LEIJNSE, D.; and MOENS, C. 1972.   Permeability distribution in a Holocene distributary channel-fill near Leerdam (the Netherlands). *Geol. Mijnb., 51,* 53–62.

WILLIAMS, J. J. 1972.   Augila field, Libya: depositional environment and diagenesis of sedimentary reservoir and description of igneous reservoir. In: *Stratigraphic Oil and Gas Fields.* R. E. King (ed.). Am. Assoc. Petrol. Geol., Mem. No. 16, 623–632.

WILSON, J. L. 1975.   *Carbonate Facies in Geologic History.* Berlin: Springer-Verlag, 471 pp.

WILSON, J. L. 1981.   A review of carbonate reservoirs. *Can. Petrol. Geol. Bull., 29,* 95–117.

WOOD, G. V. and WOLFE, M. J. 1969.   Sabkha cycles in the Arab/Darb Formation off the Trucial Coast of Arabia. *Sedimentology, 12,* 165–191.

The Trap

In the early days of oil exploration in the United States, no specific legislation governed the exploration and exploitation of petroleum. Initially, the courts applied the game laws, which stated that oil and gas were *fugacious* (likely to flee away), moving from property to property and ultimately owned by the man on whose land they were trapped (Dott and Reynolds, 1969).

The term *trap* was first applied to a hydrocarbon accumulation by Orton (1889): ". . . stocks of oil and gas might be trapped in the summits of folds or arches found along their way to higher ground." A detailed historical account of the subsequent evolution of the concept and etymology of the term *trap* is found in Dott and Reynolds (1969).

As discussed in Chapter 1, a trap is one of the five essential requisites for a commercial accumulation of oil or gas. Levorsen (1967) gave a concise definition of a trap as "the place where oil and gas are barred from further movement." This definition needs some qualification, however. Explorationists in general and geophysicists in particular search for hydrocarbon traps. Perhaps it would be more accurate to say that they search for potential traps. Only after drilling and testing is it known whether the trap contains oil or gas. In other words a trap is still a trap whether it is barren or productive.

NOMENCLATURE OF A TRAP

Many terms are used to describe the various parameters of a trap. These terms are defined as follows and illustrated with reference to an anticlinal trap, the simplest type (Fig. 7.1). The highest point of the trap is the *crest*, or *culmination*. The lowest point at which hydrocarbons may be contained in the trap is the *spill point*: this lies on a horizontal contour, the *spill plane*. The vertical distance from crest to spill plane is the *closure* of the trap. A trap may or may not be full to the spill plane, a point of both local and regional significance (p. 375). Note that in areas of monoclinal dip the closure of a trap may not be the same as its structural relief (Fig. 7.2). This situation is particularly significant in hydrodynamic traps (p. 319).

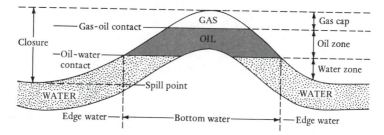

FIGURE 7.1   *Cross-section through a simple anticlinal trap.*

Within the trap the productive reservoir is termed the *pay*. The vertical distance from the top of the reservoir to the petroleum-water contact is termed *gross pay*. This thickness may vary from only one or two meters in Texas to several hundred meters in the North Sea and Middle East. All of the gross pay does not necessarily consist of productive reservoir, however, so gross pay is usually differentiated from net pay. The net pay is the cumulative vertical thickness of a reservoir from which petroleum may be produced. Development of a reservoir necessitates mapping the gross-net pay ratio across the field.

FIGURE 7.2 *Cross-section through a trap illustrating the difference between closure and structural relief.*

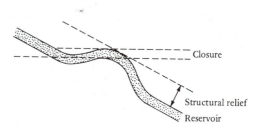

A trap may contain oil, gas, or both. The oil-water contact (commonly referred to as OWC) is the deepest level of producible oil. Similarly, the gas-oil contact (GOC) or gas-water contact (GWC), as the case may be, is the lower limit of producible gas. The accurate evaluation of these surfaces is essential before the reserves of a field can be calculated (p. 258), and their establishment is one of the main objectives of well logging and testing. Where oil and gas occur together in the same trap, the gas overlies the oil because the gas has a lower density. Whether a trap contains oil and/or gas depends both on the chemistry and level of maturation of the source rock (see Chapter 5) and on the pressure and temperature of the reservoir itself. Not only does a gross gravity separation of gas and oil occur within a reservoir but more subtle variations may also exist.

Some oil fields (e.g., Sarir in Libya) have a mat of heavy tar at the oil-water contact. This mat is produced by the degradation of oil as bottom waters move beneath the oil-water contact. Tar mats cause considerable production problems by inhibiting water from displacing oil. Fields with thick oil columns may show a more subtle gravity variation through the pay zone. Boundaries between oil, gas, and water may be sharp or transitional. Abrupt fluid contacts indicate a permeable reservoir; gradational ones indicate a low permeability with a high capillary pressure (p. 232). The zone immediately beneath the hydrocarbons is referred to as the bottom water, and the zone of the reservoir laterally adjacent to the trap as the edge zone (Fig. 7.1).

Fluid contacts in a trap are generally planar, but are by no means always horizontal. Early recognition of a tilted fluid contact is essential for the correct evaluation of a reserve. Correct identification of the cause of the tilt is necessary for the efficient production of the field.

There are several causes of tilted fluid contacts. They may occur where a hydrodynamic flow of the bottom waters leads to a displacement of the hydrocarbons from a crestal to a flank position. This displacement can happen with varying degrees of severity (Fig. 7.3). The presence of this type of tilted oil-water contact can be established from pressure data, which will show a slope in the potentiometric surface. Many, but by no means all, hydrodynamically tilted fields occur above sea level. Where these fields are shallow, the occurrence of tar mats is not unusual because of the degradation of oil due to the movement of water beneath the oil zone.

In some fields the oil-water contact has tilted as a result of production, presumably because of fluid movement initiated by the production of oil from an adjacent field. This phenomenon has been recorded, for example, from the Cairo field of Arkansas (Goebel, 1950). An alternative explanation for a sloping fluid contact is that a trap has been tilted after hydrocarbon invasion, and the contact has not moved. Considered on its own, this theory is unlikely because, within the geological time scale, the oil and/or gas have ample time to adjust to a new horizontal level.

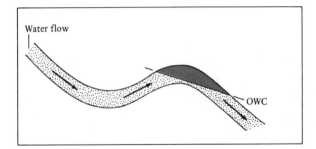

FIGURE 7.3   *Cross-section through a trap showing tilted oil-water contact due to hydrodynamic flow.*

A tar mat may decrease permeability to such an extent that if a trap is tilted, the oil-water contact may be unable to adjust to the new horizontal datum. Alternatively, cementation can continue in the reservoir beneath an oil zone while it is halted in the trap itself (see p. 247). This cementation seldom provides a seat seal sufficiently tight to restrain the hydrocarbons from later movement. It may, however, diminish permeability sufficiently to provide some interesting production problems: the up-dip edge of the field having anomalously low permeability, and the water zone on the down-dip side having anomalously high permeability (Fig. 7.4).

A third possible cause of a tilted oil-water contact may be a change in facies. Theoretically, a change in grainsize across the reservoir causes a tilted contact. With declining grainsize, capillary pressure increases, allowing a rise in the oil-water contact (p. 232). In practice the effect of this rise is likely to be negligible, at the most only a few meters (Yuster, 1953).

On the other hand, where the change is actually lithological, the lower contact of the field may be tilted. In this situation the underlying lithology, either a shale or basement, is generally impermeable. In such cases the

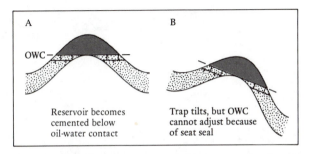

FIGURE 7.4   *Cross-sections through a trap showing how a tilted oil-water contact may be caused by cementation of the water zone (A) followed by tilting (B).*

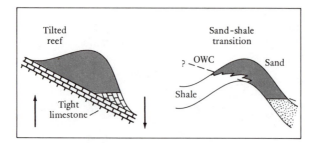

FIGURE 7.5   *Cross-sections through traps showing how apparently tilted oil-water contacts may be caused by facies changes.*

tilted lower surface of the reservoir is not truly an oil-water contact, but a seat seal (Fig. 7.5).

Within the geographic limits of an oil or gas field there may be one or more *pools*, each with its own fluid contact. *Pool* is an inaccurate term, dating back to journalistic fantasies of vast underground lakes of oil; nonetheless, it is widely used. Each individual pool may contain one or more pay zones (Fig. 7.6).

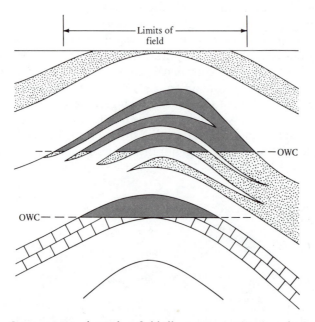

FIGURE 7.6   *Cross-section through a field illustrating various geological terms. This field contains two pools, that is, two separate accumulations with different oil-water contacts. In the upper pool the net pay is much less than the gross pay because of nonproductive shale layers. In the lower pool the net pay is equal to the gross pay.*

CLASSIFICATION OF TRAPS

Hydrocarbons may be trapped in many different ways. Several schemes have been drawn up to attempt to classify traps (e.g., Clapp, 1910, 1929; Lovely, 1943; and Hobson and Tiratsoo, 1975). Two major genetic groups of trap have been agreed upon: structural and stratigraphic. A third group, combination traps, is caused by a combination of processes. Agreement breaks down, however, when attempts are made to subdivide these groups.

Table 7.1 presents a classification of hydrocarbon traps. The table is based on information previously cited in this chapter, and can only be regarded as a crude attempt to pigeonhole such truly fugacious entities as traps. The table has no intrinsic merit other than to provide a framework for the following descriptions of the various types of hydrocarbon trap.

Structural traps are those traps whose geometry was formed by tectonic processes after the deposition of the beds involved. According to Levorsen (1967), a structural trap is "one whose upper boundary has been made concave, as viewed from below, by some local deformation, such as folding, or faulting, or both, of the reservoir rock. The edges of a pool occurring in a structural trap are determined wholly, or in part, by the intersection of the underlying water table with the roof rock overlying the deformed reservoir rock." Basically, therefore, structural traps are caused by folding and faulting.

A second group of traps is caused by diapirs, where salt or mud have moved upward and domed the overlying strata, causing many individual types of trap. Arguably, diapiric traps are a variety of structural trap; but since they are caused by local lithostatic movement, not regional tectonic forces, they should perhaps be differentiated.

Stratigraphic traps are those traps whose geometry is formed by changes in lithology. The lithological variations may be depositional (e.g., channels, reefs, and bars) or postdepositional (e.g., truncations and diage-

TABLE 7.1   A crude classification of hydrocarbon traps based on previous schemes cited in the text

---

I   Structural traps—caused by tectonic processes

  Fold traps $\begin{cases} \text{Compressional anticlines} \\ \text{Compactional anticlines} \end{cases}$

  Fault traps

II   Diapiric traps—caused by flow due to density contrasts between strata

  Salt diapirs

  Mud diapirs

III   Stratigraphic traps—caused by depositional morphology or diagenesis
          (For detailed classification see Table 7.3.)

IV   Hydrodynamic traps—caused by water flow

V   Combination traps—caused by a combination of two or more of the above processes

netic changes). Hydrodynamic traps occur where the downward movement of water prevents the upward movement of oil, thus trapping the oil
without normal structural or stratigraphic closure. Such traps are rare.
The final group, combination traps, are formed by a combination of two or
more of the previously defined genetic processes.

The various types of trap—structural, diapiric, stratigraphic, hydrodynamic, and combination—are described at greater length and illustrated
with examples in the following sections.

STRUCTURAL TRAPS

As previously stated, the geometry of structural traps is formed by postdepositional tectonic modification of the reservoir. Table 7.1 divides structural traps into those caused by folding and those caused by faulting. These
two classifications are now considered in turn.

Anticlinal Traps

Anticlinal, or fold, traps may be subdivided into two classes: compressional anticlines (caused by crustal shortening) and compactional anticlines (developed in response to crustal tension).

Compressional Anticlines

Anticlinal traps caused by compression are most likely to be found in, or
adjacent to, subductive troughs, where there is a net shortening of the
earth's crust. Thus fields in such traps are found within, and adjacent to,
mountain chains in many parts of the world.

One of the best-known oil provinces with production from compressional anticlines occurs in Iran (Fig. 7.7). Here, in the foothills of the
Zagros Mountains, many such fields occur. Sixteen of these fields are in
the "giant" category, with reserves of over 500 million barrels of recoverable oil or 3.5 trillion ft³ of recoverable gas (Halbouty et al., 1970). These
fields have been described in considerable detail over the years (Lees,
1952; Falcon, 1958, 1969; Slinger and Crichton, 1959; Hull and Warman,
1970; Colmann-Sadd, 1978). The main producing horizon is the Asmari
limestone (L. Miocene), a reservoir with extensive fracture porosity. Flow
rates and productivity are immense, with some individual wells having
flowed 50 million barrels (No. 7-7, Masjid-i-Suleiman field). The cap rock is
provided by evaporites of the lower Fars Group (Miocene), whose disharmonic folding makes it difficult to extrapolate from the surface to the
reservoirs. The traps themselves lie to the southwest of the main Zagros
Mountain thrust belt. Individual anticlines are up to 60 km in length and
some 10 to 15 km wide (Fig. 7.8). The axial planes of the folds pass down-

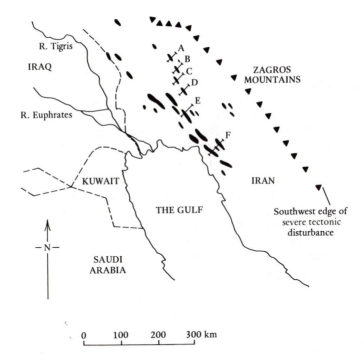

FIGURE 7.7 *Map showing the location of the folded anticlinal traps of Iran. For cross-section see Figure 7.8.*

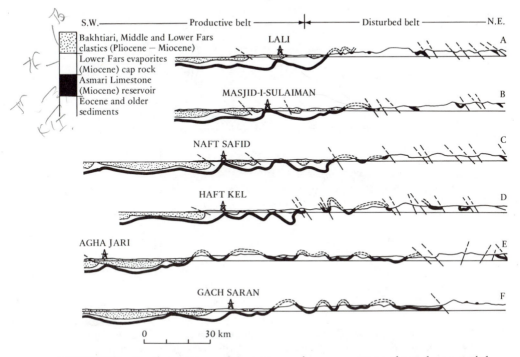

FIGURE 7.8 *Southwest to northeast true-scale cross-sections through some of the folded structures of Iran. Locations shown in Figure 7.7. (After Falcon, 1958.)*

ward into thrust faults, which die out in a zone of decollement within the underlying Hormuz salt (Precambrian?).

A second major hydrocarbon province that contains compressional anticlinal traps occurs in the Tertiary basins of California. Here, a number of fault-bounded troughs are infilled by thick regressive sequences in which organic-rich basinal muds are overlain by turbidites capped by younger continental beds. These sediments have locally undergone tight compressive folding associated with the transcurrent movement of the San Andreas fault system (Barbat, 1958; Schwade et al., 1958; Simonson, 1958). Many of the fields are associated with faulting: normal, reversed, and strike slip (Fig. 7.9). The Long Beach-Wilmington field of the Los Angeles basin is a giant field in a compressional anticline, cross-cut by normal faults perpendicular to the fold axis (Mayuga, 1970).

The folds are often involved in thrusting within the mountain chains themselves. Hydrocarbons may be trapped in anticlines above thrust

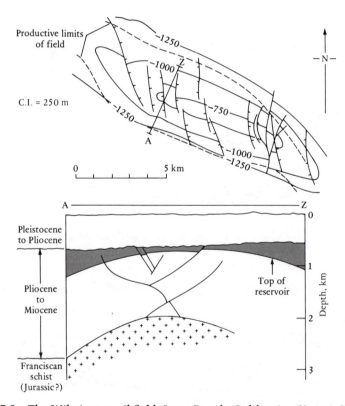

FIGURE 7.9 *The Wilmington oil field, Long Beach, California. (Upper) Structure contour map on top of the Ranger Zone, just below the crest of the reservoir. (Lower) Southwest–northeast cross-section along the line A–Z. (After Mayuga, 1970.)*

planes and in reservoirs sealed beneath the thrust. A major play of this type
occurs in the eastern Rocky Mountains, including the Turner Valley field
of Alberta and the Painter Valley Reservoir field of Wyoming (Fig. 7.10).
Such fields are extremely difficult to find and develop because of the problems of seismic interpretation due to complex faulting and steeply dipping
beds. With recent improvements in seismic technology, however, this task
is becoming easier, opening up previously neglected areas to exploration.

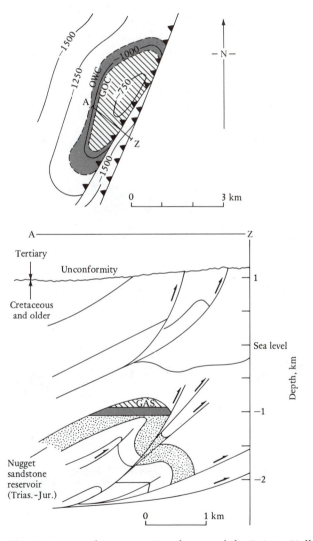

FIGURE 7.10 *Map (upper) and cross-section (lower) of the Painter Valley field of*
Wyoming. Thrust-associated compressional anticline traps such as this are
becoming easier to find because of improving seismic data. (After Lamb, 1980.)

Compactional Anticlines

A second major group of anticlinal traps is formed not by compression but by crustal tension. Where crustal tension causes a sedimentary basin to form, the floor is commonly split into a mosaic of basement horsts and grabens. The initial phase of deposition infills this irregular topography. Throughout the history of the basin the initial structural architecture usually persists, controlling subsequent sedimentation. Thus anticlines may occur in the sediment cover above deep-seated horsts (Fig. 7.11). Closure may be enhanced both by compaction and sedimentation. Differential sedimentation of clays increases the amplitude of the fold because, although the percentage of compaction is constant for crest and trough, the actual amount of compaction is greater for the thicker flank sediment (Fig. 7.12).

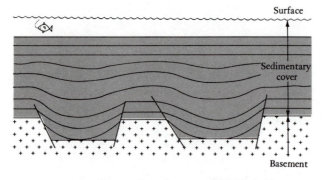

FIGURE 7.11   *Cross-section showing how basement block faulting causes anticlinal structures in sediments; closure decreases upward. These drape anticlines are caused by tension rather than compression.*

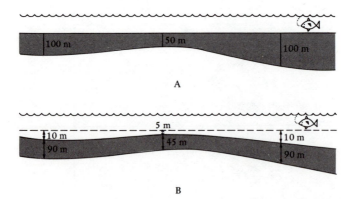

FIGURE 7.12   *Cross-sections showing how burial compaction enhances closure on drape anticlines. (A) Deposition; (B) after a uniform compaction of 10 percent.*

Differential depositional rates also enhance structural closure. Carbonate sedimentation tends to be higher in shallow, rather than deep, water; so shoal and reefal facies may pass off-structure into thinner increments of basinal lime mud. Similarly, terrigenous shoal sands may develop on the crests of structures and pass down flank into deeper water muds. Thus reservoir quality often diminishes down the flank of such structures.

Good examples of oil fields trapped in compactional anticlines occur in the North Sea. Here, Paleocene deep-sea sands are draped over Mesozoic horsts (Blair, 1975). These fields include the Forties, Montrose, Maureen, and East Frigg fields (Fig. 7.13).

The traps of compactional and compressional anticlines are very different. As just discussed, compactional folds may have considerable variations in reservoir facies across structure. Not only may there be a primary depositional control of reservoir quality, but later diagenetic changes may be extensive, since such structures are prone to subaerial exposure, leaching, and, in extreme cases, truncation.

Whereas compressional folds are generally elongated perpendicular to the axis of crustal shortening, compactional folds are irregularly shaped, reflecting the intersection of fault trends in the basement. Compressional folds generally form in one major tectonic event, whereas compactional folds may have had a lengthy history due to rejuvenation of basement faults as the basin floor subsided.

Fault and Fault-Related Traps

Faulting plays an indirect but essential role in the entrapment of many fields. Relatively few discovered fields are caused solely by faulting. A very important question in both exploration and development is whether a fault acts as a barrier to fluid movement (not only hydrocarbons but also water, which may be necessary to drive production) or whether it is permeable. The simple answer is that some faults seal, others do not. A few guidelines are available, but they are by no means foolproof. Where the throw of the fault is less than the thickness of the reservoir, it is unlikely to seal. Faults in brittle rocks are less likely to seal than those in plastic rocks. In lithified rocks faults may be accompanied by extensive fracturing, which may be permeable; indeed, some fields are caused solely by fracture porosity adjacent to a fault (Fig. 6.7). Sometimes, however, fractures may have undergone later cementation.

In unlithified sands and shales faults tend to seal, particularly where the throw exceeds reservoir thickness. Examples are known, however, where clay caught up in a fault plane can act as a seal even when two permeable sands are faulted against each other. This phenomenon is known from areas of overpressured sediments, such as the Gulf of Mexico and the Niger delta, where Smith (1980) and Weber and Daukoru (1975) have stud-

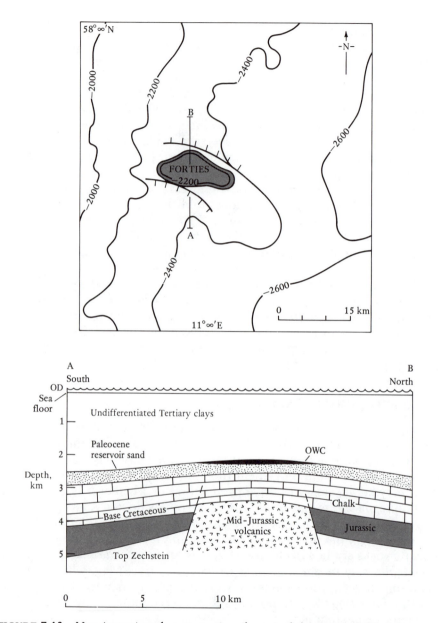

FIGURE 7.13 *Map (upper) and cross-section (lower) of the Forties field of the North Sea. This field is essentially a compactional anticline draped over an old basement high.*

ied the role of faults as seals and barriers. Smith (1980) noted in the Gulf Coast that where sands were faulted against each other, the probability of the fault sealing increased with the age difference of the two sands. Figure 7.14 shows cross-sections of the West Lake Verret field in which hydrocarbon- and water-bearing sands are faulted against one another. The

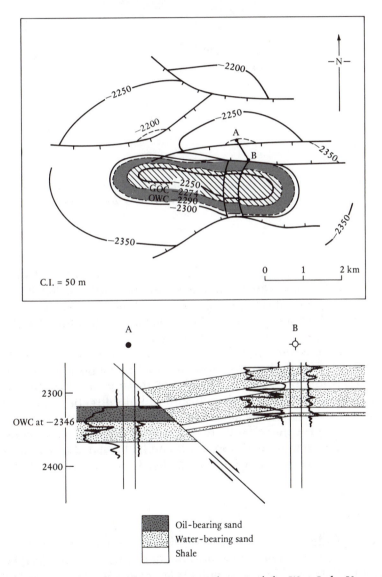

FIGURE 7.14 *Map (upper) and cross-section (lower) of the West Lake Verret field, Tertiary, Louisiana. This field provides an example of a sealing fault in which oil has not moved across the fault plane, even though permeable sands are juxtaposed. (After Smith, 1980.)*

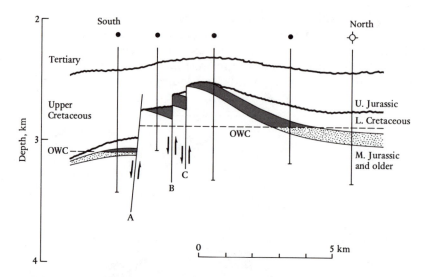

FIGURE 7.15   *Cross-section through the Piper field of the North Sea showing how,*
within the same field, some faults seal (A) and others do not (B and C). (After
Wiliams et al., 1975.)

faults do not separate sands of different facies or capillary displacement
pressure, so Smith (1980) concluded that the faults sealed by virtue of
impermeable material smeared along the fault planes. Figure 7.15 shows a
complexly faulted field in which the oil-water contacts indicate that some
faults are conduits, whereas others are seals.

Bailey and Stoneley (1981) have shown that there are eight theoretical geometries for fault traps, assuming that faults do not separate juxtaposed permeable beds (Fig. 7.16). Six of these geometries may be valid
traps provided that there is also closure in both directions parallel to the
fault plane. Although many fields are trapped by a combination of faulting
and other features, pure fault traps are rare. Figure 7.17 illustrates one
example of a simple faulted trap. Fault and fault-related traps may be conveniently categorized according to whether transverse or tensional forces
operate.

Traps Related to Transverse Faults

Transverse faults give rise to several distinctive types of petroleum trap
(Wilcox et al., 1973). Transverse movement of basement blocks takes place
along wrench faults. These movements are expressed in the overlying sediment cover in a trend of *en echelon* folds. Fold axes are oblique to the
wrench fault and indicate its direction of movement. In some instances

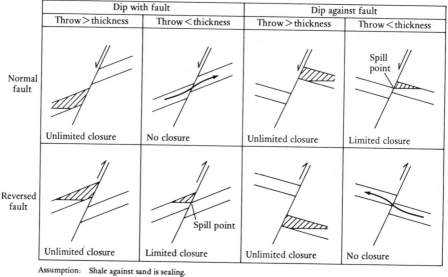

	Dip with fault		Dip against fault	
	Throw > thickness	Throw < thickness	Throw > thickness	Throw < thickness
Normal fault	Unlimited closure	No closure	Unlimited closure	Spill point Limited closure
Reversed fault	Unlimited closure	Spill point Limited closure	Unlimited closure	No closure

Assumption: Shale against sand is sealing.
Sand against sand is nonsealing.

FIGURE 7.16   *The eight theoretical configurations of petroleum traps associated with faulting. These configurations are drawn on the assumption that oil can move across, but not up, the fault plane when permeable sands are juxtaposed. (After Bailey and Stoneley, 1981.)*

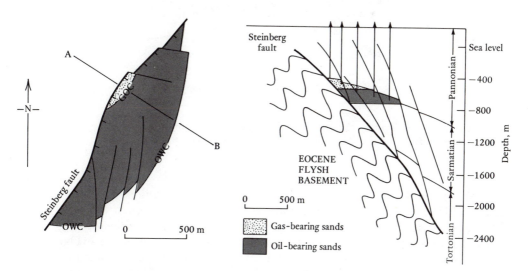

FIGURE 7.17   *Map (left) and cross-section (right) through a fault trap in the Gaiselberg field, Austria. (After Janoschek, 1958.)*

fold axes may be offset by faults (Fig. 7.18). In cross-section, wrench faults split upward into low-angle faults. This phenomenon is sometimes referred to as *flower structure* (Gregory, in Harding and Lowell, 1979). The New-port-Inglewood fault in the Ventura basin of California provides a classic example of petroleum entrapment in *en echelon* folds developed along a wrench fault (Harding, 1973).

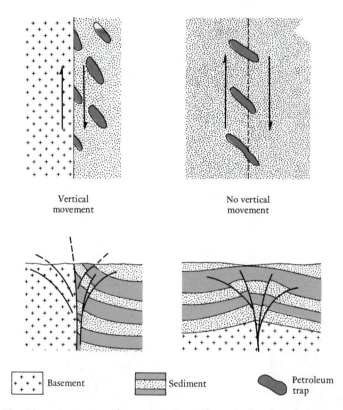

FIGURE 7.18   *Maps (upper) and cross-sections (lower) showing the types of petroleum trap associated with wrench faults. For examples and explanation see text.*

Few wrench faults have no vertical movement. Indeed, sedimentary basins often form by rapid subsidence of one side of a wrench fault (p. 371). In such settings the folds may be present only on the basinward side of the fault and may form structural noses where they are truncated (Fig. 7.18). Examples of this type of trap occur where the southwestern flank of the San Joaquin basin is truncated by the San Andreas fault, California (Harding, 1974).

Traps Related to Tensional Faults

A particularly important group of traps are associated with tensional faults. In some faults the throw increases incrementally downward. This demonstrates that fault movement was synchronous with deposition. Such faults are thus termed *growth faults*. In plan view the fault trace is frequently curved; when viewed from the downthrown side, concave. Similarly, in profile the fault angle diminishes downward.

There are two types of growth fault. One type is basement-related and tends to be long in plan view and to have had a lengthy history controlling facies and sedimentation rate (Shelton, 1968). A good example of this type occurs on the flank of the Sarir field in Libya (Fig. 7.19). One of the best-known large-scale growth faults is the Vicksburg flexure, which extends for some 500 km around the Gulf Coast of Texas. The maximum increase in sediment across the fault is in the order of 1500 m near the Mexican border. Most of this thickening occurs in the Vicksburg Group (Oligocene); considerably less thickening occurs in the overlying Frio Group (Miocene). Not only do formations thicken across the fault toward the Gulf but the percentage of net sand increases too. This suggests that subsidence on the downthrown side of the fault formed a natural sediment trap. Characteristically, there is a local reversal of the easterly regional dip of strata adjacent to the fault plane, with rollover anticlines developed on its downthrown side (Fig. 7.20). The strata dip toward the fault to fill the space caused by separation along the plane of the fault. The dip reversal of the anticlines is often enhanced by antithetic faults downthrown and dipping in toward the major fault.

Oil and gas are trapped both in the rollover anticlines and in sand pinchout stratigraphic traps on both upthrown and downthrown sides of the Vicksburg fault (Halbouty, 1972). It has been estimated that some 3 billion barrels of oil and 20 trillion ft^3 of gas are trapped adjacent to the Vicks-

FIGURE 7.19 *Cross-section through a basement-related growth fault in the Sarir field, Libya. The numbers show the throw (in meters) of the fault for various markers and demonstrate an incremental increase downward. (After Sanford, 1970.)*

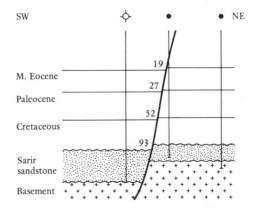

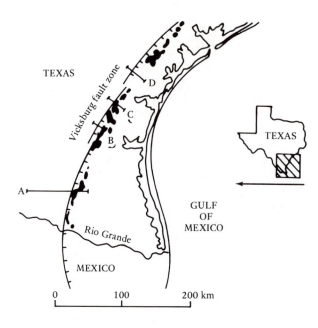

TEXAS

Vicksburg fault zone

D

C

B

A

TEXAS

GULF
OF
MEXICO

Rio Grande

MEXICO

0   100   200 km

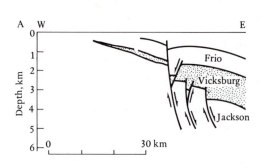

A W   E

Depth, km

0
1
2
3
4
5
6

Frio

Vicksburg

Jackson

0   30 km

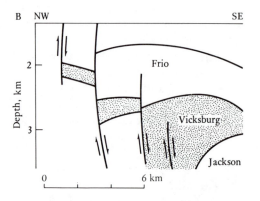

B NW   SE

Depth, km

2

3

Frio

Vicksburg

Jackson

0   6 km

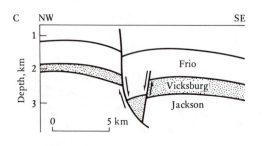

C NW   SE

Depth, km

1

2

3

Frio

Vicksburg

Jackson

0   5 km

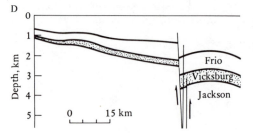

D

0
1
2
3
4
5

Depth, km

Frio

Vicksburg

Jackson

0   15 km

294

burg flexure, partly in pure fault traps, but more commonly in rollover anticlines and pinchouts.

Traced eastward into Louisiana, the Vicksburg fault disappears beneath the clastic wedge of the Mississippi delta. Growth-fault-related traps dominate this hydrocarbon province too, but they are different in scale and genesis. Individual faults are seldom more than a few kilometers in length; although with their curved traces, scalloped fault patterns can also occur. These faults are not basement-related, but pass downward to die out as horizontal shear planes either in overpressured shales or in the deeper Louann Salt (Triassic–Jurassic).

Similarly, in the Tertiary sediments of the Niger delta, growth faults play a major role in the migration and entrapment of oil and gas (Evamy et al., 1978). A detailed analysis by Weber et al. (1980) was based on observation of the fault-associated fields and laboratory experiments. They concluded that in most instances growth faults were not sealing, because they were associated not only with clay gangue but also with slivers of permeable sand. Migration of hydrocarbons appears to occur both up the fault plane and across it, from overpressured mature source shales into normally pressured (or at least lower-pressured) sands. In the upper part of a rollover anticline, traps may be filled to the spill point; but where sands are faulted against overpressured shale, traps may be filled to below the spill point (Fig. 7.21).

The Relationship Between Structural Traps and Tectonic Setting

The classification and account of structural traps just given are essentially descriptive and static; that is, the traps were not examined in their tectonic context. Structural traps do not occur at random. The type and distribution of structural traps are closely related to the regional tectonic setting and history of the region in which they are found.

A genetic classification of structural traps has been developed by Harding and Lowell (1979). They note that structural traps can be grouped according to the tectonic forces operating, according to whether the basement or only the detached cover is involved, and according to their habitat with respect to tectonic plates. This last classification is covered in Chapter

FIGURE 7.20 *Map of and cross-sections through the Vicksburg growth fault of South Texas. Traps associated with this structure contain estimated reserves of 3 billion barrels of oil and 20 trillion ft³ of gas. Note how formations thicken basinward across the fault, the rollover anticlines and antithetic minor faults dipping toward the major fault. Production comes from sands in both the Vicksburg (Oligocene) and Frio (Eocene) Groups. (Map after Stanley, 1970.)*

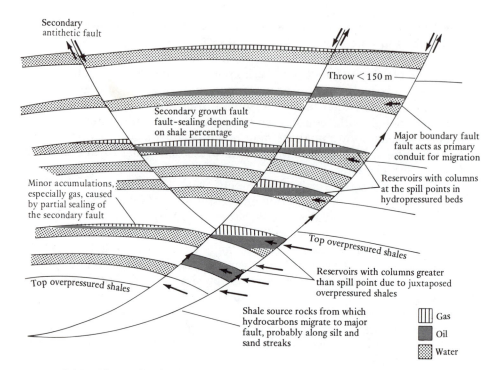

Secondary
antithetic fault

Throw < 150 m

Secondary growth fault
fault-sealing depending
on shale percentage

Major boundary fault
fault acts as primary
conduit for migration

Reservoirs with columns
at the spill points in
hydropressured beds

Minor accumulations,
especially gas, caused
by partial sealing of
the secondary fault

Top overpressured shales

Top overpressured shales

Reservoirs with columns greater
than spill point due to juxtaposed
overpressured shales

Shale source rocks from which
hydrocarbons migrate to major
fault, probably along silt and
sand streaks

☐☐☐ Gas

▨ Oil

▦ Water

FIGURE 7.21   *The mode of accumulation of oil and gas in Niger delta growth fault traps. Unlike basement-related growth faults, this type shears out horizontally into overpressured shales. (After Weber et al., 1978.)*

8. Table 7.2 presents a genetic grouping of structural traps. It shows that a major distinction is made between regions where basement and cover are attached and regions where basement and cover are detached (generally because of the presence of intervening evaporites or overpressured clays). Compressive and tensional forces can operate in both situations; the former occur at convergent plate boundaries, the latter at divergent ones. This type of genetic grouping of structural traps is more useful than a purely descriptive one. It is an exploration tool that predicts the type of structural trap to be anticipated in any given tectonic setting.

DIAPIRIC TRAPS

Diapiric traps are produced by the upward movement of sediments that are less dense than those overlying them. In this situation the sediments tend to move upward diapirically and, in so doing, may form diverse hydrocarbon traps. Such traps cannot be regarded as true structural traps,

TABLE 7.2 A genetic grouping of structural traps based on basement-cover relationship, structural style, and dominant force

	Structural style	Dominant force
Basement involved	Wrench fault	Couple
	Regional paleohigh	Mantle processes
	Thrust blocks and reversed faults	Compression
	Extensional fault blocks and drape anticlines	Tension
Cover detached from basement	Growth faults and rollover anticlines	
	Decollement-related structures	Compression
	Diapirs (salt and clay)	Density contrast

From Harding and Lowell, 1979. Reprinted with permission.

since tectonic forces are not required to initiate them (although in some cases they may do so). Similarly, diapirically related traps are not initiated by stratigraphic processes, although in some cases they may be caused by depositional changes across the structure.

Diapiric traps are generally caused by the upward movement of salt or, less frequently, overpressured clay. Salt has a density of about 2.03 g/cm^3. Recently deposited clay and sand have densities less than that of salt. As the clay and sand are buried, however, they compact, losing porosity and gaining density. Ultimately, a burial depth is reached when sediments are denser than salt. Depending on a number of variables, this point may occur between about 800 and 1200 m (Fig. 7.22). When this point is reached, the salt will tend to flow up through the denser overburden. This movement may be triggered tectonically, and the resultant structures may show some structural alignment. In other instances, however, the salt movement is apparently random. The exact mechanics of diapiric movement have been studied by observation, experiment, and mathematical calculation over many years (e.g., Halbouty, 1967; Berner et al., 1972; and Bishop, 1978).

In some salt structures the overlying strata are only up-domed, whereas in others the salt actually intrudes its way upward; the latter are referred to as piercement structures. In some instances the salt may actually reach the surface, forming solution sinks in humid climates and salt

FIGURE 7.22 *Density-depth curves for*
sand, clay, and salt. The graph shows
that salt is less dense than other
sediments below about 800 m, and salt
movement may therefore be anticipated
once this burial depth has been
reached.

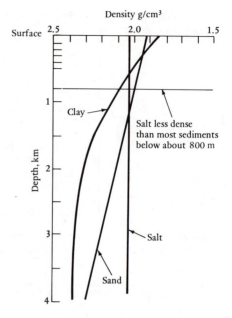

glaciers in arid climates (like Iran) (Kent, 1979). Salt movement, or *halo-kinesis*, plays an important role in the entrapment of oil and gas in the United States Gulf Coast, Iran, and the Arabian Gulf and the North Sea.

Oil and gas may be trapped by salt movement in many ways (Fig. 7.23). In the simplest cases subcircular anticlines may trap hydrocarbons

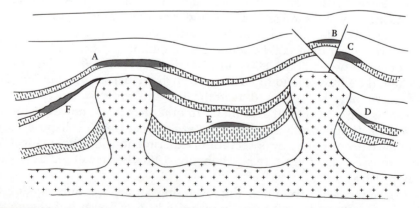

FIGURE 7.23 *Crustal cross-section illustrating the various types of trap that*
may be associated with salt movement. (A) Domal trap, (B and C) fault traps,
(D) pinchout trap, (E) turtle-back, or sedimentary anticline, (F) truncation trap.

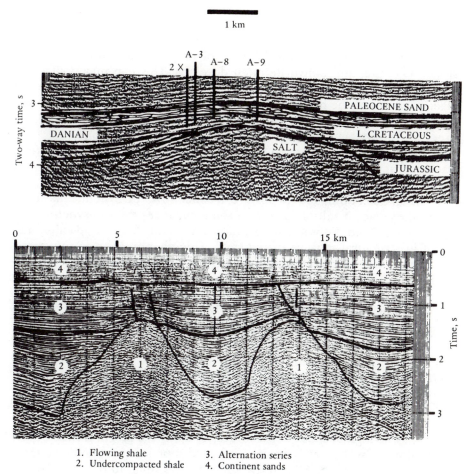

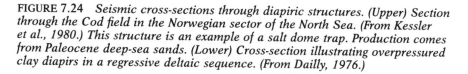

1. Flowing shale   3. Alternation series
2. Undercompacted shale 4. Continent sands

Vertical exaggeration 2:1 ±

FIGURE 7.24 *Seismic cross-sections through diapiric structures. (Upper) Section through the Cod field in the Norwegian sector of the North Sea. (From Kessler et al., 1980.) This structure is an example of a salt dome trap. Production comes from Paleocene deep-sea sands. (Lower) Cross-section illustrating overpressured clay diapirs in a regressive deltaic sequence. (From Dailly, 1976.)*

over the crest of a salt dome. Notable examples of this type include Ekofisk and associated fields of offshore Norway and Denmark (Fig. 7.24). The crestal dome may be complicated by radial faults or a central graben. Around the flank of the dome, oil or gas can be trapped by faults, both sediment against sediment and sediment against salt, and by stratigraphic truncation, pinchout, and onlap. Some salt domes are pear- or mushroom-shaped in cross-section, and petroleum is trapped beneath the peripheral overhang zone.

As the salt moves upward, a cap of diagenetically produced lime-stone, dolomite, and anhydrite often develops. This cap may contain oil and gas in fractures. When a salt dome moves up close to the earth's sur-face, its crest may be dissolved by groundwater. Overlying sediments may collapse into the space thus formed, so giving rise to solution collapse breccias. Ultimately, these too may act as petroleum reservoirs.

In the Arabian Gulf today coral atolls grow above salt domes (e.g., Das Island), and analogues containing oil and gas are present in the sub-surface.

When a salt dome moves upward, a concentric rim-syncline may form where the adjacent salt once was. In some instances the rim-syncline may be infilled by sand. Subsequent salt movement and shale compaction may result in structural closure of the sand. This phenomenon is sometimes referred to as a sedimentary anticline (i.e., of atectonic origin) or turtle-back. These structures are essentially residual highs caused by adjacent salt moving into domes. The Bryan field of Mississippi is one example of a turtle-back trap (Oxley and Herliny, 1972).

From the preceding account it is apparent that traps associated with salt domes may be very complex indeed. Not only are there many different trap situations but any one salt dome may host many separate traps of dif-ferent types. Although the total reserves pertaining to one diapir may be vast, they may be contained in many separate accumulations. Each accu-mulation may have its own fluid and pressure characteristics. Individual oil columns may be high because of steeply dipping beds; individual pres-sures may be high because of the forces set up by the salt movement itself. A salt dome can be easily located by gravity or seismic methods. Once found, however, the associated traps may be too small to be delineated by present day seismic resolution.

The foregoing account deals largely with diapirs formed by salt. Dia-piric mud structures also exist, and they too may generate hydrocarbon traps (Fig. 7.24). Overpressure and overpressured shales were discussed earlier (p. 152). By its very nature an overpressured shale has a higher porosity and therefore a lower density than does normally compacted clay.

As already discussed, pro-delta clays may be overpressured because of rapid burial beneath an advancing prism of deltaic sediment. This sedi-ment thus becomes unstable, tending to slump seaward over the clays be-neath it. This slumping is associated with growth faulting.

Sometimes diapirs of overpressured clay intrude the younger, denser cover, and, like salt domes, these mud lumps may even reach the surface. Mud diapirs are known from the Mississippi, Niger, Mackenzie, and other Recent deltas and are less common, but not absent, from pre-Tertiary del-tas. Many of these mud diapirs have hydrocarbons trapped in analogous ways to those previously described for salt domes (e.g., the Beaufort Sea of Arctic Canada).

STRATIGRAPHIC TRAPS

Another major group of traps to be considered are the stratigraphic traps, whose geometry is due to changes in lithology. Such changes may be caused by the original deposition of the rock, as with a reef or channel. Alternatively, the change in lithology may be postdepositional, as with a truncation or diagenetic trap.

The concept of the stratigraphic trap was first enunciated by Carll (1880) when he realized that the entrapment of oil in the Venango sands of Pennsylvania could not be explained by the then-prevailing anticlinal theory. Clapp (1917), in his second classification of oil fields, included lenticular sand and pinchout traps. The term *stratigraphic trap* was first coined by Levorsen in his presidential address to the American Association of Petroleum Geologists at Tulsa in 1936. He defined a stratigraphic trap as "one in which the chief trap-making element is some variation in the stratigraphy, or lithology, or both, of the reservoir rock, such as a facies change, variable local porosity and permeability, or an up-structure termination of the reservoir rock, irrespective of the cause" (Levorsen, 1967). For reviews of the evolution of the term and concept of the stratigraphic trap, see Dott and Reynolds (1969) and Rittenhouse (1972).

Stratigraphic traps are less well known and harder to locate than structural traps; their formation processes are even more complex. Nonetheless, as with structural traps a broad classification of different types can be made. Table 7.3 shows a scheme based on Rittenhouse (1972). Like most classifications it has its limitations, since many fields represent transitional steps between clearly defined types. It does, however, provide a convenient framework for the following account of stratigraphic traps. Major sources of data on stratigraphic traps are found in King (1972), Busch (1974), and Conybeare (1976).

TABLE 7.3   A classification of stratigraphic traps

I. Unassociated with unconformities	Depositional	Pinchouts
		Channels
		Bars
		Reefs
	Diagenetic	Porosity and/or permeability transition
II. Associated with unconformities	Supraunconformity	On lap
		Strike valley
		Channel
	Subunconformity	Truncation

From Rittenhouse, 1972. Reprinted with permission.

Stratigraphic Traps Unrelated to Unconformities

Table 7.3 shows that a major distinction can be made between stratigraphic traps associated with unconformities and those occurring within normal conformable sequences. This distinction itself is somewhat arbitrary, since some types of trap, such as channels and reefs, occur both at unconformities and within conformable sequences. Those traps that *do not necessarily* require an unconformity will be considered in this section. In this group the major distinction is between traps due to deposition and traps due to diagenesis. The depositional traps (the facies change traps of Rittenhouse) include channels, bars, and reefs. Diagenetic traps are due to porosity and permeability changes caused by solution and cementation. These various types of traps are discussed and illustrated as follows.

Channel Traps

Many oil and gas fields are trapped within different types of channels. Before examining examples of these fields, some background information is necessary. A channel is an environment for the transportation of sand, which may or may not include sand deposition. Thus whereas a barrier island will always be made of sand, channels are frequently clay plugged. This situation is not necessarily bad, because the channel fill may act as a permeability barrier and thus trap hydrocarbons in adjacent porous beds (e.g., p. 318). Therefore, finding a channel is not a guarantee of finding a reservoir. Also, channels occur both cut into unconformities and within conformable sequences (although it could be argued that, by definition, a channel is a *prima facie* case for the existence of some kind of depositional break).

　　　Many good examples of channel stratigraphic traps occur in the Cretaceous basins along the eastern flanks of the Rocky Mountains, from Alberta down through Montana, Wyoming, Colorado, and New Mexico. These traps occur both cut into a major sub-Cretaceous unconformity and within the Cretaceous beds. The South Glenrock oil field of Wyoming serves as a typical example illustrating many of the characteristics of these fields (Curry and Curry, 1972).

　　　The South Glenrock field occurs in a syncline, which partly underlies a thrust block of Precambrian basement. The field contains oil trapped in both barrier bar and fluvial channel reservoirs. Only the latter is considered here. The channel has a width of some 1500 m and a maximum depth of some 15 m. It has been mapped for a distance of over 15 km and shows a meandering shape (Fig. 7.25). The channel is partly infilled by sand and partly clay plugged. The S.P. curve on some wells suggest upward-fining point bar sequences (see p. 79). The channel is cut into the marine Skull Creek shale and is overlain by the marine Mowry shale. Immediately above the channel sand are thin, marine bar sand reservoirs.

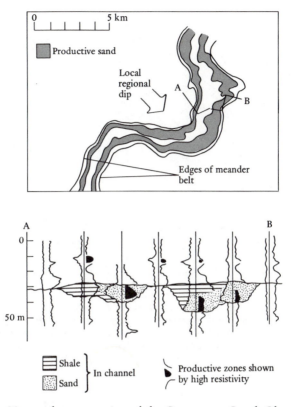

FIGURE 7.25   *Map and cross-section of the Cretaceous South Glenrock oil field, Powder River basin, Wyoming. Note the small dimensions of the reservoir and that not all of the channel contains sand. (After Curry and Curry, 1972.)*

There can be no doubt about the channel origin of the lower reservoir, and, with its meandering geometry, there can be little argument as to its fluvial environment. This example demonstrates two main points. First, note the scale: with a thickness of 15 m and a width of 1500 m, such reservoirs are not the hosts to giant oil accumulations. Second, because of the partial clay plug, only part of the channel is actually reservoir.

Oil is also trapped in channels other than fluvial ones. Perhaps one of the best examples is Busch's (1960, 1971) illustration of oil fields in deltaic distributary channels of Oklahoma (Fig. 7.26).

Barrier Bar Traps

Marine barrier bar sands often make excellent reservoirs because of their clean, well-sorted texture. Coalesced barrier sands may form blanket sands

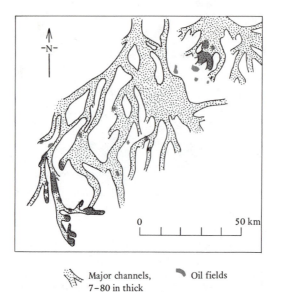

Major channels,   Oil fields
7–80 in thick

FIGURE 7.26 *Map of the Booch delta, Seminole Country, Oklahoma. This celebrated study shows how oil is stratigraphically trapped in the axes of a major Pennsylvanian (Upper Carboniferous) delta distributary channel system. This type of map can only be constructed with imagination and ample well control. (After Busch, 1960.)*

within which oil may be structurally trapped. Sometimes, however, isolated barrier bars may be totally enclosed in marine and lagoonal shales. These barrier bars may then form shoestring stratigraphic traps parallel to the paleoshoreline.

Many examples of this type of trap are known, but one of the classics is the Bisti field described by Sabins (1963, 1977). This field occurs in Cretaceous rocks of the San Juan basin, New Mexico. Three stacked sand bars, with an aggregate thickness of only about 15 m, occur totally enclosed in the marine Mancos shale. The field is about 65 km long and 7 km wide. In some wells the three sands merge totally and cannot be separated. S.P. logs show typical upward-coarsening bar sand motifs in some of the wells (Fig. 7.27).

Barrier bar sand traps occur in many of the Rocky Mountain Cretaceous basins, ranging from the Viking sands of Alberta and Saskatchewan in the north (Evans, 1970), via the Powder River basin of Montana and Wyoming (Asquith, 1970; Woncik, 1972), to the Denver basin, Colorado (Tobison, 1972; Weimer and Davis, 1977) and the San Juan basin of New Mexico (Hollenshead and Pritchard, 1961; Sabins, 1963, 1972).

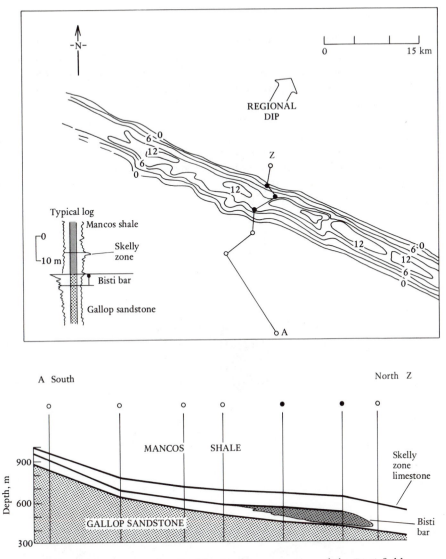

FIGURE 7.27   *Isopach map of typical log and cross-section of the Bisti field (Cretaceous) of the San Juan basin, New Mexico. This field is a classic example of a barrier bar stratigraphic trap. Note the regressive upward-coarsening grainsize motif shown on the S.P. curve. (After Sabins, 1972.)*

Pinchout Traps

Isolated barrier bar shoestring stratigraphic traps like the Bisti field are rare. Generally, a regressing barrier island deposits a sheet of sand. This sand may form a continuous reservoir, although in some instances shale permeability barriers may separate successive progradational events. Where these sheet sands pass up-dip into lagoonal or intertidal shales, they may give rise to pinchout, or feather edge, traps. Note that for these traps to be valid, they also need some closure in both directions along the paleo-strike. This closure may be stratigraphic (where, as shown in the example in Figure 7.28, the shoreline has an embayment) or structural, in which case the field should more properly be classified as a combination, rather than a stratigraphic, trap (p. 320).

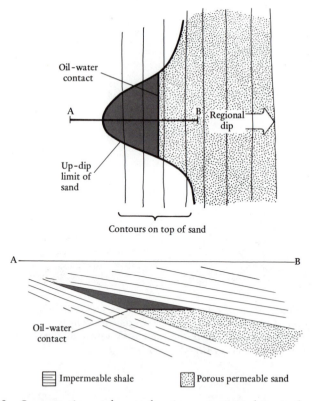

FIGURE 7.28 *Cross-section and map showing a stratigraphic pinchout trap. Note that this example is a pure stratigraphic trap because of the embayment of the coast. Usually, some structural closure on top of the sand forms a combination trap.*

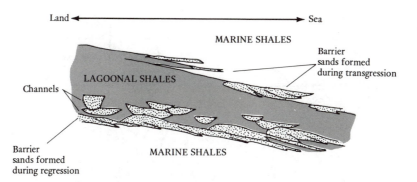

FIGURE 7.29 *Cross-section of regressive-transgressive shoreline deposits showing how barrier sands deposited during transgression can form stratigraphic traps due to up-dip lagoonal shale seal. Barrier sands formed during regressions lack up-dip seals, being in communication with channel sands. (After Mackenzie, 1972.)*

Barrier bar sands, sheets, and shoestrings often occur as an integral part of major regressive-transgressive cycles. Sands are thickest and best developed in the regressive phase, and tend to occur only as discrete shoestrings during the transgressions. Examination of the Rocky Mountain Cretaceous basins shows that the transgressive sands tend to make the best traps. The regressive sands tend to lack up-dip seals as they pass shoreward into channel sands. The transgressive sands, by contrast, pass up-dip into sealing lagoonal and tidal flat shales (Mackenzie, 1972) (Fig. 7.29).

More complex pinchout traps can occur where both barrier bar and channel sands are in fluid communication with one another. The Bell Creek field (Cretaceous) of the Powder River basin is an example of this type of trap (Berg and Davies, 1968; McGregor and Biggs, 1972). It lies on a basin margin where stratigraphic traps can be mapped in fairways of fluvial channels, deltaic and bar sands (Fig. 7.30). The Bell Creek field is one of the few stratigraphic traps of "giant" status, with reserves of over 200 million barrels. It occurs at the mouth of the major river estuary in a complex of productive channel and barrier bar sands that interfinger with nonproductive marsh and lagoon muds and sands (Fig. 7.31). This complexity of reservoir geometry has lead to the occurrence of several gas-oil and oil-water contacts.

Reefs

Reefs, or carbonate buildups (to use a nongenetic term), have long been recognized as one of the most important types of stratigraphic traps. Modern reefs have been intensely studied by biologists and geologists (e.g., Maxwell, 1968; Jones and Endean, 1973), and accounts of ancient reefs are found in most sedimentology textbooks (see also Laporte, 1974).

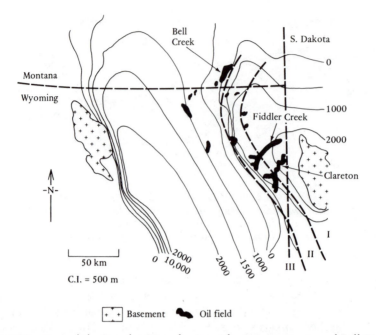

FIGURE 7.30 *Map of the Powder River basin with contours on top of Fall River Sandstone (Cretaceous) showing productive fairways of stratigraphic traps: (I) fluvial channels, (II) deltaic distributaries, and (III) marine bar sands. (After Woncik, 1972, and McGregor and Biggs, 1970.)*

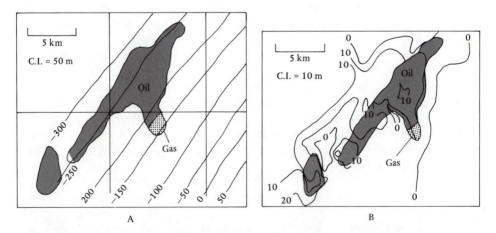

FIGURE 7.31 *Maps of the Bell Creek giant stratigraphic trap (for location see Figure 7.30). (A) Structure contour map on top of the Muddy sand (the main reservoir) showing lack of structural control on the field boundaries. (After McGregor and Biggs, 1970.) (B) Gross sand isopach. Basically an up-dip pinchout trap, the Bell Creek reservoir consists of a complex of marine shoreface and fluviodeltaic channel sands with multiple fluid contacts.*

Reefs develop as domal (pinnacle) and elongated (barrier) antiforms. They grow a rigid stony framework with high primary porosity, and they are frequently transgressed by marine shales, which may act as hydrocarbon source rocks. Small wonder, then, that few identified reef traps have been left undrilled. To show the characteristics of this type of stratigraphic trap, some examples of reef fields are examined as follows.

Many reef fields occur in the Sirte basin of Libya, although not all of these have been published. One group, the Intisar (formerly Idris) fields, has been described by Terry and Williams (1969) and Brady et al. (1980). Five pinnacle reefs were located in a concession of 1880 km². Each reef is only about 5 km in diameter, but up to 400 m thick; that is, the reefs build up in thickness from 0 to 400 m in a distance of about 2500 m. Of the five reefs located and tested, only two were found to contain oil (Fig. 7.32). Figure 7.33 shows the "A" reef in cross-section, demonstrating how the reef began to form as a biostrome of algal-foraminiferal wackestone on a platform of tight nonreef limestone. A reef (largely made of corals and encrusting algae) began to grow at one point on the biostrome. It first grew upward and then prograded over a coralline biomicrite, which had formed as a forereef talus of its own detritus. Now compare the upper and lower halves of Figure 7.33. There is little obvious correlation between facies and petrophysics. As discussed in Chapter 6, extensive diagenesis is characteristic of carbonate reservoirs. Therefore it is common to find secondary porosity whose distribution is unrelated to the primary porosity with which the sediment was deposited.

Thus the Intisar reefs demonstrate the two main problems of reef stratigraphic traps: (1) not all reefs contain hydrocarbons and (2) those that do may have reservoir characteristics unrelated to depositional facies. Nonetheless, many reef hydrocarbon provinces exist around the world, notably in the Arabian Gulf, Western Canada, and Mexico.

Diagenetic Traps

Diagenesis plays a considerable role in controlling the quality of a reservoir within a trap. As discussed in Chapter 6, solution can enhance reservoir quality by generating secondary porosity, whereas cementation can destroy it. In some situations diagenesis can actually generate a hydrocarbon trap (Rittenhouse, 1972). Oil or gas moving up a permeable carrier bed may reach a cemented zone, which inhibits further migration (Fig. 7.34A). Conversely, oil may be trapped in zones where solution porosity has locally developed in a cemented rock (Fig. 7.34B). Secondary dolomitization can generate irregular diagenetic traps as the dolomite takes up less space than the original volume of limestone (see p. 252).

Diagenetic traps are not only formed by the solution or precipitation of mineral cements. As oil migrates to the surface, it may be degraded and oxidized by bacterial action if it reaches the shallow zone of meteoric

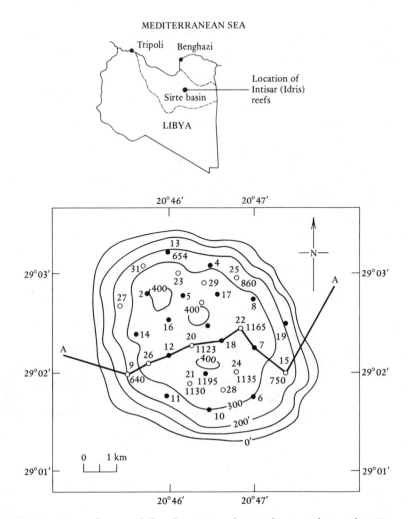

FIGURE 7.32 *Isopach map of the Idris "A" reef, Sirte basin, Libya. (After Terry and Williams, 1969.)*

water. Cases are known where this tarry residue acts as a seal, inhibiting further up-dip oil migration (Fig. 7.34C). The Shuguang oil field in the Liaohe basin is an example of a diagenetic trap caused by shallow-oil degradation (Ma Li et al., 1982).

Traps that owe their origin purely to diagenesis are rare, although there are probably a number of diagenetic traps around the world whose origin has gone unrecognized, and many are yet to be found (Wilson, 1977). Many traps, however, are due to a combination of diagenesis and one or more other causes. This type of origin is particularly true of the sub-unconformity traps discussed in the next section.

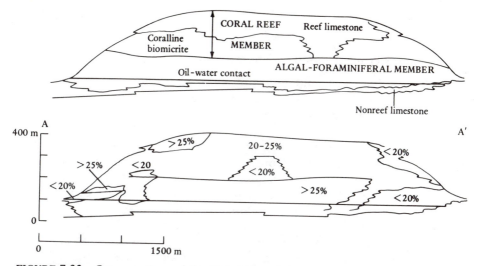

FIGURE 7.33 *Cross-sections of the Idris "A" reef showing facies (upper) and porosity distribution (lower). Note the lack of correlation between the cross-sections, a common problem of carbonate reservoirs. (After Terry and Williams, 1969.)*

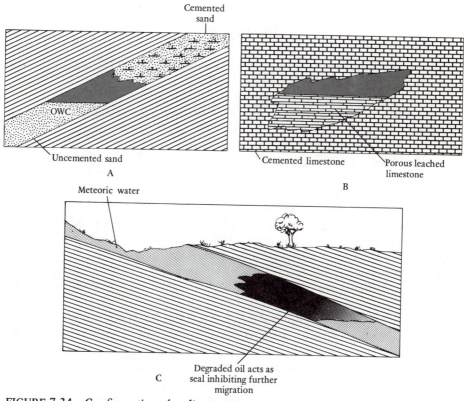

FIGURE 7.34 *Configurations for diagenetic traps caused by (A) cementation, (B) solution, and (C) shallow-oil degradation.*

Stratigraphic Traps Related to Unconformities

The channel, bar, reef, and diagenetic traps just described can occur both in conformable sequences and unconformities. The first three types often overlie unconformities, whereas diagenetically assisted traps underlie them.

The role of unconformities in the entrapment of hydrocarbons has been remarked by geologists from Levorsen (1934, 1964) to Chenoweth (1972) and Bushnell (1981). Unconformities facilitate the juxtaposition of porous reservoirs and impermeable shales that may act as source and seal. A large percentage of the known global petroleum reserves are trapped adjacent to the worldwide unconformities and source rocks of late Jurassic to mid-Cretaceous age. Many of these reserves are held in structural and combination traps, as well as the pure stratigraphic traps, which are described as follows. As already shown (Table 7.2), unconformity-related traps can be divided into those that occur above the unconformity and those that occur below it. These two types of unconformity trap are now described in turn.

Supraunconformity Traps

Stratigraphic traps that overlie unconformities include reefs and various types of terrigenous sands. These traps may be divided into three classes according to their geometry: sheet, channel, and strike valley.

Shallow marine or fluvial sands may onlap a planar unconformity. A stratigraphic trap may occur where these sands are overlain by shale and where the subunconformity rocks are also impermeable. In many ways, therefore, these onlap traps are similar to the pinchout traps described previously. Note, in particular, that both traps require stratigraphic permeability changes or structural closure in both directions along the paleostrike for the trap to be valid. The Cut Bank field of Montana is an example of an onlap stratigraphic trap (Fig. 7.35). Here the Lower Cretaceous Cutback sand unconformably onlaps Jurassic shales and is itself onlapped by the Kootenai Shale (Shelton, 1967). It contains recoverable reserves of over 200 million barrels of oil and 300 billion ft^3 of gas. In this case the reservoir is a fluvial sand, but in other onlap traps the reservoir is of marine origin.

Where an unconformity is irregular, sand often infills valleys cut into the old land surface. The location and trend of these valleys may be related to the resistance to weathering of the various strata that once cropped out at the old land surface. This process gives rise to paleogeomorphic traps (Martin, 1966). The two main groups of paleogeomorphic trap are channel and strike valley. Rivers draining a land surface may incise valleys into the bedrock, and these valleys may then be infilled with alluvium (both porous

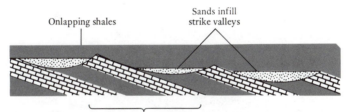

Onlapping shales

Sands infill
strike valleys

Subunconformity scarp and dip slope topography

FIGURE 7.35 *Cross-section showing the occurrence of strike valley sands. Examples of oil trapped in these sands are known from the Pennsylvanian of Oklahoma and the Cretaceous of New Mexico.*

sand and impermeable shale). The Glen Rock field, Wyoming, has already been described as an example of a fluvial channel stratigraphic trap. There is little difference between valleys cut in unconformities and those within conformable sequences. The Fiddler Creek and Clareton fields (Fig. 7.30) are examples of paleogeomorphic channel traps. There are many other such traps, ranging from fluvial channels to deep submarine ones like those of Miocene age in California (Martin, 1963).

Where alternating beds of hard and soft rocks are weathered and eroded, the soft strata form strike valleys between the resistant ridges of harder rock. Fluvial, and occasionally marine, sands within the strike valleys may be blanketed by a transgressive marine shale. Oil and gas may be stratigraphically trapped in the strike valley sands (Fig. 7.35). This type of trap was first described from the sub-Pennsylvanian unconformity of Oklahoma. Some of these sands are 60 to 70 km in length, yet only 1 or 2 km wide. Local thickening of the reservoir occurs where primary consequent valleys intersect strike valleys (Busch, 1959, 1961, 1974; Andresen, 1962, 1974). Other examples have been identified in the basal Niobrara (Upper Cretaceous) sands of the San Juan basin, New Mexico (McCubbin, 1969). These sands are shorter and wider than the Oklahoma sands. In both cases, however, the sand bodies are generally less than 30 m thick.

Subunconformity Traps

Stratigraphic traps also occur beneath unconformities where porous permeable beds have been truncated and overlain by impermeable clay. In many instances a seat seal is also provided by impermeable strata beneath the reservoir. As with pinchouts and onlaps, some closure is needed in both directions along the paleostrike. This closure may be structural or stratigraphic, but for many truncation traps it will be provided by the irregular topography of the unconformity (Fig. 7.36).

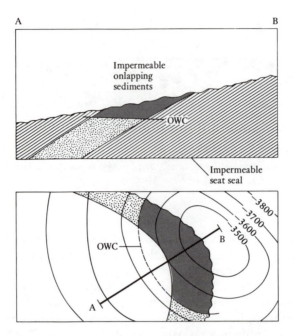

FIGURE 7.36   *Cross-section (upper) and map (lower) illustrating the geometry of a truncation trap. The contours drawn on the unconformity surface of the map define a buried hill.*

Most, if not all, truncation traps have had their reservoir quality enhanced by epidiagenesis (see p. 246). This secondary solution porosity induced by weathering is particularly well known in limestones, but also occurs in sands and basement.

Weathering of limestones ranges from minor moldic and vuggy porosity formation to the generation of karstic and collapse breccia zones of great reservoir potential. Examples of karstic weathering are known from the pre-Pennsylvanian weathering of the Cambro-Ordovician of Kansas (Walters and Price, 1948) and the Castellan field of offshore Spain. Figure 7.37 shows a Zechstein collapse breccia due to pre-Cretaceous weathering of the Auk field in the North Sea (Brennand and Van Veen, 1965). Subunconformity solution porosity is important in many sandstone reservoirs, such as the Sarir Group of Libya and the Brent Sand of the North Sea (see p. 369).

As discussed earlier (p. 172), a number of fields produce from basement rocks. In almost every case the reservoir is unconformably overlain by shales that have acted as source and seal. Production comes from fractures and solution pores where unstable minerals (generally mafics and

FIGURE 7.37 *Collapse breccia of Permian Zechstein carbonate caused by pre-Cretaceous weathering of the Auk field, North Sea. (Courtesy of Shell, U.K.)*

occasionally feldspar) have weathered out. A notable example of this type of trap is the Augila field of Libya, where one well produced 40,000 BOPD from granite (Williams, 1968, 1972). This remarkable stratigraphic trap also produces from sands and carbonates that onlap a buried granite hill (Fig. 7.38).

Unconformity-Related Traps: Conclusion

The preceding sections describe the various ways in which unconformities may trap hydrocarbons. It must be stressed that few traps are simple and monocausal. The number of combination traps in which an unconformity is combined with folding or faulting far outweighs simple unconformity traps. Many unconformity traps produce from both onlapping and sub-cropping reservoirs, as in the Augila field just noted. Similarly, as unconformities converge and merge up-structure, reservoirs that are both onlap and subcrop traps form. The East Texas field is one such example (Fig. 7.39). This giant field, with estimated recoverable reserves of 5600 million barrels, has a length of some 75 km and a maximum width of 25 km. It produces from the Cretaceous Woodbine sand, which unconformably overlies the Washita Group and is itself truncated by the Austin Chalk.

There has been considerable debate as to whether shales above or below an unconformity provide the oil for adjacent reservoirs. With modern geochemical techniques of matching crudes and their parent kerogen, this problem can generally be solved. In some cases the source rock is clearly beneath the unconformity. A notable example of this condition is the Hassi Massaoud field, in which Cambro-Ordovician sands contain oil

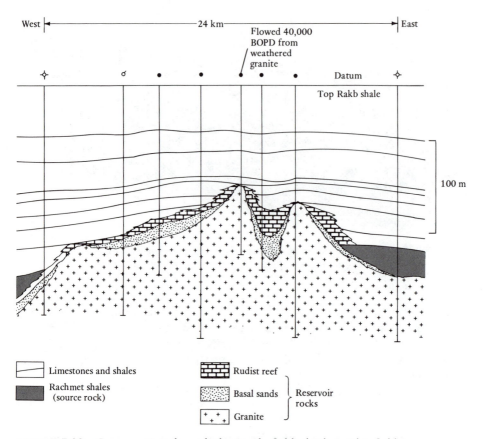

FIGURE 7.38   *Cross-section through the Augila field of Libya. This field is a complex trap that produces partly from sands and reefal carbonates and partly from fractured and weathered granite. (After Williams, 1968.)*

derived from the Tannezuft Shale (Silurian) on a large paleohigh sealed by Triassic evaporites. Another example is the Brent field and its associates in the North Sea. Here the early Cretaceous Cimmerian unconformity seals many traps ranging in age from Jurassic to Devonian. All are sourced by the Upper Jurassic Kimmeridge Clay (see also p. 370).

Conversely, in many fields shales above an unconformity act as source and seal. This situation occurs, for example, in Prudhoe Bay, Alaska (Morgridge and Smith, 1972; Jones and Speers, 1976) and Sarir field, Libya (Sanford, 1970). That an impermeable shale should act as both source and seal may seem incomprehensible at first. Actually, two explanations can account for this phenomenon. The oil trapped in a structural culmination need not necessarily have moved downward in a physical sense, but may have moved up the flank of the structure from adjacent lows, and in so

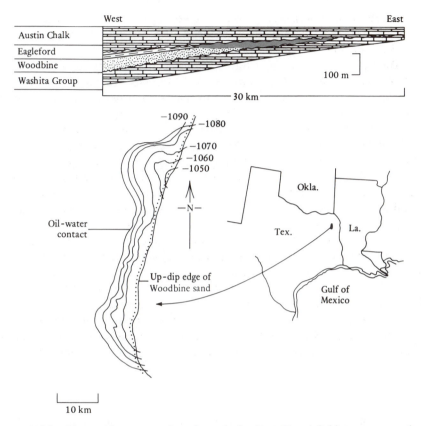

FIGURE 7.39 *Map and cross-section through the East Texas field, a supra- and subunconformity giant stratigraphic trap. Contours in meters. (After Minor and Hanna, 1941.)*

doing moved downward only in a stratigraphic sense. Also, any good engineer can prove mathematically that fluid can move downward if there is a sufficient pressure differential.

Relationship Between Stratigraphic Traps and Sedimentary Facies

In the same way that structural traps are related to tectonic style, stratigraphic traps are often related to sedimentary environments. A genetic grouping of stratigraphic traps and facies may be more useful as an aid to prospect prediction than the formal classification used for the preceding descriptive section. This approach needs to be qualified. As discussed in Chapter 6, the distribution of porosity is often unrelated to facies in carbonate reservoirs. Thus, apart from the obvious environmental control of reefs, carbonate stratigraphic traps are seldom facies related. Diagenetic processes and the position of unconformities are generally more significant than earlier environmental controls.

Turning to sandstones, Chapter 6 noted that reservoir quality is generally facies related and the effects of diagenesis are usually less significant. Pinchout traps, whether due to onlap or truncation, can occur in a blanket sand deposited in any environment. All they require is the necessary structural tilt and an adequate seal. Many varieties of sandstone stratigraphic trap have distributions that are largely facies related, as shown in Table 7.4. Channel traps may, of course, be found in almost any environment from fluvial to deep marine, as already noted. Strike valley sands are generally of continental origin. Growth-fault-related traps, such as rollover anticlines, and clay diapir traps are not stratigraphic in origin, but their occurrence is closely related to environment. They are restricted to rapidly deposited regressive wedges of deltaic sands and overpressured clay. Diapiric traps are commonly present to the basinward side of growth faults. In deep basinal settings stratigraphic traps may occur in submarine channels and in submarine fans. The latter include up-dip pinchouts as well as closed structures where fan paleotopography is preserved.

This brief review shows that many stratigraphic traps are related to depositional environment. This correlation obviously aids the prediction of the type of trap to be anticipated in a particular sedimentary facies.

TABLE 7.4   Relationship between stratigraphic traps and sedimentary environments of sandstone reservoirs

Environment		Trap type
Continental	Eolian	Pinchout
	Fluvial	Channel
		Strike valley
Coastal	Barrier bar	Shoestring
	Delta	Channel Crevasse-splay and mouth bar Growth-fault-related Diapir-related
Deep marine		Submarine fan pinchout or paleotopographic closure Submarine channel

HYDRODYNAMIC TRAPS

The third group of traps to consider, in addition to structural and stratigraphic ones, are the hydrodynamic traps. In these traps hydrodynamic
movement of water is essential to prevent the upward movement of oil or
gas. The concept was first formulated by Hubbert (1953) and embellished
by Levorsen (1966). The basic argument is that oil or gas will generally
move upward along permeable carrier beds to the earth's surface except
where they encounter a permeability barrier, structural or stratigraphic,
beneath which they may be trapped.

Where water is moving hydrodynamically down permeable beds, it
may encounter upward-moving oil. When the hydrodynamic force of the
water is greater than the force due to the buoyancy of the oil droplets, the
oil will be restrained from upward movement and will be trapped within
the bed without any permeability barrier. Levorsen illustrated this concept
by using the analogy of corks in a glass tube down which water was flowing. The corks would tend to accumulate where the tube was restricted,
causing a local increase in the fluid potential gradient. In the real world
this increase might occur where there was a local reversal of dip or facies
change, or even a local fluctuation in the potentiometric gradient (Fig.
7.40). Hubbert (1953) expounded the theory behind the concept of the
hydrodynamic trap.

The role of the hydrodynamic flow in causing tilted oil-water contacts
was discussed earlier in this chapter (p. 278). Although this phenomenon
occurs in many fields, pure hydrodynamic traps are very rare. One such

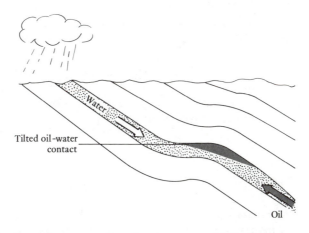

FIGURE 7.40 *Crustal cross-section showing a pure hydrodynamic trap. There is
no vertical structural relief. Oil migrating upward is trapped in a monocline by the
downward flow of water.*

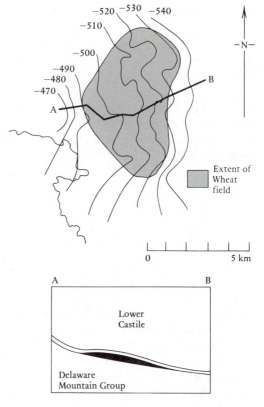

FIGURE 7.41 *Map (upper) and cross-section (lower) through the Wheat field, Delaware basin, Texas. This field is an example of a hydrodynamic trap. The structure contours on the map are drawn on top of the Wheat field reservoir. (After Adams, 1936.)*

example, however, is the Wheat field of the Delaware basin, West Texas, described by Adams (1936). This field occurs in a gentle flexure in mono-clinally dipping beds, which have closure but not vertical relief (Fig. 7.41). An obvious tilted oil-water contact points to the role of hydrodynamic flow. If there was no flow and pressures were hydrostatic, oil could not be trapped because the trap lacks four-way structural closure. Traps like the Wheat field are very rare and are not regarded as a prime target for exploration, although the significance of hydrodynamic flow in shifting fields down the flank of structures must always be remembered.

COMBINATION TRAPS

Many oil and gas fields around the world are not due *solely* to structure *or* stratigraphy *or* hydrodynamic flow, but to a combination of two or more of these forces. Such fields may properly be termed *combination traps*. Most of these traps are caused by a combination of structural and stratigraphical

processes. Structural-hydrodynamic and stratigraphic-hydrodynamic traps
are rare. Because of the multiplicity of different types of combination traps,
discussing them in groups or any logical order is impractical. One or two
examples must suffice.

On a small scale, oil may be trapped in shoestring sands (channels or
bars) that cross-cut anticlines (Fig. 7.42). As pointed out earlier, pinchout,
onlap, and truncation traps all require closure, which is very often structural, along the strike. Likewise, folded and/or faulted beds may be sealed
by unconformities to form another group of traps. Examples of folded and
faulted unconformity traps are now discussed in turn. The Prudhoe Bay
field on the North Slope of Alaska is a fine example of a combination trap
(Morgridge and Smith, 1972; Jones and Spears, 1976; Jamison et al., 1980;
Bushnell, 1981). Here a series of Carboniferous, Permian, Triassic, Jurassic, and basal Cretaceous sediments were folded into a westerly plunging
anticlinal nose. This structure was truncated progressively to the northeast
and overlain by Cretaceous shales, which act as source and seal to the trap.
Oil and gas are trapped in the older beds, which subcrop the unconformity.
The main reservoir is the fluvial Sadlerochit Sandstone (Triassic). Additional closure is provided by major faults on the northern and southwestern side of the structure (Fig. 7.43).

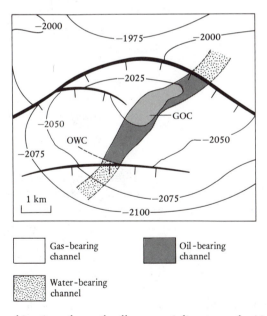

FIGURE 7.42 *Combination channel-rollover anticline trap, the Main Pass Block 35
field, offshore Louisiana. Contours are drawn on top of "G" sand (C.I. = 25 m).
(After Hartman, 1972.)*

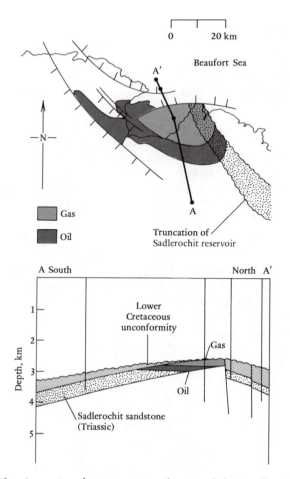

FIGURE 7.43 *Map (upper) and cross-section (lower) of the Prudhoe Bay field of Alaska. (After Jamison et al., 1980.)*

Fault-unconformity traps characterize the Jurassic Brent province of the northern North Sea. The reservoir is the Middle Jurassic deltaic Brent Sandstone Group, which is overlain by Upper Jurassic shales. These include the Kimmeridge Clay Formation, which is the source rock. Late Jurassic–early Cretaceous Cimmerian movement created numerous fault blocks, which were tilted, truncated, and unconformably overlain by Cretaceous shales. The resultant traps, which include fields such as Brent, Statfjord, Murchison, Hutton, Thistle, and Piper (Fig. 7.16), are thus all combination fault-unconformity traps with varying degrees of truncation of the reservoir.

The list of combination traps is endless. The preceding examples illustrate some of their variety and complexity.

TRAPS: CONCLUSION

Timing of Trap Development Relative to Petroleum Migration and Reservoir Deposition

The time of trap formation relative to petroleum migration is obviously extremely important. If traps predate migration, they will be productive. If they postdate migration, they will be barren. Questions concerning a prospect should include: Which horizons are known or presumed to be source rocks? Can time-burial depth curves be used to determine the time of petroleum generation? If the prospect is structural, did the fold or fault form before or after migration? In the case of truncation traps the source rock may underlie or overlie the unconformity (as in Hassi Messaoud and Prudhoe Bay, discussed on pages 247 and 321, respectively). It is important, though, to establish that a truncation trap was thoroughly sealed before petroleum generation began.

Postmigration structural movement may also be relevant. Faults can open to allow petroleum to undergo further migration. Structural closure may tighten. This tightening is not of itself harmful, unless it is accompanied by crestal fractures, which increase the permeability of the seal. Uplift and erosion may breach the crests of traps. Regional tilting may trigger extensive secondary migration of petroleum because the spill points of traps may be altered.

The time factor is also important, although less so, when considering the relationship between the deposition of the reservoir and the formation of structural traps. At its simplest level syndepositional structural growth will affect sedimentation and hence reservoir characteristics; postdepositional structures, on the other hand, may not correlate with facies variations within the reservoir. In a terrigenous basin with synchronous structural growth, channels and fans of fluvial, deltaic, or submarine origin will develop in the lows. Simultaneously, winnowed marine shoal sands may be present on the structural highs, although, where too high, truncation may be present. In carbonate basins oolite shoals and reefs may be anticipated on structural highs, but, for reasons already discussed, porosity is often unrelated to facies in carbonate sediments.

The situation may be rather different in areas where structural traps postdate sedimentation. Once sediments are lithified, they respond to stress by fracturing. Thus in both sandstone and carbonate reservoirs, porosity and permeability due to fractures may be closely related to structure. Fracture intensity, and thus reservoir performance, may be enhanced over the crests of folds and the apices of diapirs, as well as adjacent to faults.

This review of the timing of trap formation relative to reservoir deposition and petroleum generation emphasizes their importance. Generally, the relationship between structural movement and petroleum migration is more important than whether structures are syn- or postdepositional.

Relative Frequency of the Different Types of Trap

This chapter has covered a great deal of ground. Therefore it is appropriate to close with an attempt to arrange the various types of traps in some sort of order of importance. A great aid to this classification has been a global analysis of the trapping mechanisms of known giant oil fields by Moody (1975). His study showed that by far the majority of giant oil fields are anticlines, followed, a long way behind and in order of decreasing importance, by combination traps, reefs, pinchouts, truncations, salt domes, and faults (Fig. 7.44).

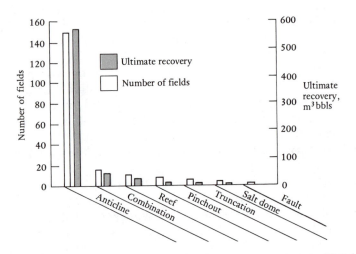

FIGURE 7.44   *Histograms showing the mode of entrapment of giant oil fields around the world. (After Moody, 1975.)*

These interesting data deserve careful analysis. They only concern giant fields, which are defined as those with over 500 million barrels of recoverable reserves. This chapter has shown that stratigraphically trapped oil tends to occur in small fields because of the limited extent of channel, bar, and reef reservoirs. Also, these data pertain to oil fields, not gas fields, although, intuitively, they would probably be similar.

Note also that these figures were compiled some time ago. With the steady improvement in seismic geophysics, the percentage of known subtle stratigraphic traps may now have increased in proportion to anticlines. Most importantly, these figures record the entrapment of *known* giant oil fields. The various percentages of the different types of giant trap that actually exist under the earth might be completely different.

These figures, therefore, reflect man's ability to find oil, not the total number of fields in the world. Finding oil in anticlines is easy. They may be mapped at the surface or detected seismically in the subsurface. The concept of the anticlinal trap is a simple one to grasp for the managers, engineers, and farmers who may actually make the decision to drill, or not, based on a geologist's recommendation.

Stratigraphic traps, on the other hand, are harder to locate. Few can simply be picked off a brightly colored seismic section; most require an integration of seismic, log, and real rock data with sophisticated geological concepts. An explorationist would therefore find it harder to develop a stratigraphic trap prospect, and harder still to explain it to the lay audience, which may hold the purse strings. Pratt (1942) spoke very truly when he said that "oil is found in the minds of men."

SELECTED BIBLIOGRAPHY

For major compilations of oil field case histories, see:

HALBOUTY, M. T. (ed.). 1970. *Geology of Giant Petroleum Fields.* Am. Assoc. Petrol. Geol., Mem. No. 14, 575 pp.

HALBOUTY, M. T. (ed.). 1980. *Giant Oil and Gas Fields of the Decade: 1968–1978.* Am. Assoc. Petrol. Geol., Mem. No. 30, 596 pp.

KING, R. E. (ed.). 1972. *Stratigraphic Oil and Gas Fields.* Am. Assoc. Petrol. Geol., Mem. No. 16, 685 pp.

For an elegant review of the relationship between trap types, sedimentary basins, and plate tectonics, see:

HARDING, T. P. and LOWELL, J. D. 1979. Structural styles, their plate tectonic habitats, and hydrocarbon traps in petroleum provinces. *Am. Assoc. Petrol. Geol. Bull.,* 63, 1016–1058.

REFERENCES

ADAMS, J. E. 1936. Oil pool of open reservoir type. *Am. Assoc. Petrol. Geol. Bull.,* 20, 780–796.

ANDRESEN, M. J. 1962. Paleodrainage patterns: their mapping from subsurface data, and their paleogeographic value. *Am. Assoc. Petrol. Geol. Bull.,* 46, 398–405.

ASQUITH, D. O. 1970. Depositional topography and major marine environments, late Cretaceous, Wyoming. *Am. Assoc. Petrol. Geol. Bull.,* 54, 1184–1224.

BAILEY, R. J. and STONELEY, R. 1981.   Petroleum: entrapment and conclusion. In: *Economic Geology and Geotectonics*. D. H. Tarling (ed.). Oxford: Blackwell, 73–97.

BALDUCCHI, A. and POMMIER, G. 1970.   Cambrian oil field of Hassi Massaoud, Algeria. In: *Geology of Giant Petroleum Fields*. M. T. Halbouty (ed.). Am. Assoc. Petrol. Geol., Mem. No. 14, 477–488.

BARBAT, F. W. 1958.   The Los Angeles basin area, California. In: *The Habitat of Oil*. L. G. Weeks (ed.). Tulsa: Am. Assoc. Petrol. Geol., 62–77.

BERG, R. R. and DAVIES, D. K. 1968.   Origin of Lower Cretaceous Muddy sandstone at Bell Creek field, Montana. *Am. Assoc. Petrol. Geol. Bull.*, *52*, 1888–1898.

BERNER, H.; RAMSBERG, H.; and STEPHANSSON, O. 1972.   Diapirism in theory and experiment. *Tectonophysics*, *15*, 197–218.

BISHOP, R. S. 1978.   Mechanism for emplacement of piercement diapirs. *Am. Assoc. Petrol. Geol. Bull.*, *62*, 1561–1584.

BLAIR, D. G. 1975.   Structural styles in North Sea oil and gas fields. In: *Petroleum and the Continental Shelf of Northwest Europe*, vol. 1. A. W. Woodland (ed.). London: Applied Science Publishers, 327–338.

BRADY, T. J.; CAMPBELL, N. D. H.; and MAHER, C. E. 1980.   Intisar 'D' Oil Field, Libya. In: *Giant Oil and Gas Fields of the Decade 1968–78*. M. T. Halbouty (ed.). Am. Assoc. Petrol. Geol., Mem. No. 30, 543–564.

BRENNAND, T. P. and VAN VEEN, F. R. 1975.   The Auk oil field. In: *Petroleum and the Continental Shelf of Northwest Europe*. A. W. Woodland (ed.). London: Applied Science Publishers, 275–284.

BUSCH, D. A. 1959.   Prospecting for stratigraphic traps. *Am. Assoc. Petrol. Geol. Bull.*, *43*, 2829–2843.

BUSCH, D. A. 1961.   Prospecting for stratigraphic traps. In: *Geometry of Sandstone Bodies*. J. A. Peterson and J. C. Osmond (eds.). Tulsa: Am. Assoc. Petrol. Geol., 220–232.

BUSCH, D. A. 1971.   Genetic units in delta prospecting. *Am. Assoc. Petrol. Geol. Bull.*, *55*, 1137–1154.

BUSCH, D. A. 1974.   *Stratigraphic Traps in Sandstones—Exploration Techniques*. Tulsa: Am. Assoc. Petrol. Geol., Mem. No. 21, 174 pp.

BUSHNELL, H. 1981.   Unconformities—key to N. slope oil. *Oil & Gas J.*, 12 Jan., 112–118.

CARLL, J. F. 1880.   The geology of the oil regions of Warren, Venango, Clarion and Butler Counties. *Penns. Geol. Surv.*, *3*, 482 pp.

CHENOWITH, P. A. 1972.   Unconformity traps. In: *Stratigraphic Oil and Gas Fields*. R. E. King (ed.). Tulsa: Am. Assoc. Petrol. Geol., Mem. No. 15, 42–46.

CLAPP, F. G. 1910.   A proposed classification of petroleum and natural gas fields based on structure. *Econ. Geol.*, *5*, 503–521.

CLAPP, F. G. 1917.   Revision of the structural classification of the petroleum and natural gas fields. *Geol. Soc. Am. Bull.*, *28*, 553–602.

CLAPP, F. G. 1929.   Role of geologic structure in the accumulation of petroleum. In: *Structure of Typical American Oil Fields*, vol. 2. Tulsa: Am. Assoc. Petrol. Geol., 667–716.

COLMANN-SADD, S. P. 1978.   Fold development in Zagros simply-folded belt, southwest Iran. *Am. Assoc. Petrol. Geol. Bull.*, *62*, 984–1003.

CONYBEARE, C. E. B. 1976.   *Geomorphology of Oil and Gas Fields in Sandstone Bodies*. Amsterdam: Elsevier, 341 pp.

CURRY, W. H. and CURRY, W. H. III. 1972. South Glenrock oil field, Wyoming: pre-discovery thinking and postdiscovery description. In: *Stratigraphic Oil and Gas Fields*. R. E. King (ed.). Tulsa: Am. Assoc. Petrol. Geol., Mem. No. 16, 415–427.

DAILLY, G. C. 1976. A possible mechanism relating progradation, growth faulting, clay diapirism and overthrusting in a regressive sequence of sediments. *Bull. Can. Petrol. Geol.*, 24, 92–116.

DOTT, R. H. and REYNOLDS, M. J. 1969. *Sourcebook for Petroleum Geology*. Tulsa: Am. Assoc. Petrol. Geol., Mem. No. 5, 471 pp.

EMERY, K. O. 1980. Continental margins—classification and petroleum prospects. *Am. Assoc. Petrol. Geol. Bull.*, 64, 297–315.

EREMAKO, N. A. and MICHAILOV, I. 1974. Hydrodynamic pools at faults. *Bull. Can. Petrol. Geol.*, 22, 106–118.

EVAMY, D. D.; HAREMBOURE, J.; KAMERLING, P.; KNAPP, W. A.; MOLLOY, F. A.; and ROWLANDS, P. H. 1978. Hydrocarbon habitat of Tertiary Niger delta. *Am. Assoc. Petrol. Geol. Bull.*, 62, 1–39.

FALCON, N. L. 1958. Position of oil fields of southwest Iran with respect to relevant sedimentary basins. In: *The Habitat of Oil*. L. G. Weeks (ed.). Tulsa: Am. Assoc. Petrol. Geol., 1279–1293.

FALCON, N. L. 1969. Problems of the relationship between surface structure and deep displacements illustrated by the Zagros Range. In: *Time and Place in Orogeny*. P. E. Kent (ed.). Geol. Soc. Lond., Sp. Pub. No. 3, 9–22.

GOEBEL, L. A. 1950. Cairo field, Union County, Arkansas. *Am. Assoc. Petrol. Geol. Bull.*, 34, 1954–1980.

HALBOUTY, M. T. 1967. *Salt Domes, Gulf Region, United States and Mexico*. Houston: Gulf Publishing Co., 425 pp.

HALBOUTY, M. T. 1972. Rationale for deliberate pursuit of stratigraphic, uncon-formity and paleogeomorphic traps. *Am. Assoc. Petrol. Geol. Bull.*, 56, 537–541.

HALBOUTY, M. T.; MEYERHOFF, A. A.; KING, R. E.; DOTT, R. H.; KLEMME, H. D.; and SHABAD, T. 1970. World's giant oil and gas fields, geologic factors affecting their formation, and basin classification. Part I, Giant Oil and Gas Fields. In: *Geology of Giant Petroleum Fields*. M. T. Halbouty (ed.). Tulsa: Am. Assoc. Petrol. Geol., Mem. No. 14, 502–556.

HARDING, T. P. 1973. Newport-Inglewood fault zone, Los Angeles basin, Califor-nia. *Am. Assoc. Petrol. Geol. Bull.*, 57, 97–116.

HARDING, T. P. 1974. Petroleum traps associated with wrench faults. *Am. Assoc. Petrol. Geol. Bull.*, 58, 1290–1304.

HARDING, T. P. and LOWELL, J. D. 1979. Structural styles, their plate-tectonic habi-tats, and hydrocarbon traps in petroleum provinces. *Am. Assoc. Petrol. Geol. Bull.*, 63, 1016–1058.

HARMS, J. C. 1966. Valley Fill, Western Nebraska. *Am. Assoc. Petrol. Geol. Bull.*, 50, 2119–2149.

HARTMAN, J. A. 1972. "G_2", channel sandstone, Mainpass Block 35 field, offshore Louisiana. *Am. Assoc. Petrol. Geol. Bull.*, 56, 554–558.

HAY, J. T. C. 1977. The Thistle oilfield. In: *Mesozoic Northern North Sea Sympo-sium*. K. Finstad and R. C. Selley (eds.). Oslo: Norweg. Petrol. Soc., *11*, 1–20.

HOBSON, G. D. and TIRATSOO, E. N. 1975. *Introduction to Petroleum Geology*. Bea-consfield: Scientific Press, 300 pp.

HOC, A. 1979. *Continental Margins Geological and Geophysical Research Needs and Problems.* Washington: Nat. Acad. Sci., 302 pp.

HOLLENSHEAD, C. T. and PRITCHARD, R. L. 1961. Geometry of producing Mesaverde Sandstones, San Juan basin. In: *Geometry of Sandstone Bodies.* J. A. Peterson and J. C. Osmond (eds.). Tulsa: Am. Assoc. Petrol. Geol., 98–118.

HUBBERT, M. K. 1953. Entrapment of petroleum under hydrodynamic conditions. *Am. Assoc. Petrol. Geol. Bull.*, 37, 1454–2026.

HULL, C. E. and WARMAN, H. R. 1970. Asmari oil fields of Iran. In: *Geology of Giant Petroleum Fields.* M. T. Halbouty (ed.). Tulsa: Am. Assoc. Petrol. Geol., Mem. No. 14, 428–437.

JAMISON, H. C.; BROCKETT, L. D.; and McINTOSH, R. A. 1980. Prudhoe Bay: a 10-year perspective. In: *Giant Oil and Gas Fields of the Decade: 1968–78.* M. T. Halbouty (ed.). Tulsa: Am. Assoc. Petrol. Geol., 289–314.

JANOSCHEK, R. 1958. The Inner Alpine Vienna Basin, an example of a small sedimentary area with rich oil accumulation. In: *The Habitat of Oil.* L. G. Weeks (ed.). Tulsa: Am. Assoc. Petrol. Geol., 1134–1152.

JONES, H. P. and SPEERS, R. G. 1976. Permo-Triassic reservoirs on Prudhoe Bay Field, North Slope, Alaska. In: *North American Oil and Gas Fields.* J. Braunstein (ed.). Tulsa: Am. Assoc. Petrol. Geol., Mem. No. 24, 23–50.

JONES, O. A. and ENDEAN, R. 1973. *Biology and Geology of Coral Reefs* (2 vols). London: Academic Press, 410, 435.

KENT, P. 1979. The emergent Hormuz salt plugs of southern Iran. *J. Petrol. Geol.*, 2, 117–144.

KESSLER, L. G.; ZANG, R. D.; ENGLEHORN, J. A.; and EGER, J. D. 1980. Stratigraphy and sedimentology of a Palaeocene submarine fan complex, Cod Field, Norwegian North Sea. In: *The Sedimentation of North Sea Reservoir Rocks.* Oslo: Norweg. Petrol. Soc., VII, 19 pp.

KING, R. E. (ed.). 1972. *Stratigraphic oil and gas fields—classification, exploration methods, and case histories.* Am. Assoc. Petrol. Geol., Mem. No. 16, 687 pp.

LAMB, C. F. 1980. Painter reservoir field—giant in Wyoming thrust belt. *Am. Assoc. Petrol. Geol. Bull.*, 64, 638–644.

LAPORTE, L. F. (ed.). 1974. *Reefs in Time and Space.* Tulsa: Soc. Econ. Pal. and Min., Spec. Pub. No. 18, 255 pp.

LEES, G. M. 1952. Foreland folding. *Quart. J. Geol. Soc. Lond.*, 108, 1–34.

LEHNER, P. and DE RUITER, P. A. C. 1977. Structural history of Atlantic margin of Africa. *Am. Assoc. Petrol. Geol. Bull.*, 61, 961–981.

LEVORSEN, A. I. 1934. Relation of oil and gas pools to unconformities in the midContinent region. In: *Problems of Petroleum Geology.* R. E. Wrather and F. H. Lahee (eds.). Tulsa: Am. Assoc. Petrol. Geol., 761–784.

LEVORSEN, A. I. 1964. Big geology for big needs. *Am. Assoc. Petrol. Geol. Bull.*, 48, 141–156.

LEVORSEN, A. I. 1966. The obscure and subtle trap. *Am. Assoc. Petrol. Geol. Bull.*, 50, 2058–2067.

LEVORSEN, A. I. 1967. Geology of Petroleum, Second Edition. San Francisco: Freeman, 724 pp.

LOVELY, H. R. 1943. Classification of oil reservoirs. *Am. Assoc. Petrol. Geol. Bull.*, 27, 224–237.

MACKENZIE, D. B. 1972. Primary stratigraphic traps in sandstone. In: *Stratigraphic Oil and Gas Fields*. R. E. King (ed.). Tulsa: Am. Assoc. Petrol. Geol., Mem. No. 16, 47–63.

MA LI; GE TAISHENG; ZHAO XUEPING; ZIE TAIJUN; GE RONG; and DANG ZHENRONG. 1982. Oil basins and subtle traps in the eastern part of China. In: *The Deliberate Search for the Subtle Trap*. M. T. Halbouty (ed.). Tulsa: Am. Assoc. Petrol. Geol., 287–316.

MARTIN, D. B. 1963. Rosedale channel: evidence for late Miocene submarine erosion in Great Valley of California. *Am. Assoc. Petrol. Geol. Bull.*, 47, 441–456.

MARTIN, R. 1966. Paleogeomorphology and its application to exploration for oil and gas (with examples from Western Canada). *Am. Assoc. Petrol. Geol. Bull.*, 50, 2277–2311.

MAXWELL, W. G. H. 1968. *Atlas of the Great Barrier Reef*. Amsterdam: Elsevier, 242 pp.

MAYUGA, M. N. 1970. California's giant—Wilmington oil field. In: *Geology of Giant Petroleum Fields*. Tulsa: Am. Assoc. Petrol. Geol., Mem. No. 14, 158–184.

McCUBBIN, D. G. 1969. Cretaceous strike-valley sandstone reservoirs, northwestern New Mexico. *Am. Assoc. Petrol. Geol. Bull.*, 53, 2114–2140.

McGREGOR, A. A. and BIGGS, C. A. 1970. Bell Creek field, Montana: a rich stratigraphic trap. In: *Geology of Giant Petroleum Fields*. M. T. Halbouty (ed.). Tulsa: Am. Assoc. Petrol. Geol., Mem. No. 14, 128–146.

McGREGOR, A. A. and BIGGS, C. A. 1972. Bell Creek field, Montana: a rich stratigraphic trap. In: *Stratigraphic Oil and Gas Fields*. R. E. King (ed.). Tulsa: Am. Assoc. Petrol. Geol., Mem. No. 16, 367–375.

MINOR, H. E. and HANNA, M. A. 1941. East Texas oil field. In: *Stratigraphic Type Oil Fields*. Tulsa: Am. Assoc. Petrol. Geol., 600–640.

MOODY, J. D. 1975. Distribution and geological characteristics of giant oil fields. In: *Petroleum and Global Tectonics*. A. G. Fischer and S. Judson (eds.). Princeton: Princeton Univ. Press, 307–320.

MORGRIDGE, D. L. and SMITH, W. B. 1972. Geology and discovery of Prudhoe Bay field, eastern Arctic Slope, Alaska. In: *Stratigraphic Oil and Gas Fields*. R. E. King (ed.). Tulsa: Am. Assoc. Petrol. Geol., Mem. No. 16, 489–501.

ORTON, E. 1889. The Trenton limestone as a source for petroleum and inflammable gas in Ohio and Indiana. *U.S. Geol. Surv. Ann. Rep. 8*, 475–662.

OXLEY, M. L. and HERLING, D. E. 1972. The Bryan field—a sedimentary anticline. *Geophysics*, 37, 59–67.

PRATT, W. E. 1942. *Oil in the Earth*. Lawrence: Univ. of Kansas Press, 105 pp.

RITTENHOUSE, G. 1972. Stratigraphic trap classification. In: *Stratigraphic Oil and Gas Fields—Classification, Exploration Methods and Case Histories*. R. E. King (ed.). Tulsa: Am. Assoc. Petrol. Geol., Mem. No. 16, 14–28.

SABINS, F. F. 1963. Anatomy of a stratigraphic trap. *Am. Assoc. Petrol. Geol. Bull.*, 47, 193–228.

SABINS, F. R. 1972. Comparison of Bisti and Horseshoe Canyon stratigraphic traps, San Juan basin, New Mexico. In: *Stratigraphic Oil and Gas Fields*. R. E. King (ed.). Tulsa: Am. Assoc. Petrol. Geol., Mem. No. 10, 610–622.

SANFORD, R. M. 1970. Sarir oil field, Libya—desert surprise. In: *Geology of Giant Petroleum Fields*. M. T. Halbouty (ed.). Tulsa: Am. Assoc. Petrol. Geol., Mem. No. 14, 449–476.

SCHWADE, I. T.; CARLSON, S. A.; and O'FLYNN, J. B. 1958. Geologic environment of Cuyuma Valley oil fields, California. In: *The Habitat of Oil*. L. G. Weeks (ed.). Tulsa: Am. Assoc. Petrol. Geol., 78–98.

SHELTON, J. W. 1967. Stratigraphic models and general criteria for recognition of alluvial, barrier bar and turbidity current sand deposits. *Am. Assoc. Petrol. Geol. Bull.*, *51*, 2441–2460.

SHELTON, J. W. 1968. Role of contemporaneous faulting during basinal subsidence. *Am. Assoc. Petrol. Geol. Bull.*, *52*, 399–413.

SIMONSON, R. R. 1958. Oil in the San Joachim Valley. In: *The Habitat of Oil*. L. G. Weeks (ed.). Tulsa: Am. Assoc. Petrol. Geol., 99–112.

SLINGER, F. C. P. and CRICHTON. 1959. The geology and development of the Gachsaran field, southwest Iran. New York: *Proc. 5th World Petrol. Cong.*, Section 1, 349–375.

SMITH, D. A. 1980. Sealing and non-sealing faults in Louisiana Gulf Coast salt basin. *Am. Assoc. Petrol. Geol. Bull.*, *64*, 145–172.

STANLEY, T. B. 1970. Vicksburg fault zone, Texas. In: *Geology of Giant Petroleum Fields*. M. T. Halbouty (ed.). Tulsa: Am. Assoc. Petrol. Geol., 301–308.

TERRY, C. E. and WILLIAMS, J. J. 1969. The Idris "A" bioherm and oilfield, Sirte basin, Libya. In: *The Exploration for Petroleum in Europe and North Africa*. P. Hepple (ed.). London: Inst. Petrol., 31–48.

TOBISON, N. H. 1972. Boxer field, Morgan County, Colorado, 1972. In: *Stratigraphic Oil and Gas Fields*. R. E. King (ed.). Tulsa: Am. Assoc. Petrol. Geol., Mem. No. 10, 383–388.

TSCHOPP, R. H. 1967. Development of the Fahud field. Mexico: *Proc. 7th World Petrol. Cong.*, *2*, 243–250.

WEBER, K. J. and DAUKORU, E. 1975. Petroleum geology of the Niger delta. Tokyo: *Trans. 9th World Petrol. Cong.*, *2*, 209–221.

WEBER, K. J.; MANDL, G.; PILAAR, W. F.; LEHNER, F.; and PRECIOUS, R. G. 1980. The role of faults in hydrocarbon migration and trapping in Nigerian growth fault structures. Houston: *Offshore Technol. Conf.*, Paper 3356, 2643–2651.

WEIMER, R. J. and DAVIS, T. L. 1977. Stratigraphic and seismic evidence for late Cretaceous faulting, Denver basin, Colorado. In: *Seismic Stratigraphy: Applications to Hydrocarbon Exploration*. C. E. Payton (ed.). Tulsa: Am. Assoc. Petrol. Geol., Mem. No. 26, 277–300.

WILCOX, R. E.; HARDING, T. P.; and SEELY, D. R. 1973. Basic wrench tectonics. *Am. Assoc. Petrol. Geol. Bull.*, *57*, 74–96.

WILLIAMS, J. J. 1968. The stratigraphy and igneous reservoirs of the Augila field, Libya. In: *Geology and Archeology of Northern Cyrenaica, Libya*. T. F. Barr (ed.). 197–206.

WILLIAMS, J. J. 1972. Augila field, Libya: depositional environment and diagenesis of sedimentary reservoir and description of igneous reservoir. In: *Stratigraphic Oil and Gas Fields*. R. E. King (ed.). Tulsa: Am. Assoc. Petrol. Geol., Mem. No. 16, 623–632.

WILLIAMS, J. J.; CONNOR, D. C.; and PETERSON, K. E. 1975. The Piper oil-field, U.K. North Sea: A fault-block structure with Upper Jurassic beach-bar reservoir sands. In: *Petroleum and the Continental Shelf of North West Europe, vol. I, Geology*. A. W. Woodland (ed.). London: Applied Science Publishers, 363–378.

WILSON, H. H. 1977. 'Frozen-in' hydrocarbon accumulations in diagenetic traps—exploration targets. *Am. Assoc. Petrol. Geol. Bull.*, *61*, 483–491.

WONCIK, J. 1972. Recluse field, Campbell County, Wyoming. In: *Stratigraphic Oil & Gas Fields*. R. E. King (ed.). Am. Assoc. Petrol. Geol., Mem. No. 16, 376–382.

YUSTER, S. T. 1953. Some Theoretical Considerations of Tilted Water Tables. Tech. Paper 3564. *Trans. Amer. Inst. Min. Engrs.* 198, 149–153.

8

Sedimentary Basins

BASIC CONCEPTS AND TERMS

A sedimentary basin is an area of the earth's crust that is underlain by a thick sequence of sedimentary rocks. Hydrocarbons commonly occur in sedimentary basins and are absent from intervening areas of igneous and metamorphic rocks. This fundamental truth is one of the cornerstones of the sedimentary-organic theory for the origin of hydrocarbons. (This theory is in opposition to the cosmic-igneous theory discussed in Chapter 5.) Therefore it is important to direct our attention not only to the details of traps and reservoir rocks but also to the broader aspects of basin analysis.

Before acquiring acreage in a new area and long before attempting to locate drillable prospects, it is necessary to establish the type of basin to be evaluated and to consider what productive fairways it may contain and where they may be extensively located. This chapter describes the various types of basin with reference to examples from around the world and discusses the relationship between the genesis and evolution of a basin and its hydrocarbon potential.

First, however, some of the basic terms and concepts must be defined. A sedimentary basin is an area on the earth's surface where sediments have accumulated to a greater thickness than they have in adjacent areas. No clear boundary exists between the lower size limit of a basin and the upper limit of a syncline. Most geologists would probably take the view that a length of more than 100 km and a width of more than 10 km would be a useful dividing line. Most sedimentary basins cover tens of thousands of square kilometers and may contain over five kilometers of sedimentary fill. Note that a sedimentary basin is defined as an area of thick sediment, with no reference to its topography. A sedimentary basin may occur as part of a mountain chain, beneath a continental peneplain, or in an ocean. Conversely, a present day ocean basin need not necessarily qualify as a sedimentary basin; indeed, many are floored by igneous rocks with only a veneer of sediment.

This distinction between topographic and sedimentary basins needs further elaboration. Both types of basin have a depressed basement. Sedimentary basins may or may not have been marked topographic basins during their history. Many basins are infilled with continental and shallow marine sediments, and totally lack deep-sea deposits.

Similarly, a distinction needs to be made between syndepositional and postdepositional basins. Most sedimentary basins indicate that subsidence and deposition took place simultaneously. This simultaneous occurrence is shown by facies changes and paleocurrents that are concordant with structure. On the other hand, in some basins paleocurrent directions and facies are discordant with and clearly predate present structure (Fig. 8.1). This is particularly characteristic of intracratonic basins, as is shown later. The distinction between these two types of basin is critically important in petroleum exploration because of the need for traps to have formed before hydrocarbon generation and migration. Stratigraphic traps are generally formed before migration, except for rare diagenetic traps. Structural traps may predate or postdate migration, and establishing the chronology correctly is essential.

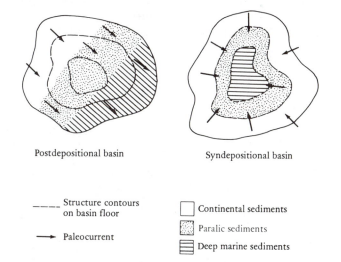

Postdepositional basin   Syndepositional basin

_ _ _ _ Structure contours
         on basin floor   ☐ Continental sediments

⟶ Paleocurrent   ▨ Paralic sediments

                                           ☰ Deep marine sediments

FIGURE 8.1 *The differences between syndepositional and postdepositional sedimentary basins.*

A further important distinction must be made between topography and sediment thickness. When examining regional isopach or isochron maps, it is tempting to assume that they are an indication of the paleo-topography of the basin. This is by no means always true. The depocenter (area of greatest sediment thickness) is not always found in the topographic nadir of the basin, but may frequently be a linear zone along the basin margin. This is true of terrigenous sediments, where maximum deposition may take place along the edge of a delta. Sediments thin out from the delta front both up the basin margin and also seawards. Similarly, in carbonate basins most deposition takes place along shelf margins, where organisms thrive in shallow, well-oxygenated conditions with abundant nutrients. Thus reefs and skeletal and oolite sands thin out toward basin margin sabkhas and basinward into condensed sequences of lime mud.

Many studies have shown that a depocenter may migrate across a basin. The topographic center of the basin need not necessarily move with it. Examples of this phenomenon have been documented from Gabon, the Maranhao basin of Brazil, and Iraq (Belmonte et al., 1965; Mesner and Woodridge, 1964; and Ibrahim, 1979, respectively). Note that the thickness of each of the formations measured at outcrop should not be added together to determine the overall thickness of sediment within a basin. This measurement can only be made from drilling or geophysical data (Fig. 8.2).

Now that basins have been considered in time and profile, they may be viewed in plan. The term *basin* has two interpretations. In the broadest sense, as already defined, a sedimentary basin is an area of the earth's surface underlain by sediments. In a narrower sense basins may be subdivided into true basins; those that are subcircular in plan and those that are elongate (troughs). Embayments, lacking centripetal closure, are basins that open out into larger basins (Fig. 8.3).

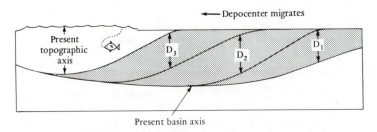

FIGURE 8.2 *Cross-section illustrating migrating basin depocenters. Note how measuring the apparent thickness of each unit at the surface leads to an erroneous overall thickness of basin fill.*

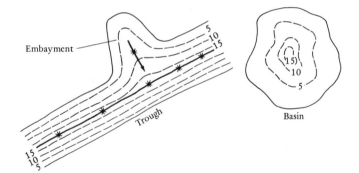

FIGURE 8.3   *Basins, defined as areas of the earth's surface underlain by sediments, may be subdivided into true basins, embayments, and troughs. Contours are in kilometers.*

MECHANISMS OF BASIN FORMATION

Sedimentary basins form part of the earth's crust, or lithosphere; they are generally distinguishable from granitic continental and basaltic oceanic crust by their lower densities and slower seismic velocities. Beneath these crustal elements is the more continuous subcrustal lithosphere. The crust is thin, dense, and topographically low across the ocean basins; but thick, of lower density, and, consequently, higher elevation over the continents (Fig. 8.4). The lithosphere is made up of a series of rigid plates, which overlie the denser, yet viscous, asthenosphere.

The lithospheric plates drift slowly across the asthenosphere. Knowledge of plate tectonics is of fundamental importance in understanding sedimentary basins. Detailed exposition of this topic is beyond the scope of this text, but a brief summary is necessary before considering the mechanics of basin formation. Further details are found in Seyfert and Sirkin (1973), Fischer and Judson (1975), Tarling and Runcorn (1973), Davies and Runcorn (1980), and Tarling (1981). A skeptical review of these ideas can be gained from Meyerhoff and Meyerhoff (1972).

The basic concept of plate tectonics may be stated as follows. Oceans are young (generally lacking rocks older than 200 m.y.), whereas continents are generally far older. Oceans are floored with basaltic volcanic rocks with a veneer of pelagic sediments. The oceans are cut by seismically active volcanic rifts, termed *midocean ridges*. Paleomagnetic reversals and age dating show that rocks become progressively older away from the ridges and toward the continental margins. The midocean ridges can be

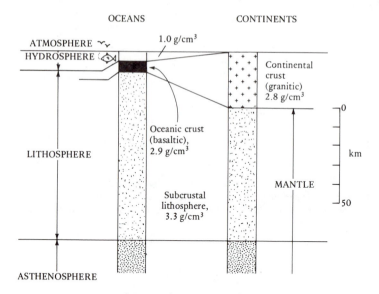

FIGURE 8.4 *Comparative columns of oceanic and continental crust showing average densities.*

traced landward into sediment-infilled rifts within continental granitic crust. The evidence suggests, therefore, that new crust, largely of basaltic composition, forms where tension and upwelling occur along these zones of sea floor spreading. Simultaneously, crust is drawn down into the asthenosphere at complementary linear features known as *zones of subduction*. These zones appear as folded troughs of sediment within or adjacent to continental masses and as volcanic island arcs within the oceans. Figure 8.5 shows the process of crustal gestation and digestion, and Figure 8.6 shows the recognized plate boundaries of the earth. Although there is general unanimity on the identification of the major plate boundaries, details of some of the smaller ones (microplates) are still somewhat unclear. Three types of plate boundary are recognized: trailing, subductive, and transcurrent. Trailing, or rift, boundaries occur where new crust forms and plates diverge. Subductive boundaries occur where plates converge. Some plate boundaries are transcurrent where two plates move past each other. Transcurrent plate boundaries are marked by extensive transform faulting accompanied by deep basins and thrust belts of local extent but great complexity. (Crowell, 1974; Dickinson and Seely, 1979).

Basins can form in four main ways (Fischer, 1975). Three of these processes are summarized in Figure 8.7. One major group of basins, the rift basins, form as a direct result of crustal tension at the zones of sea floor spreading. A second major group of basins occur as a result of crustal

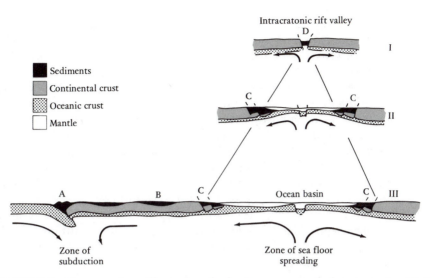

FIGURE 8.5 *Cross-sections illustrating the basic concepts of plate tectonics. (I) An axis of sea floor spreading develops, in this instance beneath continental crust. Up-doming occurs and a rift valley is formed (D). The East African rifts are a modern example.*

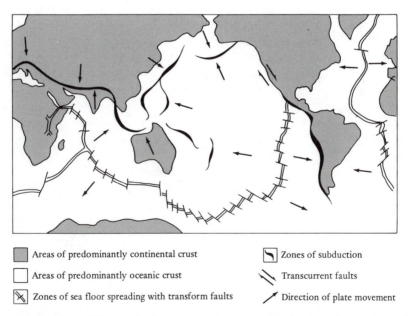

FIGURE 8.6 *Map of the earth showing approximate distribution of oceanic and continental crust and plate boundaries. (After Heirtzler, 1968; Vine, 1970; and others.)*

compression at convergent plate boundaries. A third type of basin can form in response not to lateral forces but to vertical crustal movements. For reasons not fully understood, phase changes can take place beneath the lithosphere. These changes may take the form of localized cooling, and therefore contraction, resulting in a superficial hollow, which becomes infilled by sediment. Conversely, the lithosphere may locally heat up and expand, causing an arching of the crust. Erosion of this zone will then occur. Sometimes this crustal doming is a precursor to rifting and drifting. Alternatively, subsequent cooling and subsidence result in the formation of an intracratonic hollow, which may be infilled with sediment.

A fourth mechanism of basin formation is simple crustal loading due to sedimentation. This process poses a "chicken and egg" problem, however. Basins of this type require an initial depression in the crust before deposition may begin. Thus, loaded basins characterize continental margins where a prograding delta can initiate and maintain the depression of adjacent oceanic crust.

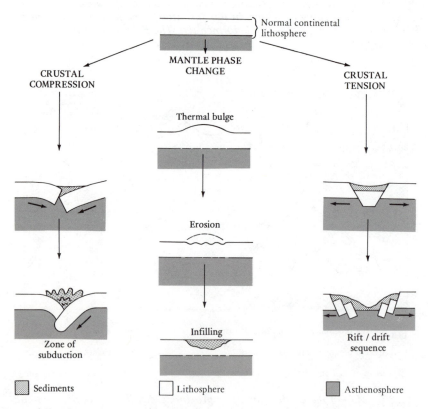

FIGURE 8.7 *Cross-sections showing the various types of basin formation discussed in the text. (After Fischer, 1975.)*

CLASSIFICATION OF SEDIMENTARY BASINS

Many schemes have been proposed to classify sedimentary basins. The early schemes were largely descriptive. Today, with the current understanding of plate tectonics, it is now possible to devise schemes that are not only descriptive but also genetic. Classifications have been proposed by Weeks (1958), Olening (1967), Uspenskaya (1967), Halbouty et al. (1970), Klemme (1975, 1980), Perrodon (1971, 1978), Selley (1975), and many others.

Table 8.1 attempts to synthesize the schemes of Halbouty, Klemme, and Selley. Sedimentary basins are difficult to classify because a basin may have had a complex history, during which it may have evolved from one type to another. Many basins could arguably be placed in more than one class. No great weight should be attached to Table 8.1. Its main merit is that it shows how basins may be linked to their genesis, providing a logical framework for the ensuing description of the various types of basin (Fig. 8.8).

CRATONIC BASINS

Cratonic basins are essentially subcircular basins that lie wholly or dominantly on granitic continental crust. The floors of such basins may be broken into a mosaic of horsts and grabens, but major rifting is absent. Cratonic basins may be subdivided into intracratonic basins, which lie wholly on continental crust, and epicratonic basins, which lie partly on continental crust and partly on oceanic crust. These classifications correspond to the type I and type II basins of Halbouty and Klemme (Table 8.1). This grouping is not based on an artificial distinction. These two types of basins differ markedly in facies, structure, and hydrocarbon potential, as the following account shows.

Intracratonic Basins

Intracratonic basins are broad, shallow, saucer-shaped basins. A major division can be made between terrigenous and carbonate intracratonic basins. The former are dominated by continental clastics, with negligible or no marine shales; the latter are more marine, although they may also be evaporitic. Examples of these two types are described and discussed as follows.

A series of intracratonic basins occur in North Africa between the Atlantic Ocean and the Red Sea. These basins are separated from one another by ridges of Precambrian igneous and metamorphic basement and

TABLE 8.1 Classifications of sedimentary basins attempting to relate basins to plate tectonics

Scheme of Selley (1975)			Scheme of Halbouty (1970) and Klemme (1975, 1980)
Cratonic suite associated with crustal stability	I Basins	Intracratonic Epicratonic	Type I Simple, saucer-shaped interior Type II Intracontinental composite foreland shelf
Geosynclinal suite at convergent plate boundaries	II Troughs	Miogeosyncline Eugeosyncline Molasse	Type VI Intermontane Type IV Extracontinental downwarp Type VII Intermontane
Transcurrent plate boundaries	III Rifts	Strike slip Intermontane (postorogenic)	Type III Rifts
Rift-drift suite at divergent plate boundaries		Intracratonic Intercratonic	Type IV Coastal graben pull-apart
	IV Ocean margin basins (Continental margin downwarp)		Type VIII Tertiary deltas

340

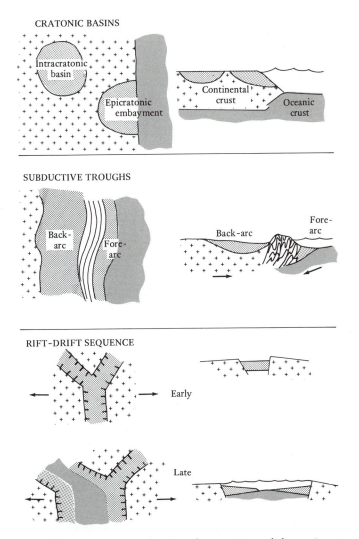

FIGURE 8.8 *Geophantasmograms showing the geometry of the various types of basin. Beware of this figure. Note that transitions occur between the different categories and that a basin may evolve from one type to another.*

tend to plunge northward toward the Mediterranean (Fig. 8.9). These basins show a remarkably uniform Paleozoic stratigraphy, but their characters become distinctly different in the Mesozoic. The Murzuk and Kufra basins of southern Libya are examples of intracratonic basins (Fig. 8.10). They are both floored with the widespread Pan-Saharan Paleozoic sequence. Names vary from basin to basin, and facies boundaries are diachronic, but the stratigraphy is remarkably uniform from the Atlantic to Arabia (see Beuf et al., 1971, and Selley, 1972, respectively). A blanket of

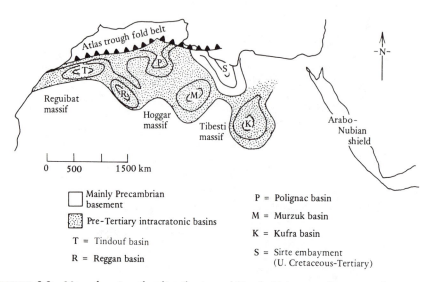

FIGURE 8.9 *Map showing the distribution of North African sedimentary basins.*

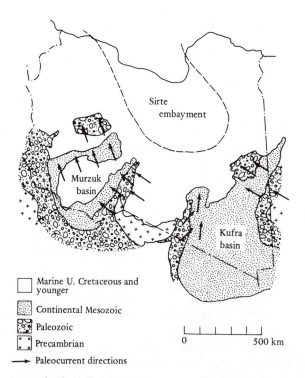

FIGURE 8.10 *Map of Libya showing the Murzuk and Kufra intracratonic basins.*
Note how paleocurrent data show that basin subsidence postdated sedimentation.

braided alluvial sands, several hundred meters thick, is overlain by a thinner, but still uniform, blanket of marine shoal sands. The precise age of these sands is uncertain because of a lack of fossils, but they are generally referred to the Cambro-Ordovician. These beds are succeeded by a marine graptolitic shale; the Tannezuft Shale (Silurian) of Algeria and Libya and the Arenig shale of the Khreim Group in Jordan. This shale is an organic-rich oil source rock in northwest Algeria, but becomes thinner, siltier, and less organic when traced south and east toward the craton. A major regression then deposited the predominantly deltaic Acacus Sandstone (Silurian) and the predominantly fluvial Tadrart Sandstone (Devonian) in North Africa, and the equivalent Khreim Group and Al Jouf Sandstone of Arabia. Overlying Carboniferous marine limestones and shales in Algeria and northwestern Libya pass southeastward into paralic sands and shales and finally to fluvial sands in the southeastern Kufra basin. After a major regression at the end of the Carboniferous, the sea has never again returned to the Murzuk and Kufra basins (Lestang, 1965; Klitzsch, 1970).

Facies analysis shows that throughout the Paleozoic the Murzuk and Kufra basins lay on a more or less uniform northerly dipping shelf. Paleocurrent analysis shows that the basins were separated by northerly plunging ridges. This morphology continued while the Continental Mesozoic Sandstones ("Nubian," Messak Sandstone) were deposited (McKee, 1965; Klitzsch, 1972; Van Houten, 1980). These sandstones include the deposits of a wide range of continental environments: dominantly fluvial, but including fanglomerate, lacustrine, and eolian deposits. Again paleocurrent analysis shows essentially a northerly paleoslope across both basins. The Murzuk and Kufra basins only became structurally enclosed basins, as opposed to embayments, after the deposition of the Continental Mesozoic. This deposit is largely barren of fossils and is considered to range in age from ?Triassic to Lower Cretaceous (Wealden). On regional grounds the closure of the basins by uplift of their northern edges would seem to have occurred toward the end of the Cretaceous period.

The Murzuk and Kufra basins thus provide good examples of intracratonic basins. Their main characteristics are subcircular shape and thin sediment fill (probably in the order of some 2 km for the Murzuk basin and 3 km for the Kufra basin). They show a remarkable intrabasinal and interbasinal uniformity of stratigraphy. Their facies are predominantly nonmarine sands, with minor volumes of marine sands, shales, and limestones. There is a shortage of organic-rich source beds, which partly reflects the shortage of marine shales, but also reflects the predominantly arid Mesozoic climate unfavorable for lacustrine oil shale deposition. The basin floors show a fairly uniform dip, and structural anomalies are largely absent. Geothermal gradients are low over the old, undisturbed granitic crust.

Intracratonic basins of this type are poor prospects for hydrocarbon exploration. They contain adequate potential reservoirs, but have a shortage of mature source rocks and structure. The hydrocarbons that may be generated may come from lacustrine source beds and be trapped stratigraphically around the basin margin. This concept is illustrated by the Green River Formation Tertiary basins of Wyoming and Utah (Picard, 1967; Eugster and Surdam, 1973; Surdam and Wolfbauer, 1975).

The second type of intracratonic basin is dominated by carbonate sedimentation. This variation is not due to an underlying structural difference, but rather to climatic and other factors, and transitions between the two types are present. The Williston and Michigan basins of North America are good examples of carbonate intracratonic basins. The Williston basin occupies parts of Saskatchewan, Montana, and North Dakota. It has a diameter of some 400 km and contains some 3 km of sediment, ranging in age from Cambrian to Tertiary (Darling and Wood, 1958; Smith et al., 1958; Dalmus, 1958; and Harding and Lowell, 1979).

The Paleozoic sequence begins with a basal Cambrian marine shoal sand, followed by largely marine shales and shallow-water limestones, with sabkha evaporites and red beds in the Devonian and Lower Carboniferous. A more or less uniform Paleozoic stratigraphy becomes regionally varied in the Mesozoic. Triassic and Jurassic sediments pinch out toward the basin margin, and a major sub-Cretaceous unconformity cuts across earlier rocks down to and including basement. This process plays a major part in the sealing of subcrop truncation traps. Over a kilometer of Cretaceous and Tertiary shales and clastics were then deposited in shallow marine and continental environments (Fig. 8.11). These beds, although not themselves significantly petroliferous, played an important part not only as a seal but as a blanket cover, which enabled the Paleozoic source shales to mature and generate oil and gas.

The Williston basin is not a major oil province and contains no giant oil or gas fields. Nonetheless, it produces oil and gas from numerous relatively small accumulations. Traps are of two main types: (1) a number of broad regional arches, such as the Miles City arch and Porcupine dome, and (2) several positive trends, that are related to basement faults, such as the Cedar Creek anticline. Neither warps nor faults shows any preferred lineation. Oil emigrating from the source beds in the deeper part of the basin has been trapped in pre-Cretaceous reservoirs in anticlinal, truncation, and combination traps (structural closure plus unconformity) beneath both the Cretaceous and Jurassic erosion surfaces.

Another good example of an intracratonic carbonate basin is the Michigan basin to the southeast of the Williston basin (Cohee and Landes, 1958; Delwig and Evans, 1969; Mesolella et al., 1974). This basin is also subcircular in plan, with a sediment thickness of some 5 km. Unlike the Williston basin, the section consists largely of Lower Paleozoic rocks, with a veneer of Devonian to Jurassic strata. Facies are largely shallow marine, with a basal Cambrian shoal sand overlain by shales, carbonates, and

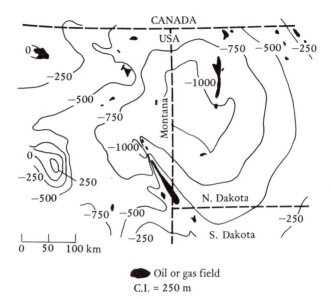

FIGURE 8.11   *Structure contour map on the top of the Cretaceous of the Williston basin (contours in 250-m intervals). This basin is a good example of a closed intracratonic basin, with a fill of over 3 km of shallow marine and continental sediments. (After Harding and Lowell, 1979.)*

evaporites (Fig. 8.12). Particular interest has centered on the Silurian rocks. These rocks consist of basinal carbonates rimmed by, in turn, pinnacle reefs on a platform slope, a barrier reef, and platform back-reef carbonates. This depositional topography has been infilled with Upper Silurian evaporites and minor carbonates.

Like the Williston basin, the Michigan basin is not a major hydrocarbon province and contains no single known giant oil or gas field. Similarly, however, it contains many hundreds of small oil and gas fields. These fields are mainly trapped in the myriad pinnacle reefs and on culminations on the concentric barrier reef. Smaller reserves have also been found, ranging from the Ordovician Trenton Limestone to the basal Upper Carboniferous (Pennsylvanian) sands.

To conclude this review of intracratonic basins the following points should be noted. Intracratonic basins generally contain abundant reservoir rocks in both terrigenous and carbonate facies. Source rocks tend to be poorly developed, except in the more marine carbonate basins. Because these basins occur on stable granitic crust, heat flow rates are low and major oil generation may not have occurred. For the same reasons there is a paucity of structural traps. Entrapment is characteristically stratigraphic because of truncation or onlap. Where mature source rocks are present, many small fields occur, rather than a few giants.

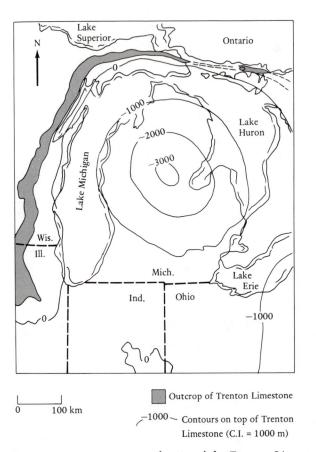

FIGURE 8.12 *Structure contour map on the top of the Trenton Limestone (Ordovician) showing the shape of the intracratonic Michigan basin of the Great Lakes region of North America. (After Cohee and Landes, 1958.)*

Epicratonic Embayments

Epicratonic embayments are basins that lie on the edge of continental crust. They are not true closed basins, but plunge toward major oceanic areas floored with basaltic crust. Epicratonic embayments correspond broadly to the type II intracontinental composite foreland shelf basins of Halbouty and Klemme (1970). As with intracratonic basins, a major distinction can be made between dominantly terrigenous and dominantly carbonate-filled embayments. The Tertiary Gulf Coast of the United States and the Niger delta embayment illustrate a terrigenous basin, and the Sirte embayment of Libya illustrates a carbonate-filled basin.

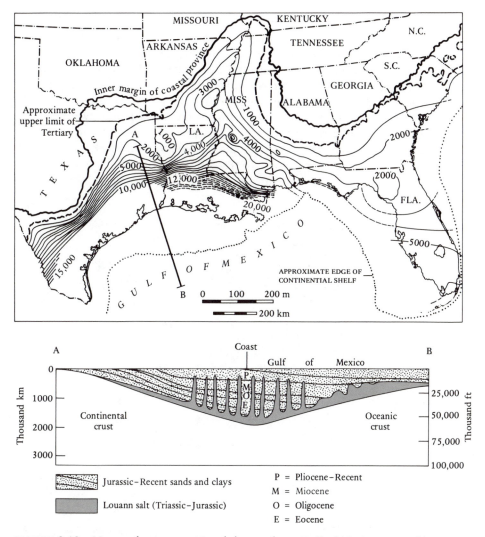

FIGURE 8.13 *Map and cross-section of the northern Gulf of Mexico coastal basin.
The map is based on Murray (1960) and shows the thickness of Cenozoic
sediments.*

The Gulf Coast embayment of the southern United States is a major
embayment containing some 15 km of sediment (Fig. 8.13). Basement is
overlain by the Louann salt of Jurassic age (?) (Murray, 1960; Wilhelm and
Ewing, 1972; Antoine, 1974; Dow, 1978). This salt is succeeded by a series
of prograding wedges, which range in age from Cretaceous to Recent.
Each wedge is composed of a thin up-dip section of fluvial sands, which
thickens seaward into deltaic sands and muds. These deltaic sands and

muds thin seaward, in turn, into deep marine clays of the Gulf of Mexico. These sediments contain major reserves of oil and gas (21.5 billion barrels of recoverable oil and 17,000 ft^3 of recoverable gas, according to Ivanhoe, 1980). These reserves occur in a series of fairways younging toward the Gulf (Fig. 8.14). Within each fairway, production occurs where the oil window intersects the *break-up zone* of interfingering slope mud source beds and deltaic sands. Hydrocarbon generation has been aided by rapid sedimentation and hence overpressuring. This process has resulted in abnormally low geothermal gradients over the Gulf depocenter as heat builds up in the "devil's kitchen" far below (Jones, 1969). Within the productive fairways hydrocarbons occur in a variety of traps. These traps include the roll-

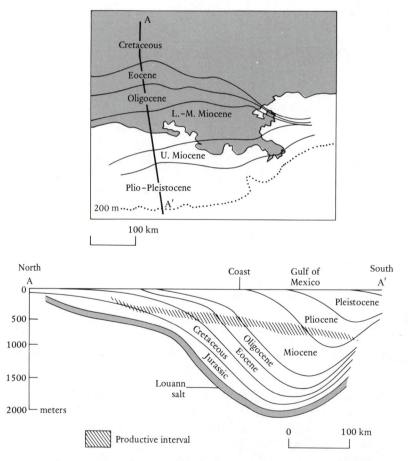

FIGURE 8.14   *(Upper) Map showing the productive fairways of the Gulf Coast embayment. (Lower) Cross-section showing the progradational nature of the basin fill and its relation to hydrocarbon production (After Dow, 1978.)*

over anticlines associated with Vicksburg flexure (p. 293), rollover anticlines associated with local growth faults originating in the overpressured clays, and diapiric traps due to both Louann salt domes and younger mud diapirs.

The Niger embayment of West Africa is in many ways analogous to the Mississippi embayment. It, too, plunges from continental to oceanic crust, but the subsidence that initiated sedimentation is clearly related to rifting of the African craton as the Atlantic ocean developed. The embayment merges up-dip into the Benue and Chari rifts, which extend into Niger and Chad (Fig. 8.15). Like the Mississippi, the Niger embayment contains a series of prograding clastic wedges, which range in age from

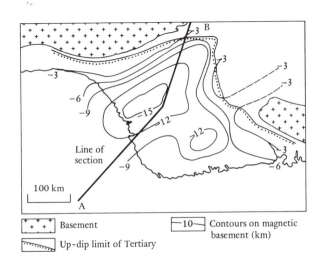

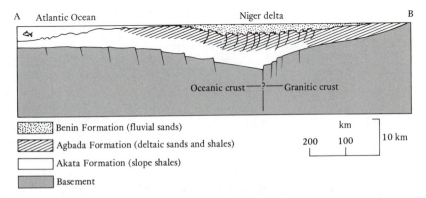

FIGURE 8.15   *(Upper) Map of the Niger delta showing depth to magnetic basement. (Lower) Cross-section along the line A–B. (After Evamy et al., 1978.)*

Upper Cretaceous to Recent. Three diachronous formations are recognized: the predominantly fluvial Benin Formation, the deltaic sands and shales of the Agbada Formation, and the slope muds of the Akata Formation (Short and Stauble, 1967; Weber and Daukoru, 1975; Evamy et al., 1978; Avbobo, 1978). Like the Mississippi delta the clays are overpressured, giving rise to growth faults and rollover anticlines, which form the major traps (page 295). The Niger delta does not show a series of seaward-younging productive fairways, although oil and gas are regularly distributed within each major growth fault structural unit. Abnormally low geothermal gradients are again encountered over the depocenter, the top of the oil window rising from some 5 km in the middle to 3 km around the edge. The Mississippi and Niger embayments are both characterized by high gas-oil ratios and waxy low-sulfur crudes. These characteristics probably reflect the high humic content of their kerogen.

Many other clastic epicratonic embayments occur around the world, especially in southeast Asia and the Canadian Arctic (Bruce and Parker, 1975). However, few are so well documented and apparently prolific as those just described.

Not all epicratonic embayments are terrigenous; some have a predominantly carbonate fill. The Sirte embayment is an example of this type (Conant and Goudarzi, 1967; Grey, 1970; Salem and Busrewil, 1981). Paleocurrent analysis shows that from the Cambrian to the early Cretaceous the area now occupied by the Sirte embayment was a northerly plunging ridge that separated the Murzuk and Kufra embayments, as they then were. This arch collapsed in the mid-Cretaceous. The Sirte unconformity directly overlies Precambrian basement in the center of the embayment and progressively younger rocks away from the basin axis. A locally developed basal sand is overlain on tilted fault blocks by Upper Cretaceous to Paleocene reefs and, in the troughs, by organic-rich shales that locally onlap and overlie the highs. The Lower Eocene consists of up to a kilometer of interbedded evaporites and carbonates. Carbonate sedimentation continued in the Middle Eocene, to be succeeded by shallow marine and fluvial sands and shales from the Oligocene to Recent (Fig. 8.16).

The Sirte embayment contains three main productive horizons. Many fields are combination traps on structural highs sealed by the Sirte unconformity. Reservoirs range in age from Precambrian granite (Augila) through various sandstones that range in age from Cambrian to Lower Cretaceous (Messla and Sarir). The second and most important play involves production from Upper Cretaceous and Paleocene reefal carbonates on the crests of fault blocks (Zelten, Waha, Dahra, etc.). A third, minor play occurs in the Oligocene sands, where the Gialo field is in the giant category (Fig. 8.17). Overall reserves of the Sirte embayment have been estimated at 30 billion barrels of recoverable oil and 32 trillion cubic feet of gas (Ivanhoe, 1980). It contains over 13 fields in the giant category (Halbouty et al., 1970).

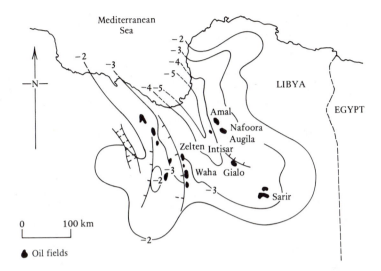

FIGURE 8.16 *Map of the Sirte embayment showing distribution of oil fields (black) and structure contours (kilometers) on the Sirte unconformity (pre-Upper Cretaceous). (After Sanford, 1970.)*

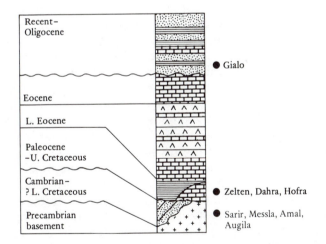

FIGURE 8.17 *Summary stratigraphy of the Sirte embayment showing the main reservoir formations. The Upper Cretaceous–Paleocene shales provide the main source rock.*

This review of epicratonic embayments shows that they are far more prospective than intracratonic basins. This statement is true of both terrigenous and carbonate embayments. Not only do they contain more marine sediments and, therefore, better source potential but they occur where continental crust is thinner and less stable. Thus heat flow is high, which favors hydrocarbon generation in areas of high geothermal gradients due to overpressure. Crustal instability also favors structural entrapment of oil, as well as stratigraphic (reefal), unconformity and growth-fault-related traps.

TROUGHS

Geosynclines and Plate Tectonics

The second major type of basin to consider are the troughs. These are linear basins far larger and far more complex in structure and facies than the basins discussed thus far. These troughs were long termed *geosynclines:* a concept defined by Hall (1859) and elaborated on by a series of workers, including Dana (1873), Haug (1900), Schuchert (1923), Kay (1944, 1947), and Glaessner and Teichert (1947), before reaching its apotheosis in the mighty work of Aubouin (1965). At that time—the dawn of the plate tectonic revolution—the concept of the geosyncline could be summarized as follows: geosynclines consist of two parallel troughs. One, the miogeosyncline, lies on continental crust and consists of an oceanward-thickening wedge of shallow marine limestones, sandstones, and shale. This trough is separated from the second trough, the eugeosyncline, by the miogeanticlinal ridge. The eugeosyncline is deeper than the miogeosyncline and is infilled largely by deep-water sediments. Initially, these sediments may be bathyal muds, but as the geosyncline evolves, turbidite sands infill it from a rising arc of islands on its oceanward side. This wedge of clastics has been referred to as *flysch* (Hsu, 1970). Compression of the flysch trough is accompanied by igneous intrusion, regional metamorphism, folding, and thrusting. Each phase of compression, or orogenesis, initiates isostatic adjustment, causing the sediments of the trough to rise and form a mountain chain. The adjacent miogeosyncline persists and is filled by a postorogenic wedge of predominantly continental sediments, referred to as *molasse* (Van Houten, 1973). This process, the classic geosynclinal cycle, was epitomized for European geologists by the Alps, in which it was easy to envisage the mountains being formed from a trough of sediments squeezed between the European and African cratons. Continental margin geosynclines, like the Appalachians, were harder to envisage, since one side of the vice was apparently absent. With the development of the concepts of sea floor spreading and crustal subduction, geologists rapidly reappraised classic geosynclinal theory (Ahmad, 1968; Mitchell and Reading, 1969; Coney, 1970; Schwab, 1971; Reading, 1972). Eugeosynclines are now interpreted as zones of subduction where plate boundaries converge.

There are three types of subduction zone. One type occurs between areas of continental crust, as, for example, in the Alps, Zagros, and Himalayas. In the second type the trough develops at the boundary between oceanic and continental crust. The Cordillera of North America, the Andes, and the Banda arc of southeast Asia illustrate this type. The third type is where subduction occurs between two plates of essentially oceanic crust, as, for example, in the Japanese and New Zealand arcs. These three types of subduction zone may form part of an evolutionary sequence in which an ocean closes as plates of oceanic crust converge, finally resulting in the juxtaposition of continental crustal blocks. Figure 8.18 illustrates these different types of subduction zone. Note the change in terminology: back-arc, arc, and fore-arc broadly correspond to the old miogeosyncline, geanticline, and eugeosyncline, respectively.

This new interpretation of sedimentary troughs has been complicated by the fact that some troughs that should theoretically be zones of subduction actually contain rifts with flat-lying sediments, which suggests a lack of compression, if not tension. The Peruvian trench is one such embarrassing example (Scholl et al., 1968). This phenomenon has lead to the idea that plate movement may occasionally reverse, so that subductive zones become divergent, albeit for intermittent periods. The Alps are now interpreted in such a way by Wilson (1966, 1968), as shown in Figure 8.19.

Back-Arc Troughs

The complexity of facies, structure, and history makes it difficult to generalize about the petroleum potential of subductive troughs. Consider first the back-arc, or miogeosynclinal, troughs. These troughs are asymmetric shelves whose sediments thicken toward the arc. This type of basin can also be regarded as an elongated epicratonic embayment and is more or less synonymous with Klemme's extracontinental downwarp. The deposits of these basins are largely shallow marine shales, carbonates (often reefal), and mature tidal shelf sands, with perhaps a feather edge of nonmarine sediments. Between major subductive phases, extensive source rocks may be deposited along the basin's seaward margin. Orogenic movements leave these back-arc basins terminated on their outer side by a thrust belt. The Arabian Gulf, Sumatra, and Western Canada illustrate this type of basin. A map and cross-section of the Arabian Gulf is shown in Figure 7.7. For further details see Kamem-Kaye (1970) and Murris (1980, 1981).

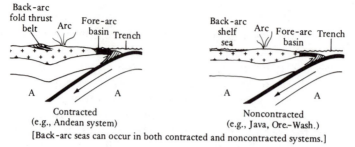

CONTINENTAL MARGIN ARC-TRENCH SYSTEMS

Contracted
(e.g., Andean system)

Noncontracted
(e.g., Java, Ore.–Wash.)

[Back-arc seas can occur in both contracted and noncontracted systems.]

INTRAOCEANIC ARC-TRENCH SYSTEMS

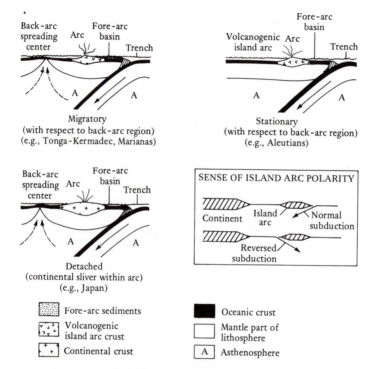

Migratory
(with respect to back-arc region)
(e.g., Tonga-Kermadec, Marianas)

Stationary
(with respect to back-arc region)
(e.g., Aleutians)

Detached
(continental sliver within arc)
(e.g., Japan)

SENSE OF ISLAND ARC POLARITY

Fore-arc sediments

Volcanogenic island arc crust

Continental crust

Oceanic crust

Mantle part of lithosphere

A   Asthenosphere

FIGURE 8.18   *Cross-sections illustrating the various types of zones of subduction and their associated sedimentary troughs. (After Dickinson and Seely, 1979.)*

Figure 8.20 illustrates the salient features of the Western Canada basin. This basin contains a wedge of sediments that thickens westward to some 5 km until it is abruptly truncated by the Rocky Mountain thrust belt. The earlier Paleozoic sediments include sands, carbonates (with spectacular Devonian reefs), organic-rich shales, and evaporites. Major oil and gas production occurs in the Devonian reefs (Illing, 1959; Barss et al., 1970; Evans, 1972; Klovan, 1974) and in the Viking and Cardium (Cretaceous) shallow marine sands. Recoverable reserves are estimated at 16.3 billion barrels of oil and 94 trillion ft^3 of gas (Ivanhoe, 1980).

354

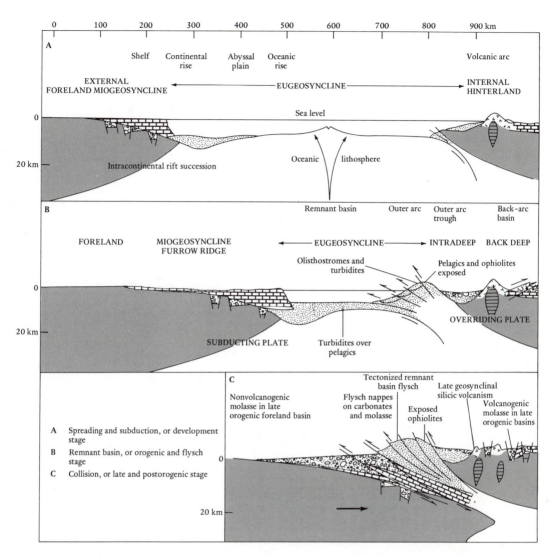

FIGURE 8.19 *Cross-sections illustrating Wilson's (1968) interpretation of the evolution of the Alps. An original sea floor spreading phase (A) reverses to become a zone of subduction (B), leading to the collision of continental plates (C). (After Mitchell and Reading, 1978.)*

The source of terrigenous detritus is important in basins such as the Western Canada basin. For most of the time the sands are produced by the slow weathering of the craton, followed by extensive reworking and deposition mainly in high-energy marine environments. These sands are thus mineralogically and texturally mature and have good porosity and permeability. As the arc rises, however, it begins to shed detritus into the back-arc basin. These later sands are often mineralogically immature, especially if derived from volcanic rocks. Rapidly deposited in fluviodeltaic environments, their mineralogical and textural immaturity may render them poorer reservoirs than the earlier shelf-derived sands.

355

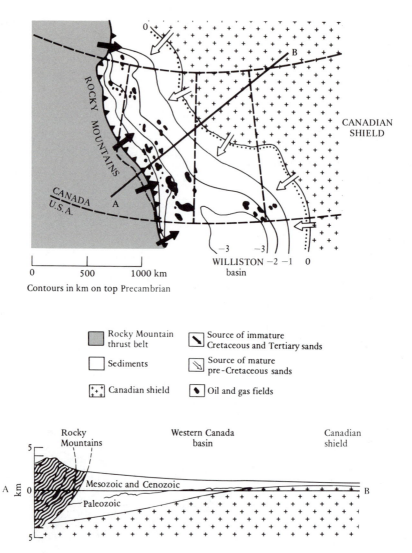

FIGURE 8.20 *Map and cross-section of the Western Canada (Alberta) trough, a major hydrocarbon province in a back-arc setting.*

Southeast Asia provides another example of back-arc basins and their relationship to fore-arc basins and plate boundaries (Schuppli, 1946; Haile, 1968; Crostella, 1977; Ranneft, 1979; Bowen et al., 1980; Wood, 1980). Here the continental crust of the Sunda Shelf moves westward toward the Indian Ocean plate, forming a convergent subductive plate boundary. Simultaneously, northward movement of the Australian plate is causing subduction on the southern margin of the Sunda plate. A number of back-arc basins separate the shelf from a volcanic arc (Fig. 8.21). These basins are infilled with a series of Eocene–Recent prograding clastic wedges in the Sumatra and Borneo basins and with carbonates over the more stable

356

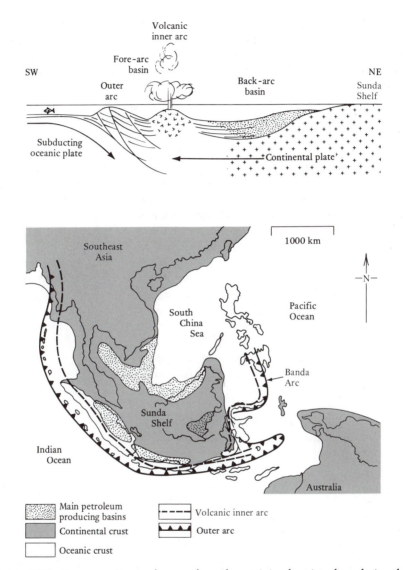

FIGURE 8.21 *Cross-section and map of southeast Asia showing the relationship between fore-arc and back-arc troughs. The back-arc troughs are the main oil-producing basins. (After Haile, 1968; Crostella, 1977; Ranneft, 1979; Bowen et al., 1980; and Wood, 1980.)*

southeastern part of the shelf. These basins contain more than 10 fields with recoverable reserves in excess of 8 billion barrels of oil and 24 trillion ft^3 of gas (St. John, 1980). The back-arc basins are separated from the fore-arc basins by a volcanic arc. The latter are infilled with marine sediment, derived largely from the volcanic arc. The fore-arc basins are gradually being subducted between the Indian and Sunda plates. Thrust slices thus appear locally as an outer nonvolcanic island arc (e.g., the Andaman and Nicober island chains). Hydrocarbon production from the fore-arc basins is negligible for several reasons, including the volcaniclastic nature of the sediments, which causes poor porosity preservation. Geothermal gradients are low and structure complex (Kenyon and Beddoes, 1977).

Back-arc basins have a good potential for favorable source rock sedimentation. An extensive marine shelf can have clay blankets deposited during marine transgressions. When these clay blankets coincide with uplift of the arc to form a barrier, restricted anaerobic conditions may occur because of poor circulation. The Cretaceous shales that were deposited in the great seaway extending from the Arctic to the Gulf of Mexico in the back-arc basins of the Rocky Mountains illustrate this point.

Traps in back-arc basins are numerous and varied. Classic anticlines may develop adjacent to the mountain front, as was already discussed in connection with Iran (p. 282). Traditionally, the thrust belts have been ignored in hydrocarbon exploration, both because of possible poor prospects and because of problems of seismic acquisition and interpretation. With improved technology, however, thrust belt exploration is increasing, as, for example, in the Appalachians, the Rockies, and the Alps (see McCaslin, 1981; Anon, 1980; and Bachman, 1979, respectively). Away from the thrust belt, entrapment may be stratigraphic, including reef and shoestring sand plays as well as onlap and truncation traps. Fairways for these prospects may parallel the basin margin or form halos around regional arches.

With this combination of favorable reservoir rocks, source rocks, and trap diversity, it is not surprising that back-arc basins are commonly major hydrocarbon provinces.

Fore-Arc Troughs

In contrast with back-arc basins, fore-arc basins are more complex in structure and facies. They are therefore more diverse in the nature and extent of their petroleum productivity. The fore-arc basins of the northeast Indian Ocean have already been noted. Dickinson and Seely (1979) have given a detailed analysis of fore-arc basins and have reviewed their petroleum potential. Figure 8.22 shows the basic structural elements of a fore-arc. Terrigenous sediments may be deposited in the trough in a wide range of environments, ranging from continental to deep marine. Initially, sands are derived from the igneous rocks of the volcanic arc. They thus tend to be mineralogically immature and lose porosity rapidly upon burial. Dickinson and Seely note that fore-arcs may have shelved, sloping, terraced, and ridged topographies. Broad shelves enable sands to mature, both mineralogically and texturally, before deposition. Narrow shelves, by

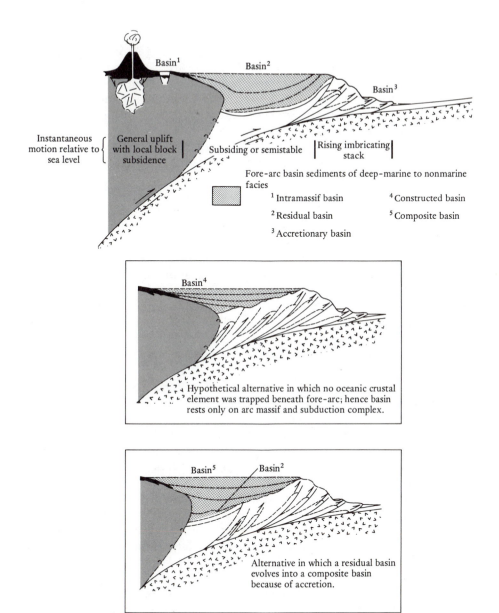

FIGURE 8.22   *Cross-section through a fore-arc trough that developed where a continental plate has overriden a subducting oceanic plate. (After Dickinson and Seely, 1979.)*

contrast, favor rapid sedimentation of immature sand with minimal re-working. As the oceanic plate subducts, wedges of fore-arc sediments are thrust up to form an outer nonvolcanic island arc. This arc will be composed of the distal edge of the fore-arc sedimentary prism, which is largely made up of ocean floor deposits. These thrust slices thus contain metamorphosed serpentinites, cherts, pelagic limestones, and turbidites. As this outer arc rises, it too will shed sediment into the fore-arc trough. The mineralogical immaturity and rapid sedimentation of these sands may diminish their reservoir quality, as was the case for those derived from the volcanic arc.

Fore-arc basins make less productive hydrocarbon provinces than do back-arc basins. As just described, sands are generally of poorer quality, lacking the polycyclic and reworked history of back-arc sediment. Carbonate reservoirs are generally absent.

Source rocks may be abundant and of good quality because of the prevalence of deep and often restricted sea floor conditions. However, locating the main area of oil generation may be difficult. Geothermal gradients are often low in fore-arcs because the cool sediment is subducted, which depresses the isotherms. Those hydrocarbons that are generated are more commonly trapped in structural fold and fault traps than in stratigraphic ones. Extensive structural deformation may cause traps to be small and hard to develop.

Those fore-arc basins that are productive tend to have fairways on their continental side and to have had broad shelves. The Cook Inlet basin of Alaska and the Peru coastal basin are examples of productive fore-arc basins with giant fields (Magoon and Claypool, 1981). The Kenai field in the Cook Inlet has over 5 million ft^3 of recoverable gas. The La Brea, Parinas, Talara fields of Peru have aggregate reserves of 1.0 billion barrels of recoverable oil (St. John, 1980).

Thrust belts, themselves, used to be avoided in petroleum exploration because of the problems of interpreting seismic data in areas of structural complexity, and also because reservoirs were not anticipated beneath metamorphic nappes. The first problem has largely been resolved with the help of modern high-resolution seismic. The second problem is now known to be a misconception in many cases. Petroleum exploration in the thrust belts of the Alps, the Rocky Mountains, and the Appalachians has been rejuvenated in recent years. In all these areas wells have penetrated metamorphic nappes and encountered sediments with potential reservoirs and remarkably low levels of kerogen maturation (Anon, 1980; Lamb, 1980; Bachmann et al., 1982). A specific example of this occurrence is the Vorderiss No. 1 well, which, after penetrating nappes with R_o values up to nearly 2.0, found R_o values of 0.5 to 0.6 at depths of some 5 km (Bachmann, 1979). Thrust belts are thus explored with more vigor today than in the past.

THE RIFT-DRIFT SUITE OF BASINS

A rift basin is bounded by a major fault system. Symmetric rifts, or grabens, are bounded by two sets of faults; asymmetric rifts, or half-grabens, are bounded by one set of faults. The introduction to this chapter discusses how rifts characteristically occur along the crests of regional arches on continental crust and along the crests of the midoceanic ridges, which are axes of sea floor spreading. Asymmetric rift, or half-graben, basins occur along the edges of many continents, notably those that border the North and South Atlantic Oceans. The concept of plate tectonics shows how all these basins are genetically related in what may be termed the rift-drift suite. This concept is described in its evolutionary sequence in the following section.

Rifts

Rifts occur today along the midoceanic ridges, which are now interpreted as zones of sea floor spreading. These rifts are formed in response to tension in the crust as the plates separate. The resultant troughs are infilled with basaltic lavas interbedded with pelagic clays, limestones, and cherts. Because of their fill and geographic location, the rift basins of midocean ridges are not attractive areas for hydrocarbon exploration.

A number of rift basins occur on continental crust, including the Rhine Valley of Germany and the Baikal rift of Russia (Illies and Mueller, 1970; Salop, 1967). Both of these basins cross-cut areas of arched crust and show a tendency to radiate into minor rifts at both ends of the main rift. They are infilled with up to 5 km of sediment and have igneous extrusives associated with them (Fig. 8.23). Equally well known, and in many ways similar, are the great rift valleys of Africa. Central Africa is now known to be crossed by rifts that, extending inland from Nigeria, Mozambique, and Somalia, intersect in Sudan, Chad, and Niger. These rifts are of Cretaceous and early Tertiary age. Largely blanketed by younger sediments, they have only recently become known as a result of petroleum exploration in these countries. More conspicuous are a series of younger rifts that occur in eastern Africa (UNESCO, 1965; Baker et al., 1972; Darrcott et al., 1973; Veevers, 1981). These rifts are very similar to the Rhine and Baikal rifts. They show crustal doming with local reversal of drainage. Major rifts bifurcate at their terminations, are volcanically and seismically active, and are infilled with volcanic, fluvial, and lacustrine sediments of Miocene to Recent age. As rifts are traced northeastward toward the Omo depression on the Ethiopian coast of the Red Sea, rift initiation began earlier (Eocene), basaltic lava outpourings are more extensive in time and space, and the sediments include extensive evaporite deposits (Fig. 8.24).

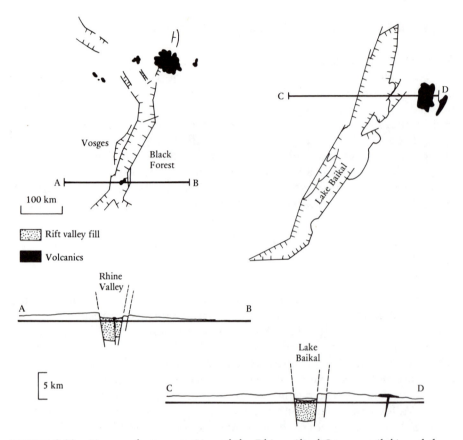

FIGURE 8.23   *Maps and cross-sections of the Rhine rift of Germany (left) and the Lake Baikal rift of the Soviet Union (right). (After Selley, 1975.)*

The Red Sea is itself a complex rift that provides a genetic link between the intracratonic rifts just discussed and the ocean margin rifts (Heybroek, 1965; Lowell and Genik, 1972; Lowell et al., 1975; Thiebaud and Robson, 1979). The margins of the Red Sea are two parallel-sided half-grabens, whose major faults downthrow seaward. On the upthrown sides of the faults the granites of the Arabo-Nubian shield crop out with occasional veneers of basaltic lavas (Fig. 8.25). Within the coastal basins Paleozoic Nubian sandstones and Cretaceous and Eocene limestones are locally truncated on numerous fault blocks. This horst and graben floor to the coastal basins is onlapped by Miocene evaporites, reefal limestones, and younger sands. Many oil fields have been discovered in Tertiary carbonates and sandstones and in Nubian sandstone reservoirs in the fault blocks of the Gulf of Suez (Fig. 8.26).

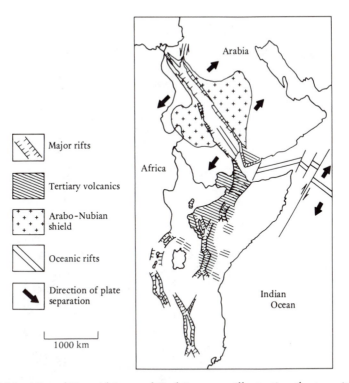

FIGURE 8.24 *Map of East Africa and Red Sea area illustrating the transition from intracratonic rifts in the south to oceanic rifts in the north.*

The deeper central part of the Red Sea consists of basaltic oceanic crust with a veneer of young deep marine deposits. It contains a longitudinal axial rift, which can be traced, offset by many transform faults, into a midoceanic rift in the Indian Ocean (Fig. 8.7). The Red Sea thus provides the link between the intracratonic rifts of East Africa and the oceanic rifts of the midocean ridges. The Red Sea may be regarded as an incipient ocean basin in which a rift in the Arabo-Nubian shield separated into two half-graben basins, between which new oceanic basaltic crust is now forming.

Armed with this concept, it is now time to examine the Atlantic Ocean. A well-defined ridge with an axial rift extends from the Arctic, through Iceland, and down through the northern and southern Atlantic Oceans before running east into the Indian Ocean (Fig. 8.6). Both sides of the Atlantic Ocean, from Labrador to the Falkland Islands and from the Barents Sea to the Agulhas Bank, are flanked by asymmetric half-graben basins, whose major bounding faults downthrow to the ocean (Drake et al.,

FIGURE 8.25 *Map of the Gulf of Suez rift basin showing location of faults and oil fields. (After Thiebaud and Robson, 1979.)*

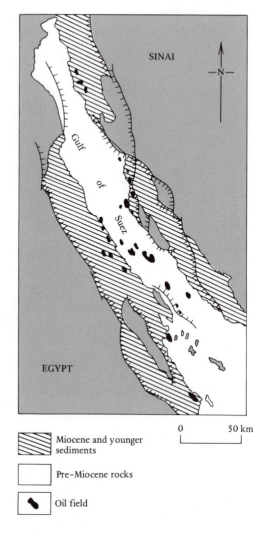

Miocene and younger sediments

Pre-Miocene rocks

Oil field

1968; Burk and Drake, 1974; Lehner and Ruiter, 1977; Pratsch, 1978; Hoc, 1979; Emery, 1980). These basins show a remarkable similarity of stratigraphy and symmetry of structure (Fig. 8.27). The Gabon basin is well known and will be used as a specific example (Belmonte et al., 1965; Brink, 1974; Vidal, 1980). It is typical of these basins geologically, but unusual in that it is a modest hydrocarbon province (Figs. 8.28 and 8.29). Seismic lines of the Atlantic coastal basins show that they are half-grabens. Two major unconformities may be discerned. One, generally the lowest mappable reflection, is considerably faulted. This marks the onset of rifting. Below this unconformity, basement and rare pre-rift sediments occur.

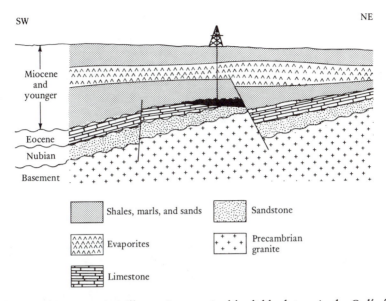

SW   NE

Miocene
and
younger

Eocene

Nubian

Basement

	Shales, marls, and sands		Sandstone
	Evaporites		Precambrian granite
	Limestone		

FIGURE 8.26   *Cross-section illustrating a typical fault block trap in the Gulf of Suez.*

The overlying sediments were laid down during active rifting. They consist largely of continental clastics, but, as the fault blocks subside progressively toward the continent, a barrier develops between the rift and new oceanic crust. Behind this barrier sapropelic lacustrine shales may form in humid climates. Within the Cocobeach Group of the Gabon basin these shales have generated oil. In arid climates, however, evaporites may form in the back basin; these evaporites are common in the Atlantic coastal basins. As the ocean opened from north to south, the onset of evaporite development youngs southward: Permian in northern Europe, Triassic in the Georges Bank and Senegal basins, Jurassic in the Gulf of Mexico, and Aptian in the Brazilian, Gabon, Cuanza, and Congo basins (Evans, 1978).

The second major unconformity overlies the evaporites. It marks the end of the rifting phase and the onset of drifting as the Atlantic Ocean widened. Rift faults generally die out at this surface, except at the basin margin. The overlying sediments start with an organic-rich shale, deposited as the sea invaded the incipient Atlantic Ocean. This shale is then overlain by a major regressive wedge. Sometimes this wedge is composed of carbonates, with marked thinning from a reefal shelf edge into basinal marls. Alternatively, there are terrigenous progrades in which basinal shales pass shoreward and upward into turbidites and paralic and continental deposits.

Basins of this half-graben type, with rift-drift sequences similar to those just described, are not unique to the Atlantic Ocean, but also occur

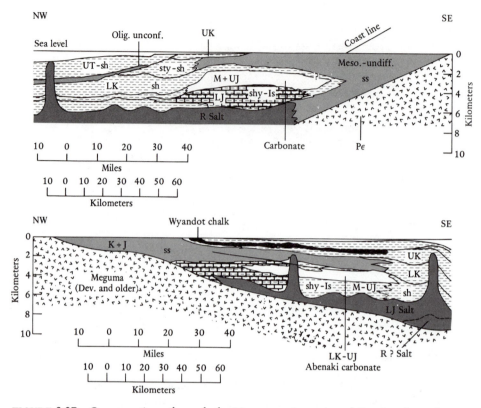

FIGURE 8.27 *Cross-sections through the Moroccan (upper) and Scotian (lower) continental shelves showing the symmetry of structure and similarity of stratigraphy of these two Atlantic margin half-graben basins. (After Bhat et al., 1975.)*

around other opening oceans. The Exmouth Plateau of northwest Australia is such an example (Exxon and Willcox, 1978). Collectively, such coasts are referred to as passive, or trailing, continental margins, in contrast to the active, or Pacific, continental margins where subductive arcs occur.

The preceding review shows that the Atlantic coastal basins evolved from intracratonic rifts, which were initiated on axes of crustal divergence and incipient sea floor spreading. For many years Atlantic-type coastal basins could only be explored along their less prospective up-dip edges. Within recent years offshore exploration has shown that many of these edges contain considerable reserves of oil and gas. Traps include tilted fault blocks (sealed by evaporites or the drift-onset unconformity), salt domes, and anticlines draped over basement horsts. For reasons still not clear, most of the productive basins appear to be in the southern hemisphere.

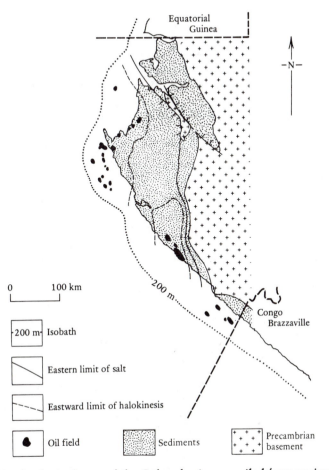

FIGURE 8.28   *Geological map of the Gabon basin, compiled from various sources.*

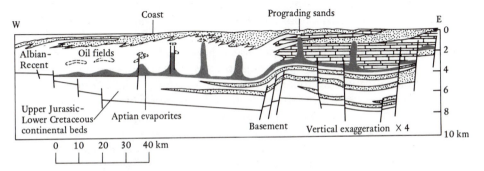

FIGURE 8.29   *Cross-section of the Gabon basin. (After Brink, 1974.)*

Failed Rift Basins: Aulacogens

There is one last type of rift basin to consider. Earlier, intracratonic rifts were noted to develop a triradiate pattern, often with triple-rift, or triple-R, junctions at the end of each major rift. As the crust draws apart, only two rifts out of each triple-R junction actually separate and become ocean margin half-grabens. One rift has failed to open. These aborted rifts, also called failed arms or aulacogens, are prime targets for petroleum exploration and deserve to be examined closely. The Benue trough of Nigeria, which has already been mentioned, is the failed arm of a triple-rift junction that developed on the site of the present Niger delta. When the old Pangean southern continent began to break up, rifts developed from the Benue trough north into Niger and Chad. Simultaneously, rifts developed along the Tibesti-Sirte arch, which separated the Murzuk and Kufra basins (p. 341). The Sirte embayment, which was introduced earlier as an epicratonic embayment, developed along this axis and could therefore be regarded as a complex failed rift.

One of the best known failed rifts occurs in the North Sea of Europe. When the European, Greenland, and North American plates began to separate, a triple-R junction developed somewhere to the northeast of Scotland. Two of these arms opened to form the Norwegian Sea and the Atlantic Ocean, which are both flanked by half-graben basins. The southeastern branch of the triple-R junction subsided but failed to open, providing the North Sea oil province. This province is described in many papers in volumes edited by Woodland (1975), Finstad and Selley (1975), Finstad and Selley (1977), and Illing and Hobson (1981). Short reviews are given by Ziegler (1975) and Selley (1976).

Figures 8.30 and 8.31 illustrate the main structural features of the North Sea as seen in plan and sections. Rifting began in the Permian and continued throughout the Triassic, with the deposition of fluvial and eolian sands and evaporites. A major transgression in the Jurassic resulted in the deposition of organic-rich shales and paralic sands. As rifting continued, submarine fault scarps along the rift margins poured conglomerates and turbidites onto the basin floor. At the end of the Jurassic a major unconformity, the Cimmerian event, marked the end of rifting and the onset of drifting as the Atlantic Ocean opened to the west. Lower Cretaceous sands and shales onlap pre-Cimmerian fault blocks. These blocks have considerable structural relief and locally underwent crestal erosion. Quiescence in the Late Cretaceous allowed the widespread deposition of coccolithic chalk. Renewed rifting in the Paleocene caused chalk turbidites and melanges to be shed into the Central Graben, following which a major delta complex prograded east and southeastward from Scotland. Submarine channel sands and turbidite fans were deposited at the foot of the delta complex. The basin continued to be infilled by marine clays and occasional shallow marine sands up to the present day.

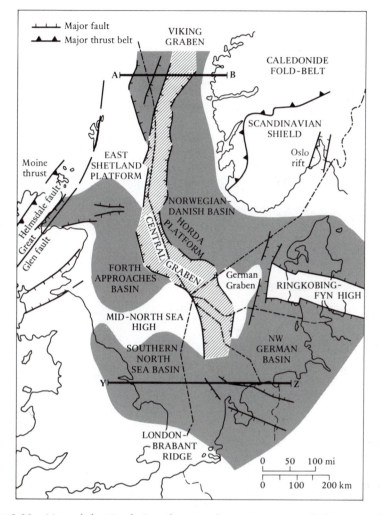

FIGURE 8.30 *Map of the North Sea showing the main structural elements. This major hydrocarbon province is a classic example of a failed rift system.*

The North Sea contains four major hydrocarbon plays. In the southern North Sea basin gas occurs in block-faulted anticlines beneath the Upper Permian Zechstein salt. Lower Permian eolian dune sands are the reservoir, and underlying Carboniferous coal beds provide the source for the gas. This play contains several gas fields in the giant category, including Groningen, Hewett, Leman, and Indefatigable. The second major play occurs in Jurassic sands in tilted fault blocks sealed by Cretaceous shales and limestones that onlap the Cimmerian unconformity. The Brent, Statfjord, Piper, and Heather fields are of this type (Fig. 7.16). The third major play occurs in southwestern offshore Norway. Here Ekofisk and as-

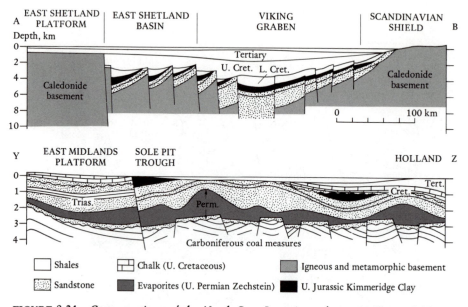

FIGURE 8.31 *Cross-sections of the North Sea. Locations shown in Figure 8.30.*

sociated fields produce from fractured, overpressured Cretaceous chalk reservoirs domed over Permian salt structures (Fig. 7.24). Finally, production comes from Paleocene deep-sea sands draped over pre-Cimmerian horsts. The Montrose, Frigg, and Forties fields are of this type (Fig. 7.13).

The Jurassic, Cretaceous, and Paleocene oil and gas fields are all believed to have been largely sourced from Jurassic shales. Estimates of North Sea reserves vary widely, but Ivanhoe (1980) cites recoverable reserves of 20 billion barrels of oil and 40 trillion ft^3 of gas.

Several conditions have made the North Sea rift basin a major hydrocarbon province. Excellent reservoirs are provided by polycyclic sands, often with subunconformity-leached porosity. Thick, rich source beds within the rift axis interfinger with and underlie the reservoirs. Traps are many and varied, including horsts, combination fault block truncations, salt domes, and compactional anticlines. Geothermal gradients, although now near average, were once abnormally high, enhancing hydrocarbon generation and migration.

We can see, therefore, why failed rift basins may be major petroleum provinces. Schneider (1972) has shown that a regular sequence of facies tends to occur within the rift-drift suite of basins. Stage one, when the rift was still above sea level, consists of continental clastics, which are often associated with volcanics. The subsiding rift floor ultimately reaches sea level. This condition favors evaporite formation as the trough surface oscillates above and below the sea. As the rift floor is finally submerged, evaporites are overlain by organic-rich marine muds deposited in the restricted

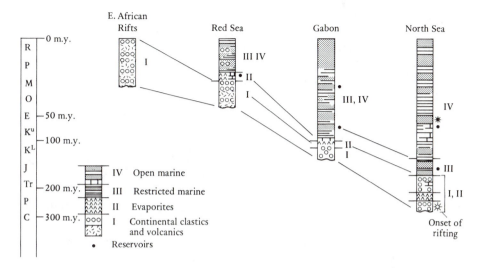

FIGURE 8.32 *Comparative sections of various rift basins showing characteristic sedimentary sequences and distribution of source rocks.*

trough. Finally, as the rift dilates into an open sea, carbonate shelves and prograding clastic wedges build out over the old rift floor onto oceanic crust. Figure 8.32 shows how this ideal sequence applies to the examples of the rift-drift sequence of basins just described.

Strike-slip Basins

We have now reviewed the different types of basin associated with convergent and divergent plate boundaries. Some plate margins are transcurrent and develop a particular type of rift basin. Transcurrent plate margins are defined by major wrench faults. Where plates move past one another, however, the movement is seldom wholly parallel; an oblique component commonly causes crustal compression. Similarly, compressive phases may alternate with phases of separation. These changes occur both in time and space. Thus, for example, the divergent plate boundary of the Red Sea can be traced northeastward to the Dead Sea rift. The transcurrent nature of the Dead Sea faults has been understood from biblical to recent times (see Zechariah 14.4 and Quennell, 1958, respectively). By contrast the San Andreas transverse fault system of California passes into compressive plate boundaries at both ends (Fig. 8.6).

Transform fault systems give rise to very distinctive types of rift basin. These rift basins are generally very deep, subside rapidly, and have rates of high heat flow. The Dead Sea rift is such an example. It is some 15 km wide and 150 km long. Left-lateral displacement has been estimated in the order of 70 to 100 km (Wilson, Kashai, and Croker, 1983). Over 6 km of subsidence has occurred since the Miocene period. The Dead Sea valley has long been noted for its petroleum seeps, occasionally giving rise to floating blocks of asphalt (Nissenbaum, 1978). Commercial quantities of oil have yet to be found within the margins of the rift.

By contrast the strike-slip basins associated with the transcurrent fault systems of California are very petroliferous (Fig. 8.33). These rifts are mainly of late Tertiary age. They are characterized by thick sequences (over 10 km) of rapidly deposited clastics in which abyssal shales (source rocks) pass up through thick submarine fans (reservoirs) into paralic and continental deposits. The basins are often asymmetric, with alluvial and submarine conglomerates developed adjacent to the active boundary faults (Crowell, 1974). Petroleum is trapped in *en echelon* and *flower-structure*-faulted anticlines (see p. 292); examples are given in Weeks (1958). Stratigraphic entrapment in submarine channels and fans is less common (p. 302). Estimated ultimate recoverable reserves for the Californian basins is in excess of 15 billion **BOE** (St. John, 1980). Additional information on strike slip basins can be found in Reading (1972).

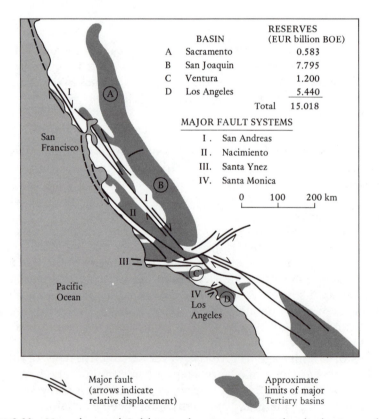

FIGURE 8.33 *Map of part of California showing major strike slip basins and their associated faults.*

CONCLUSIONS: DISTRIBUTION OF PETROLEUM BETWEEN AND WITHIN BASINS

Distribution of Hydrocarbons in Different Types of Basin

The preceding part of the chapter reviewed the various types of basin and analyzed the conditions affecting their potential for being major hydrocarbon provinces. Several geologists have quantitatively reviewed the global distribution of hydrocarbon reserves. This review may provide guidance to future exploration by establishing the relative productivity of the different types of basin (Halbouty et al., 1970; Klemme, 1975, 1980).

Figure 8.34 presents data from Klemme's study of the global distribution of hydrocarbons in different basins. His basin classification is somewhat different from the one used in this chapter, and it does not precisely correspond to his earlier version (see Table 8.1), but the figure does show many interesting features. Note particularly that the Arabian Gulf, which has 38 percent of the world's reserves (Ivanhoe, 1980), has a major effect on the type IV *Continental borderland downwarp* class. In studying these figures, remember that they do not indicate the distribution of actual reserves but only of known reserves. Thus they reflect factors of geography, economics, politics, and technology, as well as our ability as exploration-

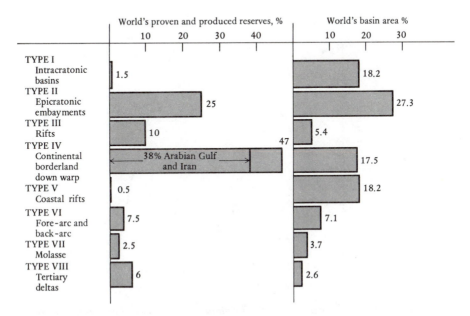

FIGURE 8.34 *Histograms comparing the surface area of different types of sedimentary basin and their currently known petroleum reserves. (Based on data from Klemme, 1980.)*

ists. Note particularly that the percentage of reserves found in Tertiary deltas and coastal basins will increase as offshore exploration extends into deeper and deeper water.

Distribution of Hydrocarbons Within Basins

Oil and gas tend to occur in sedimentary basins in a regular pattern. Considered vertically, oil gravity decreases with depth. Heavy oils tend to be shallow and, with increasing depth, pass down into light oils, condensate, and finally gas, until the point at which hydrocarbons and porosity are absent. Hunt (1979) took all the API data from the 1975 International Petroleum Encyclopedia to produce the graphs shown in Figure 8.35. These graphs bear out the statement just made. Although oil API generally increases with depth (density decreases), many local conditions may disrupt this pattern; for example, the existence of several source beds in a basin, hydrocarbon flow along faults, flushing and degradation, and uplift and erosion.

Oils tend to become lighter not only downward but also laterally toward a basin center. Typically, heavy oils occur around basin margins, and condensate and gas in the center. The cause, or causes, of these vertical and lateral variations of oil and gas are of considerable importance. An

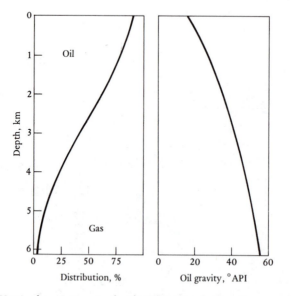

FIGURE 8.35   *Vertical variation in the distribution of oil and gas and in oil gravity. These curves were calculated from data in the 1975 International Petroleum Encyclopedia. These global trends have many local exceptions. (After Hunt, 1979.)*

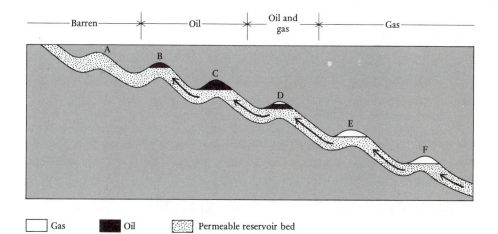

FIGURE 8.36 *Cross-section through a hypothetical sedimentary basin with a laterally continuous permeable stratum folded into a multitude of traps. This figure illustrates Gussow's principle.*

ingenious explanation was advanced by Gussow (1954) and has since gained wide acceptance as Gussow's principle. This theory may be stated briefly as follows: Consider a sedimentary basin, up whose flanks extend continuous, permeable reservoir beds that contain many traps. Consider also that the hydrocarbons, oil and gas, emigrate from the "devil's kitchen" in the deep center of the basin up the flanks. On reaching the first trap, gas may displace all the oil, which will be forced below the spill point and up to the next trap, where the process will be repeated. Finally, all the gas will be retained, so there will be a trap with a gas cap and an oil column. The trap will be full to spill point, and oil forced below the spill point will emigrate up, filling trap after trap until all the oil is contained. The final trap will not be full to spill point, and any additional traps that may be structurally higher will be barren (Fig. 8.36).

We have already pointed out that some fields show a gravity segregation of oil (p. 278). Therefore it is not hard to envisage Gussow's principle explaining not only the gross distribution of oil and gas in a basin but also the less obvious variations in oil gravity. Gussow established his principle from work on the Bonnie-Glenn-Wizard Lake Devonian reef complex of Alberta, across a distance of about 160 km. Other excellent examples of differential entrapment have been noted from the Niageran reefs of Michigan (Gill, 1979) and from the Mardin Group of Anatolia (Erdogan and Akgul, 1981).

There are problems, however, in applying Gussow's principle over large distances and in basins with discontinuous reservoirs. Thus Bailey et al. (1974) showed how oil and gas distribution can be mapped with great continuity right across the Alberta basin, where many reservoirs are isolated reefs and where unconformities provide regionally extensive permeability barriers. The zonation of gas, light oil, and heavy oil from basin center to margin may be due to a combination of thermal maturation and degradation by meteoric water.

Many other basins around the world exhibit a regular gravity zonation of hydrocarbons, yet lack continuity of reservoir from trap to trap. The chalk fields of the Ekofisk area of the North Sea, for example, show an increase in API gravity toward the Central Graben, yet lack any regional permeability. Therefore the differentiation of oil and gas found in many basins most likely reflects the levels of thermal maturation of source beds: gas being generated at higher temperatures and therefore at greater depths than oil (see p. 192). The shallowest traps around a basin rim are likely to have been flushed by meteoric water, as can be checked from their salinity. The shallowest oil is likely to be heavy where oil moving up from the basin center has been degraded by contact with meteoric water.

Gussow's principle must be utilized with care. It is particularly useful when combined with the spill point concept. This use can be illustrated by Figure 8.36. If trap C was the first to be drilled and it was full to spill point, then both B and D will be prospective. We may anticipate that D will be full to spill point, but can only speculate about the thickness of the oil column in B. Consider, on the other hand, that trap B had been drilled first and had been found not to be full to spill point. It would be sound policy to drill C, but trap A should be farmed out at the earliest opportunity, since it is unlikely to have received any oil.

SELECTED BIBLIOGRAPHY

Books and papers that describe sedimentary basins:

BURK, C. A. and DRAKE, C. L. 1974. *The Geology of Continental Margins.* Berlin: Springer-Verlag, 1009 pp.

DAVIES, P. A. and RUNCORN, S. A. 1980. *Mechanics of Continental Drift and Plate Tectonics.* London: Academic Press, 362 pp.

DICKINSON, W. R. and SEELY, D. R. 1979. Structure and stratigraphy of fore-arc regions. *Am. Assoc. Petrol. Geol. Bull., 63,* 2–31.

MITCHELL, A. H. G. and READING, H. G. 1978. Sedimentation and tectonics. In: *Sedimentary Environments and Facies.* H. G. Reading (ed.). Oxford: Blackwell, 439–476.

Books and papers that relate sedimentary basins to petroleum occurrence:

FISCHER, A. G. and JUDSON, S. (eds.). 1975. *Petroleum and Global Tectonics.* Princeton: Princeton Univ. Press, 322 pp.

HARDING, T. P. and LOWELL, J. D. 1979. Structural styles, their plate tectonic habitats, and hydrocarbon traps in petroleum provinces. *Am. Assoc. Petrol. Geol. Bull.*, *63*, 1016–1058.

NORTH, F. K. 1971. Characteristics of oil provinces: a study for students. *Can. Petrol. Geol. Bull.*, *19*, 601–658.

PRATSCH, J. C. 1978. Future hydrocarbon exploration on continental margins, and plate tectonics. *J. Petrol. Geol.*, *1*, 95–105.

REFERENCES

AHMAD, F. 1968. Orogeny, geosynclines and continental drift. *Tectonophysics*, *5*, 177–189.

ANON, 1980. Overthrust belt action hot in and out of fairway. *Oil and Gas J.*, July, 123–125.

ANTOINE, J. W. 1974. Continental margins of the Gulf of Mexico. In: *The Geology of Continental Margins*. C. A. Burk and C. L. Drake (eds.). New York: Springer-Verlag, 683–693.

AUBOUIN. 1965. *Geosynclines*. Developments in Geotectonics, vol. I. New York: Elsevier, 335 pp.

AVBOBO, A. A. 1978. Tertiary lithostratigraphy of Niger delta. *Am. Assoc. Petrol. Geol. Bull.*, *62*, 295–306.

BACHMANN, G. H. 1979. Das stratigraphische und tektonische Ergenis de Erdgestiefengufschlub—Bohring. Vorderiss 1. *Erdol-Erdgas Z. V.*, *95*, 209–214.

BACHMANN, G. H.; DOHR, G.; and MULLER, M. 1982. Exploration in a classic thrust belt and its foreland: Bavarian Alps, Germany. *Am. Assoc. Petrol. Geol. Bull.*, *66*, 2529–2542.

BAILEY, N. J. L.; EVANS, C. R.; and MILNER, W. D. 1974. Applying petroleum geochemistry to search for oil: examples from Western Canada basin. *Am. Assoc. Petrol. Geol. Bull.*, *58*, 2284–2294.

BAKER, B. H.; MOHR, P. A.; and WILLIAMS, L. A. J. 1972. Geology of the eastern rift system of Africa. *Geol. Soc. Am. Bull.*, Spec. Pub. No. 136.

BARSS, D. L.; COPLAND, A. B.; and RITCHIE, W. D. 1970. Geology of Middle Devonian reefs Rainbow area, Alberta, Canada. In: *Geology of Giant Petroleum Fields*. M. T. Halbouty (ed.). Tulsa: Am. Assoc. Petrol. Geol., Mem. No. 14, 8–19.

BELMONTE, Y.; HIRTZ, P.; and WENGER, R. 1965. The Salt basin of the Gabon and Congo (Brazzaville). In: *Salt Basins Around Africa*. London: Inst. Petrol., 55–78.

BEUF, S.; BIJU-DUVAL, B.; O de CHARPAL; and BENNACEF, A. 1971. *Les Grès du Paleozoique Inferior au Sahara*. Paris: Ed. Technip, 480 pp.

BHAT, H.; McMILLAN, N. J.; AUBERT, J.; PORTHAULT, B.; and SURIN, M. 1975. North American and African drift—the record in Mesozoic coastal plain rocks. Nova Scotia and Morocco. In: *Canada Continental Margins and Offshore Petroleum Exploration*. C. J. Yorath (ed.). Can. Soc. Petrol. Geol., Mem. No. 4, 375–390.

BOWEN, C.; PURDY, G. M.; JOHNSTON, C.; SHOR, G.; LAWVER, L.; HARTANO, H. M. S.; and JEZEK, P. 1980.   Arc-continent collision in Banda Sea region. *Am. Assoc. Petrol. Geol. Bull.*, *64*, 868–915.

BRINK, A. H. 1974.   Petroleum geology of Gabon basin. *Am. Assoc. Petrol. Geol. Bull.*, *58*, 216–235.

BRUCE, C. J. and PARKER, E. H. 1975.   Structural features and hydrocarbon deposits in the Mackenzie delta. In: *Ninth World Petrol. Cong. Tokyo, vol. 2. Geology.* London: Applied Science Publishers, 251–261.

BURK, C. A. and DRAKE, C. L. (eds.). 1974.   *The Geology of Continental Margins.* New York: Springer-Verlag, 1009 pp.

COHEE, G. V. and LANDES, K. K. 1958.   Oil in the Michigan basin. In: *The Habitat of Oil.* L. G. Weeks (ed.). Tulsa: Am. Assoc. Petrol. Geol., 473–493.

CONANT, L. C. and GOUDARZI, G. H. 1967.   Stratigraphic and tectonic framework of Libya. *Am. Assoc. Petrol. Geol. Bull.*, *51*, 719–730.

CONEY, P. J. 1970.   The geotectonic cycle and the new global tectonics. *Geol. Soc. Am. Bull.*, *81*, 739–747.

CROSTELLA, A. 1977.   Geosynclines and plate tectonics in Banda arcs, eastern Indonesia. *Am. Assoc. Petrol. Geol. Bull.*, *61*, 2063–2081.

CROWELL, J. C. 1974.   Origin of late Cenozoic basin in southern California. In: *Tectonics and Sedimentation.* W. R. Dickinson (ed.). Spec. Pub. Soc. Econ. Pal. Min., *22*, 190–204.

DALLMUS, K. F. 1958.   Mechanics of basin evolution and its relation to the habitat of oil. In: *The Habitat of Oil.* L. G. Weeks (ed.). Tulsa: Am. Assoc. Petrol. Geol., 883–931.

DANA, J. D. 1873.   On some results of the earth's contraction from cooling, including a discussion of the origin of mountains and the nature of the earth's interior. *Am. J. Sci.*, *5*, 423–443; *6*, 6–14, 104–115, 161–172.

DARLING, G. B. and WOOD, P. W. J. 1958.   Habitat of oil in the Canadian portion of the Williston basin. In: *Habitat of Oil.* L. G. Weeks (ed.). Tulsa: Am. Assoc. Petrol. Geol., 129–148.

DARRCOTT, B. W.; GIRDLER, R. W.; FAIRHEAD, J. D.; and HALL, S. A. 1973.   The East African rift system. In: *Implications of Continental Drift to the Earth Sciences.* D. H. Tarling and S. K. Runcorn (eds.). London: Academic Press, 757–766.

DAVIES, P. A. and RUNCORN, S. A. 1980.   *Mechanics of Continental Drift and Plate Tectonics.* London: Academic Press, 362 pp.

DELWIG, L. F. and EVANS, R. 1969.   Depositional processes in Salina salt of Michigan, Ohio, and New York. *Am. Assoc. Petrol. Geol. Bull.*, *53*, 949–956.

DICKINSON, W. R. and SEELY, D. R. 1979.   Structure and stratigraphy of forearc regions. *Am. Assoc. Petrol. Geol. Bull.*, *63*, 2–31.

DOW, W. C. 1978.   Petroleum source beds on continental slopes and rises. *Am. Assoc. Petrol. Geol. Bull.*, *62*, 1584–1606.

DRAKE, C. L.; EWING, J. I.; and STOKAND, H. 1968.   The continental margin of the United States. *Can. J. Earth Sci.*, *5*, 99–110.

EMERY, K. O. 1980.   Continental margins—classification and petroleum prospects. *Bull. Amer. Assoc.*

ERDOGAN, L. T. and AKGUL, A. 1981.   An oil migration and re-entrapment model for the Mardin Group reservoirs of Southeast Anatolia. *J. Petrol. Geol.*, *4*, 57–75.

EUGSTER, H. P. and SURDAM, R. C. 1973. Depositional environment of the Green River Formation of Wyoming. A preliminary report. *Geol. Soc. Am. Bull.*, *84*, 115–1120.

EVAMY, B. D.; HAREMBOURNE, J.; KAMERLING, P.; KNAPP, W. A.; MOLLOY, F. A.; and ROWLANDS, P. H. 1978. Hydrocarbon habitat of Tertiary Niger delta. *Am. Assoc. Petrol. Geol. Bull.*, *62*, 1–39.

EVANS, H. 1972. Zama—a geophysical case history. In: *Stratigraphic Oil and Gas Fields*. R. E. King (ed.). Tulsa: Am. Assoc. Petrol. Geol., Spec. Pub. No. 10, 440–452.

EVANS, R. 1978. Origin and significance of evaporites in basins around Atlantic margin. *Am. Assoc. Petrol. Geol. Bull.*, 62–234.

EXXON, N. F. and WILCOX, J. B. 1978. Geology and petroleum potential of Exmouth plateau area off western Australia. *Am. Assoc. Petrol. Geol. Bull.*, *62*, 40–72.

FINSTAD, K. G. and SELLEY, R. C. (eds.). 1975. *Proceedings: Jurassic Northern North Sea Symposium*. Stavanger: Norsk Petroleumsforening, 340 pp.

FINSTAD, K. G. and SELLEY, R. C. (eds.). 1977. *Proceedings of the Second Jurassic Northern North Sea Symposium*. Oslo: Norweg. Petrol. Soc. 270 pp.

FISCHER, A. G. 1975. Origin and growth of basins. In: *Petroleum and Global Tectonics*. A. G. Fischer and S. Judson (eds.). Princeton: Princeton Univ. Press, 47–82.

FISCHER, A. G. and JUDSON, S. (eds.). 1975. *Petroleum and Global Tectonics*. Princeton: Princeton Univ. Press, 322 pp.

GILL, D. 1979. Differential entrapment of oil and gas in Niagaran pinnacle—reef belt of Northern Michigan. *Am. Assoc. Petrol. Geol. Bull.*, *63*, 608–620.

GLAESSNER, M. F. and TEICHERT, C. 1947. Geosynclines: a fundamental concept in geology. *Am. J. Sci.*, *245*, 465–482, 571–591.

GREY, C. 1972. *The Geology of Libya*. Tripoli: Univ. of Libya, 435 pp.

GUSSOW, W. C. 1954. Differential entrapment of oil and gas: a fundamental principle. *Am. Assoc. Petrol. Geol. Bull.*, *38*, 816–853.

HAILE, N. S. 1968. Geosynclinal theory and the organizational pattern of the northwest Borneo geosyncline. *Quart. J. Geol. Soc. Lond.*, *124*, 171–194.

HALBOUTY, M. T.; MEYERHOFF, A. A.; KING, R. E.; DOTT, R. H.; KLEMME, H. D.; and SHABAD, T. 1970. World's giant oil and gas fields, geologic factors affecting their formation and basin classification. In: *Geology of Giant Petroleum Fields*. M. T. Halbouty (ed.). Tulsa: Am. Assoc. Petrol. Geol., Mem. No. 14, 502–555.

HALL, A. J. 1859. *Natural History of New York*, vol. 3. Paleontology. New York: Appleton and Coy, 1–96.

HARDING, T. P. and LOWELL, J. D. 1979. Structural styles, their plate-tectonic habitats, and hydrocarbon traps in petroleum provinces. *Am. Assoc. Petrol. Geol. Bull.*, *63*, 1016–1058.

HAUG, E. 1900. Les geosynclinaux et les aires continentales. *Geol. Soc. France Bull.*, *28*, 617–711.

HEIRTZLER, J. R. 1968. Sea floor spreading. *Sci. Am.*, *219*, 60–70.

HEYBROEK, F. 1965. The Red Sea Miocene evaporite basin. In: *The Salt Basins Around Africa*. London: Inst. Petrol., 17–40.

HOC, A. 1979. *Continental Margins Geological and Geophysical Research Needs and Problems*. Washington: Nat. Acad. Sci., 302 pp.

HSU, K. J. 1970.   The meaning of the world flysch, a short historical search. In: *Flysch Sedimentology in North America*. Geol. Assoc. Can., Spec. Paper 7, 1–11.

HUNT, J. M. 1979.   *Petroleum Geochemistry and Geology*. San Francisco: Freeman, 617 pp.

IBRAHIM, M. W. 1979.   Shifting depositional axes of Iraq: an outline of geosyncline history. *J. Petrol. Geol.*, *2*, 199–218.

ILLING, L. V. 1959.   Deposition and diagenesis of some Upper Paleozoic carbonate sediments in Western Canada. New York: *Proc. 5th World Petrol. Cong.*, 23–52.

ILLING, L. V. and HOBSON, G. D. 1981.   *Petroleum Geology of the Continental Shelf of Northwest Europe*. London: Heyden & Son, 521 pp.

ILLIES, J. H. and MUELLER, S. T. 1970.   *Graben Problems*. Stuttgart: Slucheizerbart'sche Verlagsbuchhandlung, 316 pp.

IVANHOE, L. F. 1980.   World's giant petroleum provinces. *Oil and Gas J.* 30 June, 146–147.

JAMISON, H. C.; BROCKETT, L. D.; and McINTOSH, R. A. 1980.   Prudoe Bay—a 10-year perspective. In: *Giant Oil and Gas Fields of the Decade 1968–78*. M. T. Halbouty (ed.). Tulsa: Am. Assoc. Petrol. Geol., Mem. No. 30, 289–314.

JODRY, R. L. 1969.   Growth and dolomitization of Silurian reefs, St. Clair County, Michigan. *Am. Assoc. Petrol. Geol. Bull.*, *53*, 957–981.

JONES, P. H. 1969.   Hydrodynamics of geopressure in the northern Gulf of Mexico. *J. Petrol. Tech.*, *17*, 803–810.

KAMEN-KAYE. 1970.   Geology and productivity of Persian Gulf synclinorium. *Am. Assoc. Petrol. Geol. Bull.*, *54*, 2371–2394.

KAY, M. 1944.   Geosynclines in continental development. *Sci.*, *99*, 461–462.

KAY, M. 1947.   Geosynclinal nomenclature and the craton. *Am. Assoc. Petrol. Geol. Bull.*, *31*, 1289–1293.

KENYON, C. S. and BEDDOES, K. 1977.   Geothermal gradient map of southeast Asia. Singapore: *S.E. Asia Petrol. Explor. Soc.*, 50 pp.

KLEMME, H. D. 1975.   Geothermal gradients, heat flow, and hydrocarbon recovery. In: *Petroleum and Global Tectonics*. A. F. Fischer and S. Judson (eds.). Princeton: Princeton Univ. Press, 251–306.

KLEMME, H. D. 1980.   Petroleum basins—classification and characteristics. *J. Petrol. Geol.*, *3*, 187–207.

KLITZSCH, E. 1970.   Die Strukturgeschichte der Zentral Sahara. *Geol. Rundsch.*, *59*, 459–527.

KLITZSCH, E. H. 1972.   Problems of continental Mesozoic strata of southwestern Libya. In: *African Geology*. A. J. Whiteman and T. F. J. Dessauvagie (eds.). Ibadan Univ., 483–493.

KLOVAN, J. E. 1974.   Development of Western Canadian Devonian reefs and comparison with Holocene analogues. *Am. Assoc. Petrol. Geol. Bull.*, *58*, 787–799.

LAMB, C. F. 1980.   Painter reservoir field—giant in Wyoming thrust belt. *Am. Assoc. Petrol. Geol. Bull.*, *64*, 638–644.

LEHNER, P. and DE RUITER, P. A. C. 1977.   Structural history of Atlantic margin of Africa. *Am. Assoc. Petrol. Geol. Bull.*, *61*, 961–981.

LESTANG, J. DE. 1965.   Das Palaozoikum am Rande des Afro-Arabischen Gondwanakontinentes. *Z. dtsch. Geol. Ges.*, *117*, 479–488.

LOWELL, J. D. and GENIK, G. J. 1972. Sea floor spreading and structural evolution of the southern Red Sea. *Am. Assoc. Petrol. Geol. Bull., 56*, 247–259.

LOWELL, J. D.; GENIK, G. J.; NELSON, T. H.; and TUCKER, P. M. 1975. Petroleum and plate tectonics of the southern Red Sea. In: *Petroleum and Global Tectonics*. A. G. Fischer and S. Judson (eds.). Princeton: Princeton Univ. Press, 129–156.

McCASLIN, E. 1981. Hot overthrust belt exploration continues. *Oil & Gas J.* 12 Jan., 113–114.

McKEE, E. D. 1965. Origin of Nubian and similar sandstones. *Geol. Runsch., 52*, 551–587.

MAGOON, L. B. and CLAYPOOL, G. E. 1981. Petroleum geology of Cook Inlet basin—an exploration model. *Am. Assoc. Petrol. Geol. Bull., 65*, 1043–1062.

MESNER, J. C. and WOODRIDGE, L. C. P. 1964. Maranhao Paleozoic basin and Cretaceous coastal basins. *Am. Assoc. Petrol. Geol. Bull., 48*, 1475–1512.

MESOLELLA, K. J.; ROBINSON, J. D.; McCORMICK, L. M.; and ORMISTON, A. R. 1974. Cyclic deposition of Silurian carbonates and evaporites in Michigan basin. *Am. Assoc. Petrol. Geol. Bull., 58*, 34–62.

MEYERHOFF, A. A. and MEYERHOFF, H. A. 1972. The new global tectonics: major inconsistencies. *Am. Assoc. Petrol. Geol. Bull., 56*, 269–336.

MITCHELL, A. H. and READING, H. G. 1969. Continental margins, geosynclines, and ocean floor spreading. *J. Geol., 77*, 629–646.

MITCHELL, A. H. G. and READING, H. G. 1978. Sedimentation and tectonics. In: *Sedimentary Environments and Facies*. H. G. Reading (ed.). Oxford: Blackwell, 439–476.

MURRAY, G. E. 1960. Geologic framework of Gulf Coastal Province of United States. In: *Recent Sediments, Northwestern Gulf of Mexico*. F. P. Shepard (ed.). Tulsa: Am. Assoc. Petrol. Geol., 5–33.

MURRIS, R. J. 1980. Middle East: stratigraphic evolution and oil habitat. *Am. Assoc. Petrol. Geol. Bull., 64*, 597–618.

MURRIS, R. J. 1981. Middle East: stratigraphic evolution and oil habitat. *Geol. Mijnb., 60*, 467–480.

NISSENBAUM, A. 1978. Dead Sea asphalts—historical aspects. *Am. Assoc. Petrol. Geol. Bull., 62*, 837–844.

NORTH, F. K. 1971. Characteristics of oil provinces: a study for students. *Can. Petrol. Geol. Bull., 19*, 601–658.

OLENIN, V. B. 1967. The principles of classification of oil and gas basins. *Austr. Oil and Gas J., 13*, 40–46.

PERRODON, A. 1971. Classification of sedimentary basins: an essay. *Sci. Terre., 16*, 193–227.

PERRODON, A. 1978. Coup d'oil sur les provinces geantes d'hydrocarbures. *Rev. Inst. Franc. du Petrole., 33*, 493–513.

PICARD, M. D. 1967. Paleocurrents and shoreline orientations in Green River Formation (Eocene). Raven Ridge and Red Wash areas, Northeastern Uinta Basin. Utah. *Am. Assoc. Petrol. Geol. Bull., 5*, 383–392.

PRATSCH, J. C. 1978. Future hydrocarbon exploration on continental margins and plate tectonics. *J. Petrol. Geol., 1*, 95–105.

QUENNELL, A. M. 1958. The structural and geomorphic evolution of the Dead Sea rift. *Quart. J. Geol. Soc. Lond., 114*, 1–24.

RANNEFT, T. S. M. 1979.   Segmentation of island arcs and application to petroleum geology. *J. Petrol. Geol.*, *1*, 35–53.

READING, H. G. 1972.   Global tectonics and the genesis of flysch successions. *24. Intermat. Geol. Cong. Montreal, 1972. Section 6, Stratigraphy and Sedimentology*, 59–66.

SALEM, M. J. and BUSREWIL (eds.). 1981.   *The Geology of Libya*, 3 vols. London: Academic Press, 1155 pp.

SALOP, L. I. 1967.   *Geology of the Baikal Region*, vol. 2. Moscow: Izd Nedva, 502 pp.

SANFORD, R. M. 1970.   Sarir oil field, Libya—Desert surprise. In: *Geology of Giant Petroleum Fields*. M. T. Halbouty (ed.). Tulsa: Am. Assoc. Petrol. Geol., Mem. No. 14, 449–476.

SCHNEIDER, E. D. 1972.   Sedimentary evolution of rifted continental margins. *Geol. Soc. Am.*, Mem. No. 132, 109–118.

SCHOLL, D. W.; VON HUENE, R.; and RIDLON, J. B. 1968.   Spreading of the ocean floor: undeformed sediments in the Peru-Chile trench. *Sci.*, *159*, 869–871.

SCHUCHERT, C. 1923.   Sites and nature of the North American geosynclines. *Geol. Soc. Am. Bull.*, *34*, 151–230.

SCHUPPLI, H. M. 1946.   Geology of the oil basins of East Indian archipeligo. *Am. Assoc. Petrol. Geol. Bull.*, *30*, 1–22.

SCHWAB, F. L. 1971.   Geosynclinal compositions and the new global tectonics. *J. Sedim. Petrol.*, *41*, 928–938.

SELLEY, R. C. 1972.   Diagnosis of marine and non-marine environments from the Cambro-Ordovician sandstones of Jordan. *J. Geol. Soc. Lond.*, *128*, 109–117.

SELLEY, R. C. 1975.   *Introduction to Sedimentology*. London: Academic Press, 408 pp.

SELLEY, R. C. 1976.   The habitat of North Sea oil. *Proc. Geol. Ass. Lond.*, *87*, 359–388.

SEYFERT, C. K. and SIRKIN, L. A. 1973.   *Earth History and Plate Tectonics*. New York: Harper & Row, 544 pp.

SHORT, K. C. and STAUBLE, A. J. 1967.   Outline of geology of Niger delta. *Am. Assoc. Petrol. Geol. Bull.*, *51*, 761–779.

SMITH, G. W.; SUMMERS, G. E.; WALLINGTON, D.; and LEE, J. L. 1958.   Mississippian oil reservoirs in Williston basin. In: *Habitat of Oil*. L. G. Weeks (ed.). Tulsa: Am. Assoc. Petrol. Geol., 149–177.

ST. JOHN, B. 1980.   *Sedimentary Basins of the World and Giant Hydrocarbon Accumulations*. Tulsa: Am. Assoc. Petrol. Geol., Map + 23 pp.

SURDAM, R. C. and WOLFBAUER, C. A. 1975.   Green River Formation, Wyoming: A playa-lake complex. *Geol. Soc. Am. Bull.*, *86*, 335–345.

TARLING, D. H. (ed.). 1981.   *Economic Geology and Geotectonics*. London: Blackwell, 213 pp.

TARLING, D. H. and RUNCORN, S. K. 1973.   *Implications of Continental Drift to the Earth Sciences*, vols. I, II. London: Academic Press, 1184 pp.

THIEBAUD, C. E. and ROBSON, D. A. 1979.   The geology of the area between Wadi Warden and Wadi Gharandal, East Clysmic Rift, Sinai, Egypt. *J. Petrol. Geol.*, *1*, 63–75.

UNESCO, S. 1965.   *East African Rift System, Upper Mantle*. Committee. Nairobi: University College, 91 pp.

USPENSKAYA, N. YU. 1967. Principles of oil and gas territories subdivisions and the classification of oil and gas accumulations. Mexico: Proc. *7th World Petrol. Cong.*, vol. 2. Amsterdam: Elsevier, 961–969.

VAN HOUTEN, F. B. 1973. Meaning of molasse. *Geol. Soc. Amer. Bull.*, *84*, 1973–1976.

VAN HOUTEN, F. D. 1980. Latest Jurassic—Cretaceous regressive facies, northeast African craton. *Am. Assoc. Petrol. Geol. Bull.*, *64*, 857–867.

VEEVERS, J. J. 1981. Morphotectonics of rifted continental margins in embryo (East Africa), youth (Africa-Arabia) and maturity (Australia). *J. Geol.*, *89*, 57–82.

VIDAL, J. 1980. Geology of Grondin field. In: *Giant Oil and Gas Fields of the Decade: 1968–1978*. M. T. Halbouty (ed.). Tulsa: Am. Assoc. Petrol. Geol., Mem. No. 30, 577–590.

VINE, F. J. 1970. The geophysical year. *Nature*, *227*, 1013–1017.

WALTERS, R. F. and PRICE, A. S. 1948. Kraft-Prusa oilfield, Barton County, Kansas. In: *Structure of Typical American Oil Fields*. J. V. Howell (ed.). Am. Assoc. Petrol. Geol., 249–280.

WEBER, K. J. and DAUKORU, E. 1975. Petroleum geology of the Niger delta. Tokyo: *Proc. 9th World Petrol. Cong.*, vol. 2. Geology. London: Applied Science Publishers, 209–221.

WEEKS, L. G. 1958. Factors of sedimentary basin development that control oil occurrence. *Am. Assoc. Petrol. Geol. Bull.*, *32*, 1093–1160.

WILHELM, O. and EWING, M. 1972. Geology and history of the Gulf of Mexico. *Geol. Soc. Am. Bull.*, *83*, 575–600.

WILSON, J. E.; KASHAI, E. L.; and CROKER, P. 1983. Hydrocarbon potential of Dead Sea rift valley. *Oil & Gas J.*, 20 June, 147–154.

WILSON, J. T. 1966. Did the Atlantic close and then re-open? *Nature*, London, *211*, 676–681.

WILSON, J. T. 1968. Static or mobile earth: the current scientific revolution. *Proc. Am. Phil. Soc.*, *112*, 309–320.

WOOD, P. W. J. 1980. Hydrocarbon plays in Tertiary S.E. Asia basins. *Oil & Gas J.*, 21 July, 90–96.

WOODLAND, A. W. (ed.). 1975. *Petroleum and the Continental Shelf of North West Europe*, vol. I. Geology. London: Applied Science Publishers, 501 pp.

ZIEGLER, P. A. 1975. Geologic evolution of North Sea and its tectonic framework. *Am. Assoc. Petrol. Geol. Bull.*, *59*, 1073–1097.

9

Tar Sands and Oil Shales

The petroleum geologist is largely concerned with exploring for crude oil and natural gas, and the major part of this book is devoted to this theme. Vast reserves of energy are also locked up in the tar sands and oil shales. These reserves have been hard to unlock because the relatively cheap price of oil and gas has inhibited technological research into the extraction of oil from these resources. The increasing cost of producing conventional hydrocarbons has recently given an incentive to develop methods of oil extraction from tar sands and oil shales. This chapter deals with the plastic and solid hydrocarbons (excluding coal), describing their composition, origin, and distribution, and briefly reviewing production processes.

Plastic and solid hydrocarbons are common in sedimentary rocks of diverse ages in many parts of the world. They are distinct from crude oils, which are liquid at normal temperatures and pressures, although many of these hydrocarbons are viscous, and their viscosity decreases with increasing temperature. Figure 9.1 classifies the solid hydrocarbons.

Man's first exposure to hydrocarbons was from surface occurrences of solid and viscous tars rather than from liquid crude oil, which was not encountered in large quantities until boreholes were drilled. The heavy hydrocarbons are colloquially termed *pitch, asphalt, bitumen,* or *tar,* and the words are often used interchangeably. Although the first three terms are

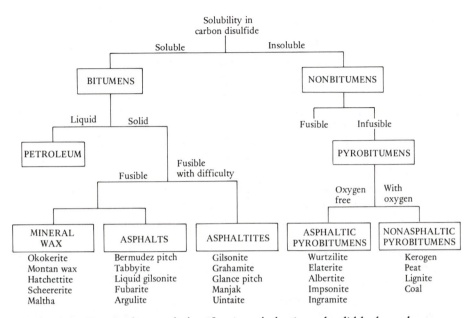

FIGURE 9.1 *Terminology and classification of plastic and solid hydrocarbons.*
(After Abraham, 1945.)

synonymous, tar is, strictly speaking, a dark brown, viscous liquid produced by the destructive distillation of organic matter—wood, coal, or
shale. This distillation may occur naturally in the subsurface or artifically
by man. Stockholm, or Archangel, tar was collected from the cavities below the *meilers* in which charcoal was burnt. This tar was used extensively
for the wooden ships of the maritime nations of northern Europe.

The following sections describe the mode of occurrence, the composition, and the economic significance of solid hydrocarbons.

PLASTIC AND SOLID HYDROCARBONS

Occurrence

The solid and heavy, viscous hydrocarbons occur as lakes or pools on the
earth's surface and disseminated in veins and pores in the subsurface. Two
genetically distinct modes of occurrence can be recognized:

1. Inspissated deposits
2. Secondary deposits

Inspissated Deposits

Inspissated deposits are heavy hydrocarbons from which the light fraction has been removed. This situation may occur where an accumulation of liquid oil has been brought to the surface of the earth by a combination of uplift and erosion. As this process takes place, the oil will be subjected to flushing by meteoric water, leading to oxidation and bacterial degradation. The lighter hydrocarbons may come out of solution as the pressure drops and may be lost to the surface through fractures as the cap rock is breached. Thus the oil accumulation is proportionally enriched in the viscous, heavy hydrocarbons.

Surface occurrences of asphalt are not only fossilized unroofed oil fields but may also be produced from seepages escaping from oil fields in the subsurface. These hydrocarbon seeps can occur in many ways, but are principally found where faults and fractures allow hydrocarbons to emigrate up through the cap rocks of reservoirs (Fig. 9.2). The seeps may be of mud, brine, gas, or oil, although oil has a tendency to become tarry for the reasons already outlined.

These various kinds of hydrocarbon seeps are known from all over the world and are characteristic of many, but by no means all, major hydrocarbon provinces. Notable examples occur in Oklahoma, Venezuela, Trinidad, Burma, Iran, Iraq, and elsewhere in the Middle East. The numerous references to pitch and bitumen in the Bible have already been noted. In the Dead Sea, blocks of pitch that are sufficiently large for men to stand on them are occasionally exuded from fault planes (Nissenbaum, 1978;

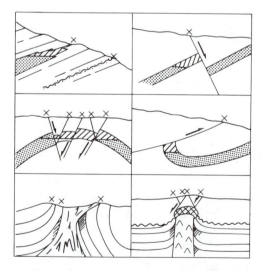

FIGURE 9.2 *Sections illustrating the various ways in which oil seeps may occur. (After Levorsen, 1967.)*

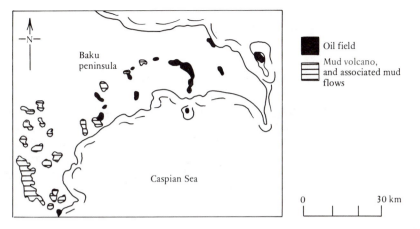

FIGURE 9.3 *Map of Aspheron peninsula, Baku province, showing association of oil fields with mud flows and mud volcanoes. (After Levorsen, 1967.)*

Spiro et al., 1983). The seepages on the shores of the Caspian Sea have long attracted attention, especially in the Baku region of what is now the USSR (Fig. 9.3). Here, not only oil but also mud and gas pour out on the earth's surface. Spontaneous ignition frequently occurs, so that this region has been called the *Land of Eternal Fires*. It has been suggested that this region, and other burning seeps in this area, formed the basis for the Zoroastrian religion, which flourished in Persia for many hundreds of years. In the thirteenth century Marco Polo wrote of one seep in the Baku area: "On the confines toward Geirgene there is a fountain from which oil springs in great abundance, in as much as a hundred shiploads might be taken from it at one time" (translated by Yule, 1871).

Hydrocarbon seeps also occur offshore, notably in the Santa Barbara Channel of California (where the oil companies have been blamed for causing pollution) and off the coasts of Trinidad, Yucatan, and Ecuador. Gas seeps, possibly of shallow biogenic origins, have been recorded on sparker surveys, notably off the edge of the Mississippi delta.

Hydrocarbon seeps are, of course, not just a modern phenomenon but are a form of nature's own pollution, which has gone on throughout Phanerozoic time. The Pleistocene seeps of California are well known for their contained vertebrate fauna. Since hydrocarbons are biodegradable, ancient inspissated oil occurs along faults and unconformities. Tarry seeps ranging from Miocene to Pliocene age occur in Iran (Elmore, 1979).

Secondary Deposits

The extensive flow of oil from an oil seep may give rise to secondary deposits. These deposits are admixtures of sediment and heavy degraded oil. Extensive deposits of secondary petroleum are extremely rare, because of the tendency for oil to be degraded and oxidized in the atmosphere and under water.

Secondary deposits are occasionally found around oil seeps and mud volcanoes. In some rare cases of intraformational conglomerates, pebbles and even boulders occur of sand cemented by dead oil. One such example is known in Lower Cretaceous fluvial channels at Mupe Bay on the southern coast of England (Lees and Cox, 1937) (Fig. 9.4). Such secondary deposits are of limited economic significance. They may, however, be extremely important because of how they date the migration of petroleum in an area. The relationship between petroleum migration and trap formation is obviously extremely important (see p. 323).

FIGURE 9.4 *Photograph and sketch of Wealden (Lower Cretaceous) oil sand, Mupe Bay, Dorset. A cross-bedded fluvial channel sand is saturated with sweet light oil. The floor of the channel is lined by boulders of sand cemented with dead oil. The channel was eroding a penecontemporaneous oil seep. This outcrop demonstrates that oil generation and migration had begun by the Lower Cretaceous. The light oil saturation testifies to present-day migration to the surface.*

Composition

As crude oil is dissipated at the surface of the earth, its composition gradually changes as it loses its lighter components and becomes a solid hydrocarbon. Although, as already shown, the terms pitch, tar, and asphalt have been used synonymously for such material, their composition can be described more specifically. This classification is related to the type of the original crude oil and its degree of inspissation.

Paraffinic oils degenerate to waxy residues, whereas the naphthenic oils move along a more complex path. These two major types of solid hydrocarbon are discussed in the following sections.

Waxy Solid Hydrocarbons

Waxy solid hydrocarbons are produced by the inspissation of paraffinic crude oils. They are generally plastic, waxy yellow to dark brown substances called *ozokerite* (from the Greek *oderiferous* wax). This substance occurs as a vein filling in the Boryslav area of Russia, in Utah, and in India. A particular variant known as Hatchettite, or mineral tallow, occurs in ironstone nodules in the Coal Measures of South Wales.

Chemically, the waxy solid hydrocarbons are paraffin derivatives, ranging from C_{22} to C_{29}. They contain 84 to 86 percent carbon, 14 to 16 percent hydrogen, and traces of sulfur and nitrogen. In terms of molecular structure, ozokerites from Utah contain 81 percent paraffins and naphthenes, 10 percent aromatics, and 9 percent oxygen, nitrogen, and sulfur compounds (Hunt et al., 1954).

Asphaltic Solid Hydrocarbons

Asphaltic solid hydrocarbons are produced by inspissation of naphthenic crude oil. In terms of their elemental chemistry these compounds are similar to the waxy solid hydrocarbons. They contain 79.5 to 87.2 percent carbon, 8.9 to 13.2 percent hydrogen, up to 8 percent sulfur, and commonly traces of nitrogen, oxygen, and inorganic compounds (generally rock-forming minerals). Their molecular composition, however, is quite different from that of the waxy solid hydrocarbons. Asphalts from Utah have only 18 percent paraffins and naphthenes, 48 percent aromatics, and 34 percent compounds of oxygen, nitrogen, and sulfur. Details of the composition of these substances are given by Hobson and Tiratsoo (1981).

As the asphalts are subjected to inspissation, they grade through the asphaltites to asphaltic pyrobitumen. Asphalt (Greek for bitumen) is a solid to semisolid, dark brown to black substance with a specific gravity of less than 1.15 and a melting point of 65 to 95°C. It is 10 to 70 percent soluble in naphtha and totally soluble in carbon disulfide.

The asphaltites have a specific gravity of less than 1.20 and a melting point of 120 to 320°C. They are up to 60 percent soluble in naphtha and 60 to 90 percent soluble in carbon disulfide. Gilsonite is an asphaltite that occurs in fissures in the Tertiary shales of the Uinta basin of Utah and Colorado. Glance pitch, manjak, and Grahamite are other local names for asphaltites.

The asphaltic pyrobitumens are rubbery, brown substances with a specific gravity of less than 1.25. They do not melt, but swell and decompose on heating. They are insoluble in naphtha and only slightly soluble in carbon disulfide. The various names for asphaltic pyrobitumens include Albertite, Elaterite, Impsonite, and Wurtzilite (Table 9.1). Some of the best-studied deposits of these minerals are associated with the Green River Shales (Hunt et al., 1954).

TABLE 9.1 Characteristics of some bitumens and waxes

Name	H-C ratio	Percent by weight S	N	Origin
Ozokerite	1.96–1.89	0.1	0.1	Paraffinic oil
Tabbyite	1.62			↑
Ingranite	1.62			
Wurtzilite	1.6 –1.59	4.0	2.2	Increasing degradation of naphthenic oil
Gilsonite	1.47–1.42	0.4	2.8	
Albertite	1.32–1.24	0.7	1.0	
Elaterite				

Chemical data from Hunt et al., 1954. Reprinted with permission.

Figure 9.5 summarizes the composition and evolution of the solid hydrocarbons.

TAR SANDS

Composition, Distribution, and Origin

Heavy, viscous oil deposits occur at or near the earth's surface in many parts of the world (Table 2.7). These deposits have API gravities in the range of 5 to 15° and typically occur within highly porous sands, generally referred to as tar sands, or oil sands. As Table 9.2 shows, vast reserves of hydrocarbons are contained in these beds. For many years tar sands have largely been a curiosity. The oil industry lacked the incentive and hence the technology to extract the oil. Global price rises in the 1970s have changed this picture. Much research is now being carried out in methods of extraction.

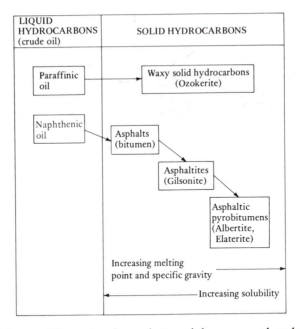

FIGURE 9.5   *Diagram illustrating the evolution of the waxy and asphaltic hydrocarbons.*

This section examines the composition, occurrence, origin, and extraction of tar sands and associated heavy oils. Further details are given by Phizackerley and Scott (1967) and Hills (1974). The change from normal crude oil, via heavy oil, into tar is a gradational one, both in terms of physical and chemical properties. Physically, the oils become heavier and more viscous. Chemically, the oils tend to contain more inorganic impurities and to be more sulfurous and aromatic. The sulfur content of oil sand varies from about 0.5 to 5.0 percent, but occasionally exceeds this range, attaining 6 to 8 percent in the La Brea deposits of Trinidad. This value compares with an average of about 1 percent for normal crudes. Figure 9.6 shows the composition of a number of tar sands compared with light crudes. As shown in the figure, compared with normal oils, tar sands tend to be enriched in resins and asphaltenes and impoverished in saturated hydrocarbons. There is little significant difference in their aromatic content.

Two possible explanations have been proposed to account for these differences between normal light crudes and tar sands. The heavy oils may be young and have yet to mature to light oil. Alternatively, the heavy oils may have been produced by the degradation of mature light oil. Many geologists have discussed the chemical evolution of mature oils as the oils emigrate into seeps on shallow traps and are subjected to flushing by meteoric water (e.g., Evans et al., 1971; Byramjee, 1983). Degradation occurs by both organic and inorganic processes. Many of the saturated hydrocarbons

TABLE 9.2   Data on some major oil sands

Location	Age	In-place reserves	API gravity	Reference
Albania				
Selenizza	Miocene	371×10^6 bbl	4.6–13.2	Iniechen, 1951
Canada				
Melville	Permian	$50–100 \times 10^6$ bbl	7	Sproule, 1963
Kindersley	Mississip.–L. Cret.	410,432,000 bbl	13–15	White, 1974
Athabaska				
Surface	L. Cret.	120 bbl	6–8	Jardine, 1974
Subsurface	L. Cret.	505 bbl	8–10	Jardine, 1974
Peace River	L. Cret.	50 bbl	8–23	Jardine, 1974
Cold Lake	L. Cret.	164 bbl	10–12	Jardine, 1974
Wabaska	L. Cret.	53 bbl	10–13	Jardine, 1974
Malagasy				
Bemolanga	Trias	1.75 bbl	?	Spraggins, 1973
Rumania				
Derna	Pliocene	25×10^6 bbl	?	Walters, 1974
Trinidad				
La Brea	Miocene	60×10^6 bbl	1–2	Kugler, 1961
United States				
California	Mioc.–Plioc.	200×10^6 bbl	4–13	Burk, 1971
Kansas	Penns.	750×10^6 bbl	15	Banks and James, 1974
Kentucky	Miss.–Penns.	75×10^6 bbl	?	Burk, 1971
New Mexico	Triassic	57×10^6 bbl	?	Walters, 1974
Utah	Cret.–Olig.	$17.7–27.5 \times 10^9$ bbl	8–12	Walters, 1974
USSR				
Kobystan	Miocene	24×10^6 bbl	?	Walters, 1974
Olenek	?	?	?	Walters, 1974
Venezuela				
Orinoco	Olig.–Mioc.	$26,064 \times 10^6$ bbl	9–12	Alayeto and Louder, 1974

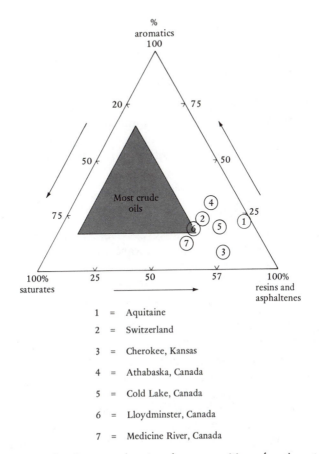

FIGURE 9.6 *Triangular diagram showing the composition of various tar sands.*
(Data from Hills, 1974.)

are soluble in water, especially the light end members of the paraffin series. Organic degradation also occurs through bacterial action. Sulfate-
reducing bacteria may also be responsible for taking sulfate ions from the
meteoric water and combining them with hydrocarbons to form benziothi-
ophenes and dibenzothiophenes.

Nonetheless, the origin of heavy oils is controversial. The Cretaceous
tar sands of Alberta are a case in point, even though they are probably the
best-studied example in the world. An early theory stated that the oil was
actually formed in the sands that now contain them (Ball, 1935). Reasons
for this theory included the flat-lying nature of the sediments and the apparent absence of normal crudes in the area. This idea is now known to be
untrue. The Athabaska hydrocarbons are now generally believed to have
migrated into their present reservoir in a conventional manner.

Many geologists and chemists have suggested that the heavy oils are mature oils that have undergone degradation by meteoric flushing. Deroo et al. (1974) show how these oils can be traced gradationally up the basin margin from deep light-oil pools. This change is accompanied by a gradual diminution in water salinity from over 80 g/l to less than 20 g/l over a distance of some 300 km. Nevertheless, this theory has been disputed by Montgomery (1974), who favors an immature origin for these deposits. Evidence for this theory includes their higher porphyrin content, the absence of insoluble benzene matter, their molecular weight distribution, and their nickel-vanadium ratio.

It is perhaps appropriate at this point to consider what light the geological setting of tar sands may throw on their origin. Kappeler (1967) has reviewed the mode of occurrence of tar sands around the world. The majority occur where major basins transgress over Precambrian shields. They occur in fluvial or deltaic sands; they rarely occur in marine sands or carbonates. Considered on a smaller scale, tar sands occur in two ways: either in situ in erosionally breached traps or migrated from deep traps into surface seepages. The Bemolanga oil sands of Malagasy are a good example of the former; the La Brea deposits of Trinidad of the latter.

The Bemolanga oil sands of Malagasy are the third largest of such deposits in the world (Fig. 9.7). They occur in the Isalo Group (Triassic) of the Karroo system, covering an area of some 388 km^2 with an overburden of up to 100 m. The tar sands are in a complex faulted anticline, where the cap rock of the Bemolanga Clay has been removed by recent erosion. The average pay is some 30 m thick, and the in-place reserves are estimated at 1.75 million barrels. The sands are cross-bedded, coarse, and fluvial. Local dykes of Cretaceous age have locally metamorphosed the tar. The underlying Sakamena Shale is believed to have provided the source for the oil (Kent, 1954; Kamen-Kaye, 1983).

The La Brea asphalt lake of Trinidad illustrates the second type of tar sand, where oil has emigrated from an accumulation in a Cretaceous anticline up through an imperfect seal to reach the surface (Fig. 9.8). The tar lies in a depression within Miocene sands. It has a gravity of 1 to 2° API bitumen saturation and 6 to 8 percent sulfur. The lake covers an area of 126 acres and has estimated reserves of 60 million barrels (Sutor, 1951; Walters, 1974).

The genesis of these two tar sands is fairly clear. It may be instructive to contrast them with the more equivocal deposits of Alberta, Canada. The basic statistical and geological data of the Lower Cretaceous tar sands of Alberta are given in Table 9.2 and Figure 9.9. They occur in fluviodeltaic sands of the Manville Group, which were derived mainly from the west, but with local deltaic spreads of sand from the east. Subsequently, the strata were tilted regionally to the west and very gently folded. Normal light oils (35–40° API) were generated in the western part of the Alberta basin, but

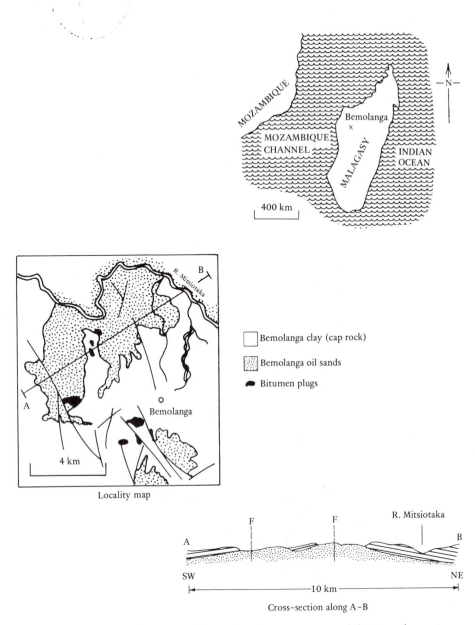

FIGURE 9.7 *Maps and sections illustrating the occurrence of the Bemolanga tar sands of Malagasy. (After Kent, 1954.)*

tar sands are now encountered in stratigraphic pinchout and combination (fold and pinchout) traps along the eastern margin.

As already noted, the origin of these tar sands is controversial. The debate revolves around whether the oil is immature, or mature and subsequently degraded. Advocates of the immature origin believe that the chemical composition of the oil shows that it has not undergone significant

395

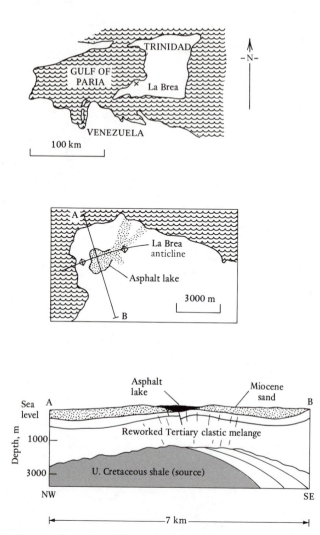

FIGURE 9.8 *Maps and sections illustrating the La Brea tar sands of Trinidad.*
(After Kugler, 1961.)

thermal maturation. Advocates of the degradation theory point to the gradational lightening of the oil basinward, accompanied by increased formation water salinities. They also find it hard to understand how a viscous,
heavy oil could have migrated into the reservoirs. The migration of a light
oil up the basin margin and its subsequent degradation is easy to explain.
The sands are highly permeable, however (1–5 darcies). So, given time,
even a heavy oil might move through them. Not only is the origin of tar
sands debatable but so is their economic worth.

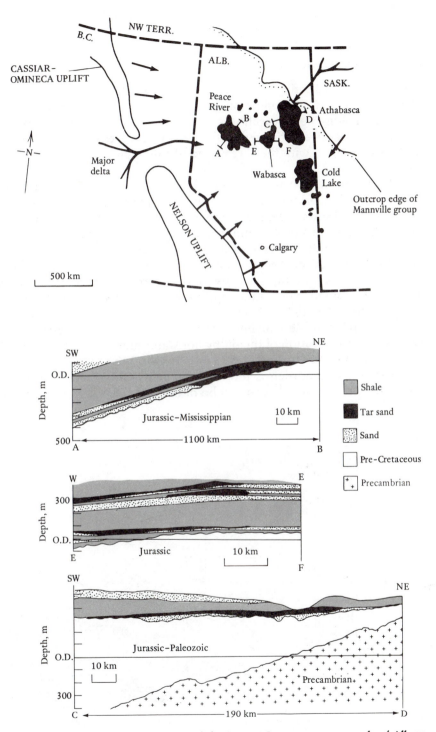

FIGURE 9.9 *Map and cross-sections of the Lower Cretaceous tar sands of Alberta. (After Jardine, 1974.)*

Extraction of Oil from Tar Sands

During the last half century, the low cost of conventional light crude inhibited interest in heavy oil production. With no incentive to produce oil from the tar sand, negligible research was put into the technology of extracting the oil. All this has been changed by the huge price rises of crude oil in the early 1970s, coupled with the threatened ultimate global shortage of conventional oil.

Many schemes have now been developed to extract oil from tar sands. Some are already working commercially or as pilot plants. More will follow. Figure 9.10 shows the major methods of extracting heavy oil. The two basic approaches are surface mining and subsurface extraction. The technology of strip mining tar sands is analogous to that already developed for the open-cast extraction of other economic rocks. It presents similar problems, namely, the economic thickness of overburden that can be tolerated and the degree of environmental opposition of the native population.

Once the tar sands have been quarried, their contained oil must be extracted. At present the conventional method consists of disaggregation of the sand and separation of the oil by hot water and/or steam. Variations on this technique are currently being employed on the Alberta tar sands.

Where the overburden is too thick for conventional open-cast mining, in situ extraction methods must be developed. Basically, two types of

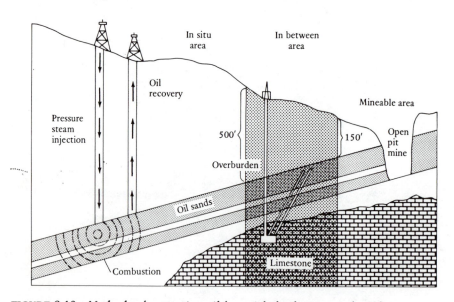

FIGURE 9.10   *Methods of extracting oil from Athabaska tar sands. (After Wennekers, 1981.)*

extraction methods are used: One involves the injection of a solvent to dissolve the oil; the other seeks to reduce the oil's viscosity by heating. Both these methods are inhibited by the tendency for the high permeability of clean sands to decrease in proportion to the level of oil saturation. Great care is thus needed to correctly assess both the petrophysical variations of the sands and their oil saturations (Towson, 1977; Kendall, 1977). For thermal extraction, heat may either be taken down into the reservoir in hot water or steam or, alternatively, be generated in situ by combustion of the tar or by a nuclear explosion. In situ combustion necessitates the introduction of oxygen. All of these methods, except for nuclear extraction, can be accomplished either by the *huff and puff* procedure or by monodirectional flooding. In the first method wells are alternately used for injection and production. The second method is akin to conventional oil or gas recovery techniques in that a continuous flood is established from peripheral injection wells to centrally located producers. A pilot steam injection plant has been considered for the Cold Lake deposits of Alberta (Winestock, 1974; Dome, 1979).

In 1956 Shell began a 10-year program of steam injection in the Athabaska tar sands (Doscher, 1967). Hot sodium hydroxide was pumped into the base of the pay along horizontal fractures connecting the injecting and producing wells. This process emulsified the oil, allowing it to be produced. While the rocks become heated, the oil viscosity declines so that it begins to flow into the fracture. Once the process is established, the sodium hydroxide solution may be removed and replaced by injections of hot water or steam.

In the same region Amoco tried out a method in which combustion was alternated with water flooding. Initially, the sands were hydraulically fractured to enhance permeability. Combustion phases were alternated with periods when air-water mixtures were injected (Anon, 1972).

A number of in situ extraction methods have been employed on the Orinoco tar sands of Venezuela (Finken and Heldau, 1972). One of the simpler methods involved the injection of refined diesel oil to dissolve and transport the heavier crude. In the United States in situ extraction projects have been carried out in Missouri and California. A reverse combustion process has been attempted at Bellamy in Missouri (Trantham and Marx, 1966). In the early 1960s a huff-puff steam injection experiment was carried out in the Vaca tar sands of California. All of these in situ extraction methods are still in the development stage. For a review of the various methods of extraction, see Ali (1974).

In conclusion, at the present time the extraction of oil from tar sands is a marginal economic proposition. This situation will gradually change because of a combination of improved technology and long-term rising global energy costs.

OIL SHALES

Oil shale is a fine-grained sedimentary rock that yields oil on heating. It differs from tar sand in more than grainsize. In tar sands the oil is free and occurs within the pores. In oil shales, however, oil seldom occurs free, but is contained within the complex structure of kerogen, from which it may be distilled. Oil shales are widely distributed around the globe and may contain locked within them more energy than in all the presently discovered conventional oil reserves. The world's oil shales may contain some 30 trillion barrels of oil, only about 2 percent of which is accessible using present day technology (Yen and Chilingarian, 1976).

Oil shales have been known to man for many centuries. The Mongol hordes used burning oil shale arrowheads to terrify mandarins on the Great Wall of China as long ago as the thirteenth century (Naasiam, 1937). In 1694 the British Crown issued a patent to Martin Eele for the extraction of oil from shale (p. 2). Subsequently, a considerable extraction industry has developed, although it is now temporarily quiescent.

The following sections review the composition, distribution, origin, and extraction of oil shales. More detailed accounts are found in Yen and Chilingarian (1976), Burger (1973), and Russell (1980).

Composition

Most oil shales are essentially claystones and siltstones, but bituminous marls and lime mudstones are also known. The basic constituents of oil shales may be grouped as follows:

1. *Inorganic components (<90%)*

    Detrital
    {
    Quartz
    Feldspar
    Mica
    Accessory minerals
    Carbonates } Authigenic
    Clays
    Pyrite
    }

2. *Organic components (10–27%)*
    Bitumens (hydrocarbons soluble in CS_2) $\simeq$ 2%
    Kerogens (insoluble in CS_2) $\simeq$ 8%

The exact amount of organic matter needed before a shale can be classified as an oil shale is arbitrary. A cut-off value of 10 percent has been cited. With increasing organic content, oil shales grade into the cannel coals. A figure of 27 percent has been taken as the critical value between oil shales and cannel coals (Shanks et al., 1976).

The inorganic component of oil shales differs little from that of conventional shales. There is a framework of detrital grains of quartz, feldspar, mica, and other accessories. Large amounts of clay are present, both detrital floccules and authigenic crystals. Various carbonate minerals are present, both calcitic skeletal fragments and authigenic calcite, dolomite, ankerite, and siderite. Authigenic pyrite is especially characteristic of oil shales because of their anaerobic origin. The presence of this sulfur causes one of the major problems of oil shale refining.

The important constituent of oil shale is kerogen. In a study of the Scottish Carboniferous oil shales, Stewart (1912) defined kerogen as "the name given to carbonaceous matter in shale which gives rise to crude oil on distillation." The term *kerogen* was suggested by Professor Crum Brown. Subsequently, more rigid definitions have been proposed (see, for example, Dott and Reynolds, 1969). Stewart's original definition of kerogen is now too broad for most authorities, since it would include the bitumen fractions. Currently, kerogen is defined and distinguished from bitumen by its insolubility in carbon disulfide. Further discussion of the chemical origin and significance of kerogen is given on page 187.

The organic precursor of the kerogens in oil shales are hard to determine. However, two particular types of oil shale whose biological origin is well known are Torbanite and Tasmanite. Torbanite comes from the village of Torban in the Midland Valley of Scotland, where this particular variety of oil shale occurs in Carboniferous coal measures. Torbanites, also termed *bog head* or *cannel coal*, are made of algal matter. They contain the remains of the freshwater blue-green algae *Botryococcus*. Torbanite appears to be synonymous with Kukuserite, an algal oil shale of Ordovician age, which occurs in Estonia.

Modern algal deposits analogous to Torbanite have been described from Lake Balkash, Siberia, and from the Coorong lagoon of South Australia. These occurrences are naturally known as Balkashite and Coorongite, respectively. Coorongite is a gelatinous, yellow-green mess, which dries to a dark, rubbery material (Broughton, 1920). Its algal composition can be seen under the microscope.

Tasmanite is a rare variety of oil shale that occurs principally in Tasmania, Australia (the type locality), and in the Brooks Range of North Alaska. It is composed of leiospheres of the algae *Tasmanites*. Although analogous in composition to torbanite, it occurs in marine deposits.

Torbanites and tasmanites are relatively rare varieties of oil shale. The source of the organic matter of most oil shales is equivocal, but generally believed to be originally admixtures of algal and terrestrial humic materials.

Distribution

The following account of the environment, global distribution, and temporal distribution of oil shales is based on surveys by Schlatter (1969), Duncan and Swanson (1965), Cane (1976), and Duncan (1976).

Although carbonaceous shales and hydrocarbon source rocks are known (p. 183), there appear to be few genuine Precambrian oil shales. In the Lower Paleozoic, however, siliceous oil shales were deposited on marine shelves throughout the northern hemisphere. The Alum shales and Kukersite deposits of the Baltic Shield are examples of these. In the Upper Paleozoic, marine-shelf oil shales were also deposited in central European Russia and in the central and eastern United States and Canada. The Chattanooga shale is an example of the latter type, and, although it covers some 500,000 km^2, it has a remarkably uniform stratigraphy.

Oil-shale-forming environments appear to have changed significantly at the Devonian–Carboniferous boundary. Paralic deltaic-lacustrine oil shales are characteristic of Carboniferous deposits around the world. The Midland Valley of Scotland oil shales, which include torbanite, are of this age and type, as are other examples in France, Spain, South Africa, Australia, and Russia. The Carboniferous Albert oil shale of Nova Scotia is totally lacustrine, however.

Both paralic and marine shelf oil shales were formed during the Permian period. The first type include the coal measures of the Sidney basin in New South Wales, Australia. Here an area of some 3,000 mi^2 contains rich torbanite deposits. The Permian coal measures of this basin are some 200 m thick. The oil shales occur interbedded with the coal-bearing strata, but are best developed in the west toward the margin of the basin (Cane, 1976).

The Irati oil shale is of similar age. This oil shale covers several hundred thousand square kilometers in Brazil, Argentina, and Uruguay, and the equivalent *White Band* of the Karroo Series in South Africa (Anderson and McLachlan, 1979) (Fig. 9.11).

The late Mesozoic era saw the formation of many oil shales in marine, deltaic, and lacustrine environments. In northwest Europe, in particular, a major phase of oil shale and organic-rich shale deposition began in northern Germany in the Liassic and moved diachronically northward along the North Sea into the East Greenland Mesozoic basin, where organic-rich mud deposition continued into the Cenomanian. These deposits include the Kimmeridge oil shale of the Dorset Coast. Here the 2-m thick "Kimmeridge Coal" was mined as an oil shale for many years. Although it was used locally as a domestic fuel, its high sulfur content was incompatible with the tourist trade. The organic matter of the Kimmeridge Clay is a type II mixed kerogen, although the oil shale itself is a type I algal kerogen (Plate 1). Kimmeridgian organic shales are the major source rock of North Sea oil, their richness increasing as they are traced from shelf areas into

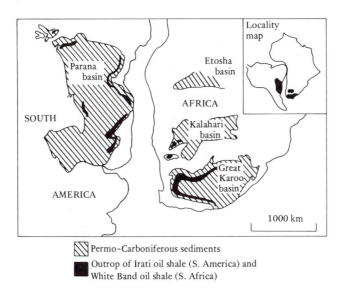

Permo-Carboniferous sediments

Outrop of Irati oil shale (S. America) and
White Band oil shale (S. Africa)

FIGURE 9.11   *Outcrop of the Permian oil shales of South America and South Africa. These oil shales were deposited in several vast lakes. (After Anderson and McLachlan, 1979.)*

the grabens of the axial rift system (Barnard and Cooper, 1981; Goff, 1983). Analogous marine platform oil shales occur in Alaska, in Saskatchewan and Manitoba, Canada, and in Nigeria.

The Cenozoic saw another return to lacustrine oil shale deposition, although marine oil shales do occur. The best-known example of lacustrine oil shale deposition is the Green River Formation (Eocene) of Colorado, Utah, and Wyoming (Fig. 9.12). In this area oil-shale-bearing rocks cover some 16,500 mi^2 in a formation up to 700 m thick (Duncan and Swanson, 1965; Knutson, 1981). Total reserves, if they can be recovered, have been estimated at 1.8 trillion barrels (Yen and Chilingarian, 1976). Analyses of Green River oil shales yield the following composition (Yen and Chilingarian, 1976).

Kerogen	11.04%	
Bitumen	2.76%	
Total organic matter:		14.16%
Dolomite and calcite	43.1%	
Feldspar	20.7%	
Micas and clays	12.9%	
Quartz	8.6%	
Pyrites	0.86%	
Total inorganic matter:		85.8%
TOTAL:		99.96%

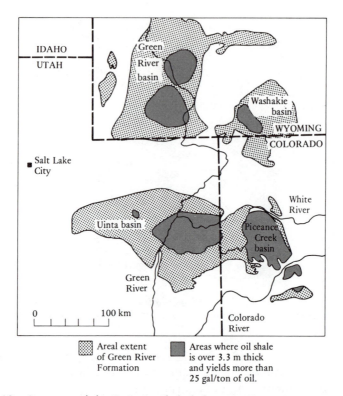

FIGURE 9.12 *Outcrops of the Eocene oil-shale-bearing Green River Formation of the central United States. This formation was deposited in a series of lakes occasionally associated with evaporites. (After Knutson, 1981.)*

This review shows that the occurrence and distribution of oil shales are widespread in place and time. Three major depositional environments are favorable for oil shale generation: lakes, fluvially dominated deltas, and certain marine shelves.

Torbanites and oil shales of mixed humic and nonmarine algal origin characterize the lacustrine and deltaic shales. The shallow marine oil shales are of mixed humic and marine algal composition, including the rarer algally dominated tasmanite variety. Oil shales of both lacustrine and marine types are occasionally associated with autochthonous chemical deposits. The carbonates and evaporites of the Green River Formation illustrate the lacustrine types, whereas the phosphates and cherts of the Phosphoria Formation illustrate the marine types.

Extraction of Oil from Oil Shale

The preceding section described the widespread occurrence of oil shales around the globe. The world's reserves of shale oil are estimated to be in the order of 30 trillion barrels, and the United States alone contains over 700 billion barrels (Yen and Chilingarian, 1976).

The commercial extraction of oil from shale was begun by James Young in Scotland in 1862, 3 years after the drilling of Drakes' well in Pennsylvania. During the next century, oil shale extraction was carried out in England, France, Spain, Sweden, Australia, and South Africa. In all of these countries the industry has collapsed.

In the state of Parana, Brazil, a 2500 ton/day plant based on the Irati oil shale began production in the mid 1970s. Petrobras is currently producing 2200 tons/day from a retort at Sao Mateus. A 4400 bbl/day plant is being built, and plans are laid for a 50,000 bbl/day plant. In the Green River Shale areas, about 10 oil extraction plants are in various stages of planning and development. These plants include both strip mining, room and pillar mining, and in situ retorting (Fig. 2.12). For the last 30 years, two plants have operated in the Grand Valley of Colorado.

At the present time only Russia and China have large-scale oil shale extraction industries. In Russia extraction plants operate in Estonia (near Leningrad) and in the Volga Basin. In China there are plants at Fushun in Manchuria and at Mowming in Kwantung Province (see Prien, 1976, for a historical review).

The reasons for the rise and fall of the oil shale extraction industry are two-fold: economics and technology. As technology for extracting oil from boreholes commenced, the rapid expansion of the conventional oil industry largely put oil shale extraction out of business. Crude oil could be found, produced, transported, refined, and marketed at a lower cost than shale oil. Today, however, the oil shale industry is in a similar situation as the tar sand industry. The cost of conventional oil is steadily increasing, while the technology of winning unconventional oil is improving. The time will soon come when the extraction cost of shale oil is comparable with that of conventional crude.

Again, like the tar sands, there are two basic methods of winning oil from shale: (1) by retorting shale quarried at the surface or (2) by underground in situ extraction (Dinneen, 1976). Most of the surface extraction methods are based on the open-cast quarrying of shale, although the Kimmeridge "coal" was mined along conventional adits. The crushed shale is then placed in a retort and heated. Several sources of heat can be used, but the most efficient is by the combustion of gas previously generated from the shale. As the shale is heated (generally in the range of 425 to 475°C), oil and gas are driven off by the pyrolysis of the kerogen. Spent shale collects at the bottom of the retort. The oil is refined, the gas reinjected, and the ash cooled and disposed of, hopefully at least partly back in the hole whence it came.

In situ methods of retorting are rather different. The two basic problems are similar to those encountered in tar sand extraction. First, an effective method of heating the shale is necessary; second, the rock must be rendered permeable. In Sweden, during the 1940s, the Ljungstrom process was used (Salomonsson, 1951). Closely spaced, shallow boreholes were drilled into the oil shale. Heating elements were inserted, and the rock temperature was raised to 400°C. The oil thus generated was pumped from the wells. The amount of energy used to extract the oil was probably greater than that recovered in the oil itself.

Sinclair Oil and Gas tried another in situ method in 1953 on the Green River Shale of the Piceance basin (Grant, 1964). A series of wells were drilled into the oil shale and artificially fractured to generate permeability. Compressed air was pumped into certain wells, which were then ignited. Combustion gas from these wells moved through the shale and retorted the oil. This oil was recovered from producing wells.

A slightly different technique was developed by Equity Oil Co. in the same area (Dougan et al., 1970). In the site selected the shales were already permeable due to the leaching of salts. Natural gas was injected and ignited in the wells. Shell Oil Co. has experimented with the injection of hot hydrocarbon solvents, such as carbon disulfide. Further details of shale oil extraction technology are found in Dinneen (1976).

Whether the energy budget is in credit in all of these methods is debatable. In other words, is the amount of energy derived from the shale oil likely to be greater than that expended in its extraction? Even if the derived energy is less than the expended energy, these processes may one day become financially viable. This event will happen when the rising cost of conventional crude oil intersects the cost of shale oil. The latter may be expected to decline, in real terms, with improved technology.

SELECTED BIBLIOGRAPHY

Solid hydrocarbons and oil seeps:

HUNT, J. M. 1979. *Petroleum Geochemistry and Geology.* San Francisco: Freeman.
    A good review of seeps and the degradation of oil is given on pages 409–433.
TISSOT, B. P. and WELTE, D. H. 1978. *Petroleum Formation and Occurrence.* Berlin: Springer-Verlag. Chapter 5, Petroleum Alteration, covers the degradation of oils to form asphalts.

Tar sands:

CHILINGARIAN, G. V. and YEN, T. F. (eds.). 1978. *Bitumen Asphalts and Tar Sands.* Amsterdam: Elsevier, 331 pp.
HILLS, L. V. (ed.). 1974. *Oil Sands: Fuel of the Future.* Can. Soc. Petrol. Geol., Mem. No. 3, 263 pp. This volume contains many papers on the occurrence of tar sands and the technology of extracting oil from them. Most of the papers concentrate on the Athabaska deposits.

Oil shales:

TISSOT, B. P. and WELTE, D. H. 1978. *Petroleum Formation and Occurrence.* Berlin: Springer-Verlag. Chapter 8, pages 225–236, deals with oil shales.

YEN, T. F. and CHILINGARIAN, G. V. (eds.). 1976. *Oil Shale.* Amsterdam: Elsevier, 292 pp. This volume contains many papers on the occurrence of oil shales and the technology of extracting oil from them. A large part of the text is concerned with the oil shales of the Green River Formation.

REFERENCES

ABRAHAM, H. 1945. *Asphalts and Allied Substances,* Sixth Edition. New York: Van Nostrand, 432 pp.

ALAYETO, M. B. E. and LOUDER, L. W. 1974. The geology and exploration potential of the heavy oil sands of Venezuela (the Orinoco Petroleum Belt). In: *Oil Sands: Fuel of the Future.* L. V. Hills (ed.). Can. Soc. Petrol. Geol., Mem. No. 3, 1–18.

ALI, S. M. F. 1974. Application of in situ methods of oil recovery to tar sands. In: *Oil Sands: Fuel of the Future.* L. V. Hills (ed.). Can. Soc. Petrol. Geol., Mem. No. 3, 199–211.

ANDERSON, A. M. and McLACHLAN, I. R. 1979. The oil-shale potential of the Early Permian White Band Formation in southern Africa. In: *Some Sedimentary and Associated Ore Deposits of South Africa.* A. M. Anderson and W. J. van Biljon (eds.). Geol. Soc. S. Africa, Spec. Pub. No. 6, 83–90.

ANON, 1972. Wet in situ combustion aid recovery of oil. *Oil & Gas J.,* 3 Apr., 50–53.

BALL, M. W. 1935. Athabaska oil sands: apparent example of local origin of oil. *Am. Assoc. Petrol. Geol. Bull., 19,* 153–171.

BANKS, W. J. and JAMES, G. W. 1974. Heavy-crude bearing sandstones of the Cherokee Group (Desmoinesian) in Southeast Kansas. In: *Oil Sands: Fuel of the Future.* L. V. Hills (ed.). Can. Soc. Petrol. Geol., Mem. No. 3, 19–34.

BARNARD, P. C. and COOPER, B. S. 1981. Oils and source rocks of the North Sea area. In: *Petroleum Geology of the Continental Shelf of North-West Europe.* L. V. Illing and G. D. Hobson (eds.). London: Inst. Petrol., 169–175.

BROUGHTON, A. C. 1920. 'Coorongite.' *Trans. Proc. Roy. Soc. S. Austral., 44,* 386–395.

BURGER, J. 1973. L'exploitation des pyroschistes on schistes bitumineux. *Rev. Instit. Francais du Petrol., 28,* 315–372.

BURK, R. O. 1971. Tar sands task group report. In: *U.S. Energy Outlook—An Initial Appraisal. 1971–5.* J. G. McLean (ed.). Natl. Petrol. Council's Comm. on U.S. Energy Outlook, Summ. Task Group Repts., *1,* 209–252.

BYRAMJEE, R. J. 1983. Heavy crudes and bitumen categorized to help assess resources, techniques. *Oil & Gas J.* 4 July, 78–82.

CANE, R. F. 1976. The origin and formation of oil shale. In: *Oil Shales.* T. F. Yen and G. V. Chilingarian (eds.). Amsterdam: Elsevier, 27–60.

CONNAN, J. and YAN DER WEIDE, B. M. 1974.   Diagenetic alteration of natural as-
    phalts. In: *Oil Sands: Fuel of the Future*. L. V. Hills (ed.). Can. Soc. Petrol.
    Geol., Mem. No. 3, 134–147.

DEROO, G.; TISSOT, B.; McCROSSAN, R. G.; and DER, F. 1974.   Geochemistry of the
    heavy oils of Alberta. In: *Oil Sands: Fuel of the Future*. L. V. Hills (ed.). Can.
    Soc. Petrol. Geol., Mem. No. 3, 148–167.

DINNEEN, G. V. 1976.   Retorting technology of oil shale. In: *Oil Shale*. T. F. Yen
    and G. V. Chilingarian (eds.). Amsterdam: Elsevier, 181–198.

DOME, K. 1979.   Cold comfort for Alberta's oil industry. London: *Financial Times*,
    8 May, 32.

DOSCHER, T. M. 1967.   Technical problems in in-situ methods for recovery of bitu-
    men from tar sands. Mexico City: *7th World Petrol. Cong.*, Session 13, Paper 6.

DOTT, R. H. and REYNOLDS, M. J. 1969.   *Sourcebook for Petroleum Geology*. Tulsa:
    Am. Assoc. Petrol. Geol., Mem. No. 5, 471 pp.

DOUGAN, P. M.; REYNOLDS, F. S.; and ROOT, P. J. 1970.   The potential for in situ
    retorting of oil shale in the Piceance Creek basin of Northwestern Colorado.
    Qu. Colorado Sch. Mines, 65 (4), 57–62.

DUNCAN, D. C. 1976.   Geological setting of oil-shale deposits and world prospects.
    In: *Oil Shale*. T. F. Yen and G. V. Chilingarian (eds.). Amsterdam: Elsevier,
    13–26.

DUNCAN, D. C. and SWANSON, V. E. 1965.   Organic-rich shale of the United States
    and world land areas. Washington: *U.S.G.S. Circular 523*, 30 pp.

ELMORE, R. D. 1979.   Miocene-Pliocene syndepositional tar deposit in Iran. *Am.
    Assoc. Petrol. Geol. Bull., 63*, 444.

EVANS, C. R.; ROGERS, M. A.; and BAILEY, N. J. L. 1971.   Evolution and alteration of
    petroleum in western Canada. *Chem. Geol., 8*, 147–170.

FINKEN, R. E. and HELDAU, R. F. 1972.   Phillips solves Venezuelan tar-belt produc-
    ing problems. *Oil & Gas J.* 17 July, 108–114.

GOFF, J. C. 1983.   Hydrocarbon generation and migration from Jurassic source
    rocks in the E. Shetland Basin and Viking Graben of the northern North Sea.
    J. Geol. Soc. Lond., 140, 445–474.

GRANT, B. F. 1964.   Retorting oil shale underground—problems and possibilities.
    Qu. Colorado Sch. Mines, 59 (3), 39–46.

HILLS, L. V. 1974.   *Oil Sands: Fuel of the Future*. Can. Soc. Petrol. Geol., Mem.
    No. 3, 263 pp.

HOBSON, G. D. and TIRATSOO, N. 1981.   *Introduction to Petroleum Geology*, Second
    Edition. Scientific Edition. London: Scientific Press, 352 pp.

HUNT, J. M.; STEWART, F.; and DICKEY, P. A. 1954.   Origins of hydrocarbons of Uinta-
    basin, Utah. *Am. Assoc. Petrol. Geol. Bull., 38*, 1671–1698.

INIECHEN, G. 1951.   Petrole, gas naturels et asphales du Geosynclinal Adriatic.
    Proc. World Petrol. Cong., 3, Section 1, 199–219.

JARDINE, D. 1974.   Cretaceous oil sands of Western Canada. In: *Oil Sands: Fuel of
    the Future*. L. V. Hills (ed.). Can. Soc. Petrol. Geol., Mem. No. 3, 50–67.

KAMEN-KAYE, M. 1983.   Mozambique-Madagascar geosyncline, vol. II. Petroleum
    Geology. *J. Petrol. Geol., 5*, 287–308.

KAPPELER, J. H. 1967.   Discussion on paper by Phizackerly and Scott. *Proc. World
    Petrol. Cong., 3*, 551–71.

KENDALL, G. H. 1977. Importance of reservoir description in evaluating *in situ* recovery methods for Cold Lake heavy oil. Part I—Reservoir Description. *Can. Petrol. Geol. Bull.*, *25*, 314–317.

KENT, P. E. 1954. A visit to Western Madagascar. *B. P. Unpublished Report.*

KNUTSON, C. F. 1981. Oil shale. *Am. Assoc. Petrol. Geol. Bull.*, *65*, 2283–2289.

KUGLER, H. G. 1961. Geological map and section of Pitch Lake Morne l'Enfer area. *Petrol. Assoc. Trinidad.*

LEES, G. M. and COX, P. T. 1937. The geological basis of the present search for oil in Great Britain by the D'Arcy Exploration Co., Ltd. *Quar. J. Geol. Soc. Lond.*, *93*, 156–194.

LEVORSEN, A. I. 1967. *Geology of Petroleum*, Second Edition. San Francisco: Freeman, 724 pp.

MONTGOMERY, D. S. 1974. Geochemistry of the heavy oils of Alberta: discussion. In: *Tar Sands: Fuels of the Future.* L. V. Hills (ed.). Can. Soc. Petrol. Geol., Mem. No. 3, 184–185.

NAASIAM, O. W. T. 1937. Early use of hydrocarbons in the Imperial Chinese Empire. In: *Professor Li Lo Memorial Volume.* Univ. of Peiping, 301–331.

PHIZACKERLY, P. H. and SCOTT, L. O. 1967. Major tar sand deposits of the world. *Proc. World Petrol. Cong. No. 7*, *3*, 551–71.

PRIEN, C. H. 1976. Survey of oil shale research in the last three decades. In: *Oil Shale.* T. F. Yen and G. V. Chilingarian (eds.). Amsterdam: Elsevier, 235–267.

RUSSELL, P. L. 1980. *History of Western Oil Shale.* Barking: Applied Science Publishers, 152 pp.

SALOMONSSON, G. 1951. The Ljungstrom in situ method for oil-shale recovery. In: *Oil Shale and Cannel Coal*, vol. 2. London: Inst. of Petrol., 260–280.

SCHLATTER, L. E. 1969. Oil shale occurrences in Western Europe. In: *The Exploration for Petroleum in Europe and North Africa.* P. Hepple (ed.). London: Inst. Petrol., 251–258.

SHANKS, W. C.; SEYFRIED, W. E.; MEYER, W. C.; and O'NEIL, T. J. 1976. Mineralogy of oil shale. In: *Oil Shale.* T. F. Yen and G. V. Chilingarian (eds.). Amsterdam: Elsevier, 81–102.

SPIRO, B.; WELTE, D. H.; RULLKOTTER, J.; and SCHAEFER, R. G. 1983. Asphalts, oils and bituminous rocks from the Dead Sea area—a geochemical correlation study. *Am. Assoc. Petrol. Geol. Bull.*, *67*, 1163–1175.

SPRAGINS, F. K. 1973. Outlook for tar sands and other bitumens. *Quar. J. Colorado Sch. of Mines*, *68*, No. 2.

SPROULE, J. C. 1963. Five factors govern the future of the oil sands on Melville Island. *Can. Oil and Gas Indust.*, *16*, No. 9.

STEWART, D. R. 1912. *The Oil Shales of the Lothians*, Part III. Scotland: Mem. Geol. Surv., 199 pp.

SUTER, H. H. 1951. The general and economic geology of Trinidad. *Col. Geol. and Min. Res.* London: H.M.S.O., *2*, 271–307.

TOWSON, D. E. 1977. Importance of reservoir description in evaluating *in situ* recovery methods for Cold Lake heavy oil, Part II. In situ application. *Can. Petrol. Geol. Bull.*, *25*, 328–340.

TRANTHAM, J. C. and MARX, J. W. 1966. Bellamy field tests: oil from tar by counterflow underground burning. *J. Petrol. Tech.*, Jan., 109–115.

WALTERS, E. J. 1974.   Review of the world's major oil sands deposits. In: *Oil Sands: Fuels of the Future*. L. V. Hills (ed.). Can. Soc. Petrol. Geol., Mem. No. 3, 240–263.

WENNEKERS, J. H. N. 1981.   Tar sands. *Am. Assoc. Petrol. Geol. Bull.*, *65*, 2290–2293.

WINESTOCK, A. G. 1974.   Developing a steam recovery technology. In: *Oil Sands: Fuel of the Future*. L. V. Hills (ed.). Can. Soc. Petrol. Geol., Mem. No. 3, 190–198.

YEN, T. F. and CHILINGARIAN, G. V. (eds.). 1976.   *Oil Shale*. Amsterdam: Elsevier, 292 pp.

10

Conclusions

PROSPECTS AND PROBABILITIES

The preceding chapters described and discussed the elements of petroleum geology: from the generation and migration of petroleum to its entrapment in reservoirs and distribution in basins. This chapter considers how these concepts can actually be applied to petroleum exploration and production and reviews the global distribution of reserves and their future supply and demand.

As noted in Chapter 1, many years passed before geology was applied to petroleum exploration, and even now its relevance is occasionally questioned. In particular, several studies purport to show that oil could be found in a new area by drilling wells at random rather than by the application of science. Harbaugh et al. (1977) have reviewed many such studies.

Regarding the random approach to petroleum exploration, three comments are appropriate. First, one should recall the statistician who was drowned trying to cross a river with an average depth of 1 m. Second, one should note that the success of random drilling is likely to increase in proportion to the percentage of productive acreage in a basin (Fig. 10.1). Thus almost anyone (even in a hurry) could find oil in parts of Texas and Oklahoma, but might be less fortunate in offshore Labrador. Petroleum exploration has been likened to betting on horses. One can either pick a name at random or study the racing form. Similarly, drilling locations may be picked at random or by the application of geology in an attempt to

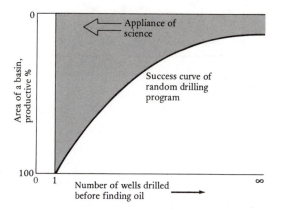

FIGURE 10.1   *Graph showing that the smaller the productive area of a basin, the greater the need for the application of science.*

reduce the odds. The third point is that no commercial enterprise could embark on a program of random drilling if geological methods might shorten the number of wells drilled before oil is found and produced and a cash flow established. Good geology is essential for a small company, for which a string of dry holes spells catastrophe. Only a state company, with an unlimited supply of taxpayers' money, could adopt a philosophy of random drilling.

Prospect Appraisal

Having established that geology is more appropriate than random drilling, we now turn to prospect evaluation. The two aspects of prospect evaluation are geology and economics. The questions that need to be answered are: Is the well likely to find oil and/or gas? If so, are the economics such that it would be commercially viable?

Geological Aspects

The probability of a well finding oil or gas ranges from 1.0 (oil or gas are certain to be present) to 0.0 (oil or gas are certain to be absent). Chapter 1 introduced the five parameters that must be present for a commercial oil or gas field to occur:

1.   A source rock
2.   A reservoir
3.   A trap
4.   A cap rock
5.   Sufficient heat to generate oil or gas, but not destroy it.

Thus instead of subjectively rating the probability of a prospect finding oil or gas on a scale of 1.0 to 0.0, the probability of each condition being fulfilled may be assessed, and the overall probability found from their product:

$$\frac{\text{Probability of a prospect}}{\text{containing oil or gas}} = (p_1 \times p_2 \times p_3 \times p_4 \times p_5)$$

where p_1 = probability of a source rock being present
p_2 = probability of a reservoir being present
p_3 = probability of a trap being present
p_4 = probability of a cap rock being present
p_5 = probability of correct thermal maturation occurring

Where several prospects have been identified, they may be analyzed individually, placed in a matrix, their probability of success calculated, and hence their likelihood of finding oil ranked, as shown in Table 10.1.

TABLE 10.1   Matrix for calculating the probability of success for four prospects

	Prospects			
Probabilities	A	B	C	D
Probability of a source rock	0.9	0.6	0.7	0.3
Probability of a reservoir	1.0	0.7	0.5	0.4
Probability of a trap	0.7	0.6	0.5	0.6
Probability of a seal	0.8	0.5	0.6	0.0
Probability of correct maturation level	0.6	0.4	0.4	0.1
PROBABILITY OF SUCCESS	0.302	0.050	0.042	0.000

Decreasing probability of success →

Economic Aspects

The probability of a well finding oil or gas is only one aspect of successful exploration. Since the object of an oil company is to make money, the financial risk involved and the potential profitability must be established, both for individual prospects and for a string of wells. According to the "gamblers' ruin" law, there is a chance of going broke because of a run of bad luck, irrespective of the long-term probability of success. Whether a gambler (or company) is ruined before being commercially successful depends both on the probability of geological success (previously discussed) and four commercial parameters:

1. Potential profitability of venture
2. Available risk investment funds
3. Total risk investment
4. Aversion to risk

Greenwalt (1982) has shown how some of these parameters may be quantified:

$$1 - S = (1 - P_s)^N$$

where S = aversion to risk
P_s = probability of geological success
N = number of ventures necessary to avoid
gamblers' ruin

Further, and much more elaborate, aids to exploration decision making involve more sophisticated quantification of these commercial parameters. Computer simulation techniques may then be used to aid the decision of whether or not to embark on an exploration venture, and, if so, to determine the amount of risk the investors' finances can tolerate. Such economic considerations lie beyond the field of geology, although they are an extremely important aspect of petroleum exploration. For further details see Newendorp (1976) and Megill (1979).

RESERVES AND RESOURCES

Now that the probability of finding petroleum in individual prospects has been considered, it is relevant to discuss the probable reserves to be found in sedimentary basins and ultimately the whole world. Chapter 6 briefly reviewed how reserves in a field could be measured, and Chapter 8 showed how the petroleum potential of different types of basin could vary. It is useful to be able to calculate the reserves that a sedimentary basin may contain. First, however, some terms must be defined.

Resources are industrially useful natural materials, not only petroleum but water, minerals, and aggregate. Resources may be defined by two criteria: economic feasibility of extraction and geological knowledge (Fig. 10.2). *Reserves* are resources that can be economically extracted from the resource base (McKelvey, 1975). Thus as the price of oil rises and falls, the oil in a marginally commercial field may be a reserve one day, but only a resource the next. Similarly, as extraction technology improves, oil shale and tar sand resources become reserves.

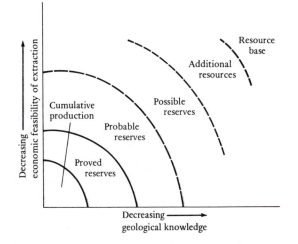

FIGURE 10.2 *The concepts of reserves and resources. (After Ion, 1979.)*

Reserves are subdivisible into several categories, as shown in Figure 10.3. Recent definitions have been proposed jointly by the Society of Petroleum Engineers, the American Petroleum Institute, and the American Association of Petroleum Geologists: "proved reserves are estimates of hydrocarbons to be recovered from a given date forward." They are divisible into proved developed reserves ("those reserves that can be expected to be recovered through existing wells") and proved undeveloped reserves ("those additional reserves that are expected to be recovered from (1) future drilling of wells, (2) deepening of existing wells to a different reservoir, or (3) the installation of an improved recovery project").

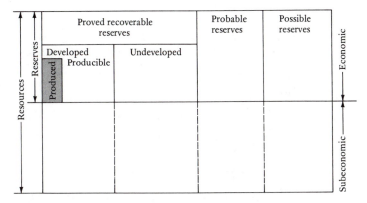

FIGURE 10.3 *The different types of reserves. For definitions see text. (After McKelvey, 1975.)*

From the preceding definitions, it is apparent that there are reserves and *reserves*. Any estimates of the reserves of a field or sedimentary basin should be qualified as to whether they are total, recoverable, proved, probable, or possible, and so forth.

Assessment of Basin Reserves

Companies and countries have paid considerable attention to estimating the total recoverable reserves that are likely to be found in a basin (Hubbert, 1967; Moody, 1977; Harbaugh et al., 1977). One popular method was described by White et al. (1975). In a study of south Louisiana, gas reserves were divided into speculative, possible, and probable categories. The mean, maximum, and minimum possible reserves were plotted on a cumulative probability curve (ranging from 0.0 = the likelihood of their estimates being totally wrong to 1.0 = their estimates being absolutely correct). The three curves were then integrated to produce a summation curve for the total reserves (Fig. 10.4). This result was achieved by a very time-consuming piece of statistical gymnastics called the Monte Carlo technique, which a computer can execute in seconds. The method has also been applied to other areas of the United States (Miller et al., 1975).

Many studies show that the size-frequency distribution of fields in a sedimentary basin approaches a lognormal distribution. That is to say, the number of fields increases with decreasing field size. When the sizes of

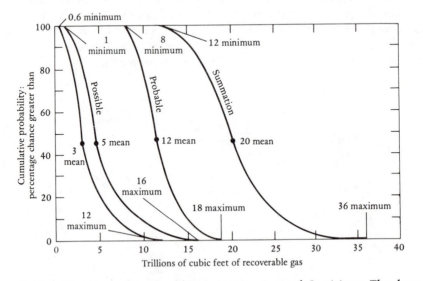

FIGURE 10.4 *Estimates of producible gas reserves in south Louisiana. The three left-hand curves show the probable, possible, and speculative gas reserves estimated by Exxon. The right-hand curve is a summation of the three curves produced by a Monte Carlo procedure. (After White et al., 1975.)*

fields are plotted on a logarithmic scale against cumulative percent, they approximate to a straight line. Lognormal distributions have been noted for field sizes in northern Louisiana (Kaufman, 1963), Kansas (Griffiths, 1966), the Denver basin (Haun, 1971), and western Canada (Kaufman et al., 1975). Table 10.2 documents reserve estimates for fields discovered in the North Sea up to 1981. Figure 10.5 plots these data as cumulative percentages against log probability. Although the North Sea is still a relatively poorly explored province, the straightish trend of the line suggests that these fields approximate to the lognormal distribution discovered in other basins. This type of information is useful because it gives the probability of discovering a field of a particular size. When the minimum size of an economically viable field is known, so is the probability of finding one. For

TABLE 10.2   Oil reserves in the northern North Sea (British Sector)

Field	Recoverable reserves, million bbl	Percent of total	Cumulative percent
Forties	1918	.176	100.0
Brent	1610	.148	81.8
Ninian	1007	.092	67.0
Piper	647	.059	57.8
Statfjord	573	.053	51.9
Beryl	485	.045	46.6
Thistle	441	.041	42.1
Magnus	441	.041	38.1
Fulmar	434	.040	34.4
Cormorant N	412	.038	30.4
Claymore	419	.038	26.6
Murchison	318	.029	22.8
Dunline	294	.027	19.9
Brae	294	.027	17.2
Beryl B	293	.027	14.5
Hutton NW	276	.025	11.8
Hutton	257	.023	9.3
Cormorant S	191	.018	7.0
Beatrice	154	.014	5.2
Maureen	154	.014	3.8
Montrose	89	.008	2.4
Heather	73	.007	1.6
Auk	59	.005	0.9
Argyll	48	.004	0.4
TOTAL	10,887		

From 1981 data published by the Department of Energy.

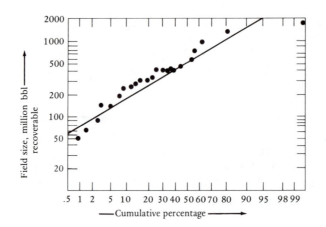

FIGURE 10.5   *North Sea, United Kingdom, reserves shown from a plot of field size against cumulative probability based on the data in Table 10.2.*

different basins the slope of the line in Figure 10.5 will vary, but once the slope has been established from a few fields, the ultimate recoverable reserves of the basin can be estimated.

Assessment of Global Reserves

Many estimates of the global oil and gas reserves have been attempted by making a basin by basin evaluation. Most oil companies are continually compiling estimates, both of the ultimate recoverable reserves of the world, and future demand for oil, gas, and other sources of energy. Needless to say, such estimates differ widely. Table 10.3 presents a breakdown of the distribution of oil and gas reserves in the major sedimentary basins of the world. The data are presented in gas-oil equivalents (GOE); 6 million ft^3 of gas are taken as equal to 1 barrel of oil. Figure 10.6 presents a distribution of known reserves by continents. Many similar statistics have been published (Fig. 10.7). Tippee (1981) analyzed similar data from some 2000 major companies. He concluded that from 1985 to 2000 noncommunist energy demand will climb from some 120 million bbl/day of oil equivalent (OE) to 140 to 170 million bbl/day OE. This demand will increase in developing countries two to four times as fast as in the developed ones. OPEC exports will remain about constant, partly because of increasing internal demand. Non-OPEC production will increase.

Bolder still are Ion's attempts to predict energy utilization beyond the year 2000 (Ion 1979, 1981). His analyses suggest that the production of oil and gas will increase until about 2010, at which point it will gradually decline (Fig. 10.8). Oil and gas are obviously finite sources. Until recently

TABLE 10.3   Known produced, proven, and probable recoverable reserves
in the largest 25 hydrocarbon basins of the world

Basin	Total petroleum, 10^9 BOE[a]	World ranking
Arabian/Iranian[b]	651	1
West Siberian	167	2
Volga-Ural	58	3
Maracaibo (Venezuela)	50.2	4
Mississippi delta	49.7	5
Permian basin (USA)	42.3	6
Texas Gulf Coast (USA and NE Mexico)	41.6	7
Reforma-Campeche (Mexico)	40.6	8
Sirte (Libya)	35.3	9
Alberta (Canada)	32.0	10
Amarillo-Anadarko-Ardmore (USA)	29.2	11
Niger delta	29.0	12
Northern North Sea (UK and Norway)	26.7	13
Triassic (Algeria)	25.8	14
E. Texas-Arkla (USA)	23.7	15
N. Caucasus-Mangyshlak (USSR)	20.8	16
Netherlands and NW Germany	17.5	17
E. Venezuela and Trinidad	16.6	18
South Caspian (USSR)	14.5	19
North Slope Alaska (USA)	14.2	20
San Joaquin (USA)	13.6	21
Tampico-Misantla (Mexico)	12.2	22
Amu Daryu (USSR)	10.5	23
Appalachian (USA)	10.4	24
Los Angeles (USA)	10.0	25
TOTAL	1700.0	

[a]BOE = barrels of oil equivalent.
[b]Note that nearly 40 percent of the world's known reserves occur in the Arabian/Iranian basin. For locations see Figure 10.7.
From Nehring; in Ivanhoe, 1980. Reprinted with permission.

many forecasts of oil and gas production showed an increase for several years from the date of the study, followed by a decline. Figure 10.9 shows that for the last two decades the ratio of proved reserves to production has declined. To maintain a 30-year lead of oil discovery over production, the world needs to find a new basin akin to the North Sea every year (Pearce, 1980). This rate is not being achieved. More and better-trained petroleum geologists will be needed in the years ahead, lest our grandchildren freeze in the dark. Go to it!

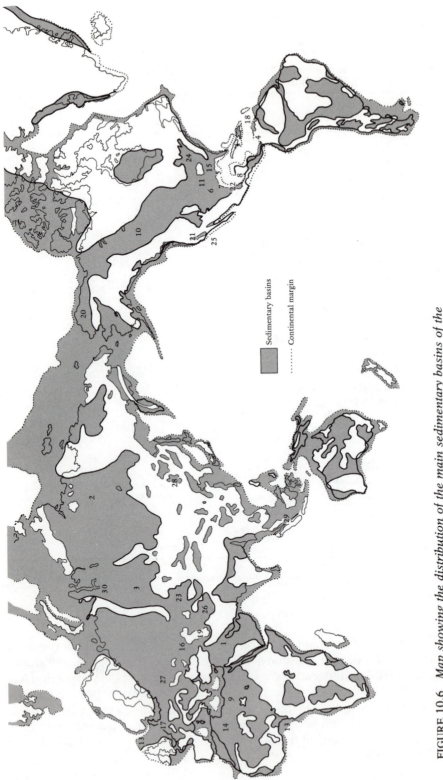

FIGURE 10.6 Map showing the distribution of the main sedimentary basins of the world. Numbers refer to basins listed in Table 10.3, arranged in decreasing magnitude.

FIGURE 10.7   *The geographic distribution of known recoverable reserves. (After St. John, 1980.)*

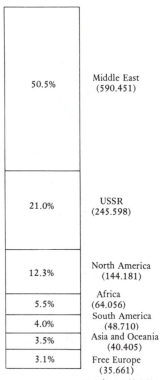

50.5%   Middle East
        (590.451)

21.0%   USSR
        (245.598)

12.3%   North America
        (144.181)

        Africa
5.5%   (64.056)

        South America
4.0%   (48.710)

3.5%   Asia and Oceania
        (40.405)

3.1%   Free Europe
        (35.661)

Total: 1,169.062

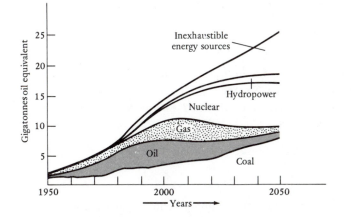

FIGURE 10.8   *Possible world energy supply and demand from 1950 to 2050. Note: A Gigatonne = 1000 million metric tons = 1 billion metric tons U.S. (After Ion, 1979.)*

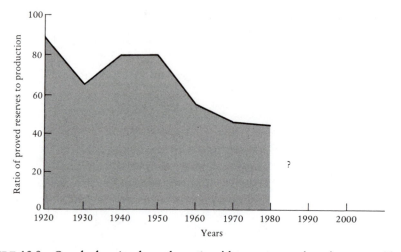

FIGURE 10.9   *Graph showing how the ratio of known to produced recoverable reserves has declined over the last 20 years. (Data from Warman, 1971, and Pearce, 1980.)*

SELECTED BIBLIOGRAPHY

For an exposé of statistical techniques applied to prospect evaluation, see:
HARBAUGH, J. W.; DOVETON, J. H.; and DAVIS, J. C. 1977.   *Probability Methods in Oil Exploration.* London: Wiley, 269 pp.

For methods of assessing the economics of oil prospects, see:
MEGILL, R. E. 1979.   *Exploration Economics.* Tulsa: Petroleum Publishing Co., 180 pp.
NEWENDORP, P. D. 1976.   *Decision Analysis for Petroleum Exploration.* Tulsa: Petroleum Publishing Co., 750 pp.

For discussions and predictions of future demand/production of oil, gas, and other energy sources, see:
ION, D. C. 1981.   *Availability of World Energy Resources.* London: Graham and Trotman, 248 pp.

REFERENCES

GREENWALT, W. A. 1982.   Determining venture participation. *J. Petrol. Tech., 33,* 2189–2195.
GRIFFITHS, J. C. 1966.   Exploration for natural resources. *Oper. Res., 14(2),* 189–209.

HARBAUGH, J. W.; DOVETON, J. H.; and DAVIS, J. C. 1977. *Probability Methods in Oil Exploration*. London: Wiley, 269 pp.

HAUN, J. D. 1971. Potential oil and gas resources. *Colorado School of Mines*, unpublished progress report. 18 pp.

HAUN, J. D. (ed.). 1975. Methods of estimating the volume of undiscovered oil and gas resources. Tulsa: Am. Assoc. Petrol. Geol., *Studies in Geology*, No. 1, 206 pp.

HUBBERT, M. K. 1967. Degree of advancement of petroleum exploration in the United States. *Am. Assoc. Petrol. Geol. Bull.*, *51*, 2207–2227.

ION, D. C. 1979. World energy supplies. London: *Proc. Geol. Assoc.*, *90*, 193–203.

ION, D. C. 1981. *Availability of World Energy Resources* (and supplements). London: Graham and Trotman, 248 pp.

IVANHOE, L. F. 1980. World's giant petroleum provinces. *Oil & Gas J.*, 30 June, 146–7.

KAUFMAN, G. M. 1963. *Statistical decision and related techniques in oil and gas exploration*. Englewood Cliffs: Prentice-Hall, 307 pp.

KAUFMAN, G. M.; BAKER, Y.; and KRUYT, D. 1975. A probabilistic model of oil and gas discovery. In: *Methods of Estimating the Volume of Undiscovered Oil and Gas Resources*. J. D. Haun (ed.). Tulsa: Am. Assoc. Petrol. Geol., *Studies in Geology*, No. 1, 113–142.

McKELVEY, V. E. 1975. Concepts of reserves and resources. In: *Methods of Estimating the Volume of Undiscovered Oil and Gas Resources*. J. D. Haun (ed.). Tulsa: Am. Assoc. Petrol. Geol., *Studies in Geology*, No. 1, 11–14.

MEGILL, R. E. 1979. *Exploration Economics*, Second Edition. Tulsa: Petroleum Publishing Co., 180 pp.

MILLER, B. M. et al. 1975. Geological estimation of undiscovered recoverable oil and gas resources in the United States. *U.S. Geol. Surv. Circ.*, *725*, 78 pp.

MOODY, J. D. 1977. Perspectives on energy problems. In: *Exploration and Economics of the Petroleum Industry*. V. S. Cameron (ed.). New York: Bender, 1–16.

NEWENDORP, P. D. 1976. *Decision Analysis for Petroleum Exploration*. Tulsa: Petroleum Publishing Co., 750 pp.

PEARCE, A. E. 1980. Oil: hydrocarbons or B.T.U.'s. *Esso Magazine*, *113*, 3–8.

ST. JOHN, B. 1980. *Sedimentary Basins of the World and Giant Hydrocarbon Accumulations*. Tulsa: Am. Assoc. Petrol. Geol., Map + 23 pp.

TIPPEE, R. T. 1981. The oil surplus—a fleeting phenomenon? *Oil & Gas J.* 9 Nov., 141–148.

WARMAN, H. R. 1971. Why explore for oil and where? *Austr. Petrol. Explor. Assoc. J.*, 9–13.

WHITE, D. A.; GARRETT, R. W.; MARSH, G. R.; BAKER, R. A.; and GEHMAN, H. M. 1975. Assessing regional oil and gas potential. In: *Methods of Estimating the Volume of Undiscovered Oil and Gas Resources*. J. D. Haun (ed.). Tulsa: Am. Assoc. Petrol. Geol., *Studies in Geology*, No. 1, 143–159.

WHITE, D. A. and GEHMAN, H. M. 1979. Methods of estimating oil and gas resources. *Am. Assoc. Petrol. Geol. Bull.*, *63*, 2183–2203.

Appendixes

APPENDIX A:
WELL CLASSIFICATION GUIDELINES

The American Association of Petroleum Geologists' Committee on Statistics of Drilling has prepared the following guidelines on well classification (quoted with permission):

1. *New-field wildcat.* A new-field wildcat is a well located on a structural feature or other type of trap that previously has not produced oil or gas. In regions where local geologic conditions have little or no control over accumulations, these wells are generally at least 2 mi from the nearest productive area. Distance, however, is not the determining factor. Of greater importance is the degree of risk assumed by the operator, and his intention to test a structure or stratigraphic conditions not previously proved productive.

2. *New-pool wildcat.* A new-pool wildcat is a well located to explore for a new pool on a structural feature or other type of trap already producing oil or gas, but outside the known limits of the producing area. In some regions where local geologic conditions exert an almost negligible control over accumulations, exploratory holes of this type may be called "near wildcats." Such wells usually will be less than 2 mi from the nearest productive area.

3. *Deeper pool test.* A deeper pool test is an exploratory hole located within the productive area of a pool, or pools, already partly or wholly developed. It is drilled below the deepest productive pool to explore for deeper unknown prospects.

4. *Shallower pool test.* A shallower pool test is an exploratory well drilled in search of a new productive reservoir that is unknown but possibly suspected from data secured from other wells and is shallower than known productive pools. This test is located within the productive area of a pool, or pools, previously developed.

5. *Outpost, or extension, test.* An outpost is a well located and drilled with the expectation of extending for a considerable distance the productive area of a partly developed pool. It is usually two or more locations distant from the nearest productive site.

6. *Development well.* A development well is a well drilled within the proven area of an oil or gas reservoir to the depth of a stratigraphic zone known to be productive. If the well is completed for production, it is classified as an oil or gas development well. If the well is not completed for production and is abandoned, it is classified as a dry development hole.

7. *Stratigraphic test.* A stratigraphic test is a drilling effort, geologically directed, to obtain information pertaining to a specific geologic condition that might lead toward the discovery of an accumulation of hydrocarbons. Such wells customarily are drilled without the intention of being completed for hydrocarbon production. This classification also includes tests identified as core tests and all types of expendable holes related to hydrocarbon exploration.

8. *Service well.* A service well is a well drilled or completed for the purpose of supporting production in an existing field. Wells of this class are drilled for the following specific purposes: gas injection (natural gas, propane, butane, or flue gas), water injection, steam injection, air injection, saltwater disposal, water supply for injection, observation, and injection for in situ combustion. An *old well drilled deeper* is a previously drilled hole that is reentered and deepened by additional drilling. Such wells are reported as (1) oil or gas wells if completed for production of oil or gas or (2) dry holes if sufficient quantities of oil or gas are not found to justify completion at the greater depth. The additional footage below the old total depth (OTD) is reported under the appropriate well classification.

Figure A.1 illustrates the various categories of wells.

OBJECTIVE OF DRILLING	INITIAL CLASSIFICATION φ WHEN DRILLING IS STARTED	FINAL CLASSIFICATION AFTER COMPLETION OR ABANDONMENT	
		SUCCESSFUL ●☼☼	UNSUCCESSFUL ✧
Drilling for a new field on a structure or in an environment never before productive	1. NEW-FIELD WILDCAT	NEW-FIELD DISCOVERY WILDCAT	DRY NEW-FIELD WILDCAT
Drilling for a new pool on a structure or in a geological environment already productive — NEW POOL TESTS — Drilling outside limits of proved area of pool	2. NEW-POOL (PAY) WILDCAT	NEW-POOL DISCOVERY WILDCAT (*Sometimes an extension well*)	DRY NEW-POOL WILDCAT
Drilling inside limits of proved area of pool — For a new pool below deepest proven pool	3. DEEPER POOL (PAY) TEST	DEEPER POOL DISCOVERY WELL — NEW-POOL DISCOVERY WELLS (*Sometimes extension wells*)	DRY DEEPER POOL TEST — DRY NEW-POOL TESTS
For a new pool above deepest proven pool	4. SHALLOWER POOL (PAY) TEST	SHALLOWER POOL DISCOVERY WELL	DRY SHALLOW POOL TEST
Drilling for long extension of a partly developed pool	5. OUTPOST OR EXTENSION TEST	EXTENSION WELL (*Sometimes a new-pool discovery well*)	DRY OUTPOST, OR DRY EXTENSION, TEST
Drilling to exploit or develop a hydrocarbon accumulation discovered by previous drilling	6. DEVELOPMENT WELL	DEVELOPMENT WELL	DRY DEVELOPMENT WELL

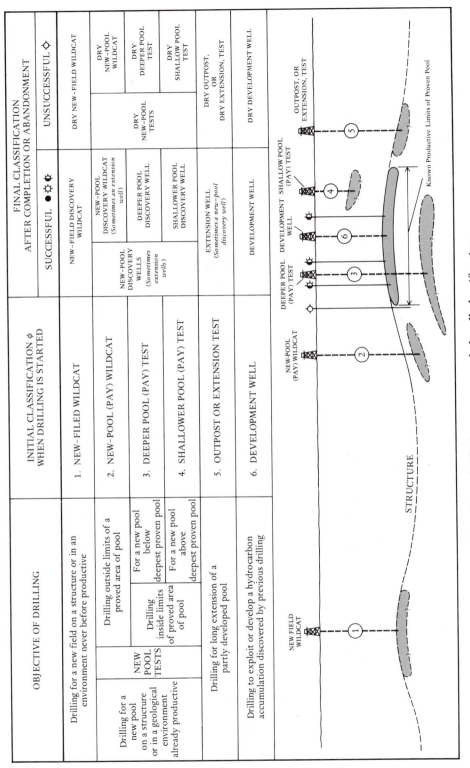

FIGURE A.1   *American Association of Petroleum Geologists recommended well classification.*

APPENDIX B: CONVERSION FACTORS

TABLE A.1   Oil field units and SI units

	To convert	into	Multiply by
Length, L	feet	inches	12.0
	feet	meters	0.3048
	meters	feet	3.280
	inches	meters	2.54×10^{-2}
	meters	inches	39.7
Area, L^2	acres	hectares	0.4046
	hectares	acres	2.471
	square meters	square feet	10.76
	square feet	square meters	9.290×10^{-2}
Volume, L^3	acre-feet	barrels	7758.4
	barrels	cubic meters	0.1590
	barrels	cubic feet	5.615
	cubic meters	barrels	6.290
	cubic meters	cubic feet	35.31
	cubic feet	barrels	0.1781
	cubic feet	cubic meters	2.832×10^2
Flow rate, L^3/T	bbl/day	m^3/s	1.8401×10^{-6}
	m^3/s	bbl/day	0.5434×10^6
Mass, M	pounds	kilograms	0.4536
	kilograms	pounds	2.205
Density, M/L^3	g/cm^3	kg/m^3	10^3
	g/cm^3	lb/ft^3	62.43
	kg/m^3	g/cm^3	10^{-3}
	kg/m^3	lb/ft^3	6.243×10^{-2}
	lb/ft^3	g/cm^3	1.602×10^{-2}
	lb/ft^3	kg/m^3	16.02
Pressure, $M/(L)(T^2)$	atm	psi	14.7
	atm	dyn/cm^2	1.0133×10^6
	atm	N/m^2	1.0133×10^5
	psi	atm	6.805×10^{-2}
	psi	dyn/cm^2	6.895×10^4
	psi	N/m^2	$6.895 \times 10^{+3}$
	N/m^2	atm	9.869×10^{-6}
	N/m^2	psi	1.450×10^{-4}
	N/m^2	dyn/cm^2	10

continued

TABLE A.1 Continued

	To convert	into	Multiply by
Absolute viscosity,	centipoise	g/(cm)(s) (poise)	1×10^{-2}
M/(L)(T)	centipoise	kg/(m)(s) (= N · s/m²)	1×10^{-3}
Permeability, L²	Darcy	square centimeters	9.869×10^{-9}
	Darcy	square meters	9.869×10^{-13}
	square centimeters	Darcy	1.013×10^{8}
	square centimeters	square meters	10^{-4}
	square meters	Darcy	1.013×10^{-2}
	square meters	square centimeters	10^{4}
Heat work energy,	gram calorie	Joule	4.186
ML²/T²	Joule	gram calorie	0.2389
	cal/(g)(mol)	J/(kg)(mol)	4186
	J/(kg)(mol)	cal/(g)(mol)	0.2389×10^{-3}

APPENDIX C: GLOSSARY

Some common abbreviations, definitions, and colloquialisms

AAPG American Association of Petroleum Geologists
ADMA Abu Dhabi Marine Areas: a consortium of BP, CFP, and ADNOC
ADNOC Abu Dhabi National Oil Company
ADPC Abu Dhabi Petroleum Company: a consortium of BP, Shell, and ADNOC
AFE Authority for expenditure: an estimate of the cost of drilling a well
API American Petroleum Institute (see also API units and gravity)
API gravity The gravity of oil is calculated according to the formula:

$$\text{API gravity} = \frac{141.5}{\text{specific gravity at } 60°F} - 131.5$$

API units Units of the American Petroleum Institute; one of the most common is the scale of radioactivity used on the gamma-ray log
ARCO The Atlantic Richfield Oil Company

Barrel One barrel = 159 liters, or 35 Imperial gallons, or 42 U. S. gallons
bbl Barrel (q.v.)
BCPD Barrels of condensate per day
BD or B/D Barrels (of oil) per day
BHT Bottom hole temperature
BNOC British National Oil Corporation
BOPD Barrels of oil per day

BOPs   Blowout preventers (on a rig)
btu   British thermal unit
BWPD   Barrels of water per day

CDP   Common depth point (see p. 98)
CFP   Companie Francaise des Petroles: 35 percent owned by the French
government; often operates under the name TOTAL
Christmas tree   The assembly of valves at a well head
Condensate   Light hydrocarbons that condense when cooled to surface
temperature; natural gas liquids
CONOCO   Continental Oil Company

DEMINEX   Deutsche Erdolversorgungs GmbH: West German oil com-
pany; largely government owned
Dry gas   Gas with no condensate; generally pure methane
Driller   Person in charge of a drill crew
Dry hole   Well containing neither oil nor gas
DST   Drill stem test (see p. 147)
Duster   Well with no commercial oil or gas

ESSO   Standard Oil of New Jersey; name changed to EXXON on 1
November, 1972
EXXON   Formerly Esso (see ESSO)

Fairway   A productive trend in a basin
FARO   Flowed at rate of . . . (e.g., FARO 8 BOPD)

GCM   Gas cut mud
GOC   Gas-oil contact
GOR   Gas-oil ratio
GUPCO   Gulf of Suez Petroleum Company; a consortium of Amoco and
the General Egyptian Oil Company
GWC   Gas-water contact

IFP   L'Institute Francais du Petrole; an institute for research and teaching
in petroleum science founded by the French Government
ISP   Initial shut in pressure
IP   Institute of Petroleum; an independent British organization founded
to promote the study of petroleum

Jack-up   Drilling rig that jacks up over the sea on steel girders

KB   Kelly bushing (see p. 40)
Kerosene   American word for paraffin
Kill (a well)   To prevent a well from flowing
KOC   Kuwait Oil Company; British Petroleum and Gulf Oil Company
joint venture

Lead Evidence suggesting a possible hydrocarbon trap; precursor to a prospect

LOM Level of organic maturation (referring to kerogen in a source rock)

LPG Liquified petroleum gases

LASMO London and Scottish Marine Oil Company

MBD (also **mb/d**) Million barrels per day

md Millidarcy; measure of permeability (see p. 229)

MFIC A tool pusher or other important person in charge

Mudlogger A geologist (generally) who monitors the drilling mud for shows of oil or gas, describes samples, and so forth as a well is drilled

NGL Natural gas liquids; e.g., propane, butane, and pentane

NIOC National Iranian Oil Company

OAPEC Organization of Arab Petroleum Exporting Countries

OASIS Oasis Oil Company of Libya; a consortium of Marathon, Amerada, Conoco, and Shell

OPEC Organization of Petroleum Exporting Countries (see p. 4)

OWC Oil-water contact

P & A Plugged and abandoned (well)

Paraffin English word for kerosene

Pay Productive interval (see p. 280)

PEMEX Petroleos Mexicanos; the national oil company of Mexico

PERTAMINA Perusahan Negara Pertambangan Minjak Dan Gas Bumi Nasional; the Indonesian state oil company

PETRONAS Malaysian national oil company

Play Petroleum traps of a particular genetic type; e.g., reef play and growth fault play

Pool Individual oil accumulation (see p. 280)

Prospect A location where a well could be drilled to locate oil or gas

PRT Petroleum revenue tax

psi Pounds per square inch (measure of pressure)

psia Pounds per square inch absolute

PVT Pressure-volume-temperature (analysis of reservoir fluids)

ROP Rate of penetration (of bit during drilling)

Roughneck A member of the drilling crew who works on the drill floor

Roustabout Drilling rig laborer

RT Rotary table (see p. 40)

Scout Company employee concerned with gathering intelligence data on the activities of rival companies

Show Indication of oil/gas during drilling

Semi Semisubmersible drilling rig

Shut-in field/well   A field/well from which oil/gas are not being produced
SL   Sea level
SOEKOR   Southern Oil Exploration Corporation: the South African state
   oil company
Sour gas/oil   Gas or oil containing hydrogen sulfide
SP   Spontaneous potential (log/curve)
SPE   Society of Petroleum Engineers
SPWLA   Society of Professional Well Log Analysts
SSL   Seismograph Services Ltd.
STATOIL   The Norwegian state oil company
Stripper well   An oil well with minimal production from a depleted field

TD   Total depth
Tool pusher   Person in charge of all drilling operations on a rig
TOTAL   A name used by Companie Francaise des Petroles
TRICENTROL   Trinidad Canadian Oil Fields Company; now with North
   Sea and other interests
TVD   True vertical depth

UKOA   United Kingdom Operators Association

Wet gas   Gas with condensate fraction (ethane, propane, butane)
Well-site geologist   Geologist on the rig employed by the operating oil
   company to safeguard their interest
Wildcat   Exploration well in a new area
Wild well   A well out of control—flowing oil, gas, or water
Workover rig   A light derrick used for maintenance work on a produc-
   ing well
WOW   Waiting on weather

Author Index

Subject Index